The Amber Lands in the Time of the Roman Empire

Vladimir I. Kulakov

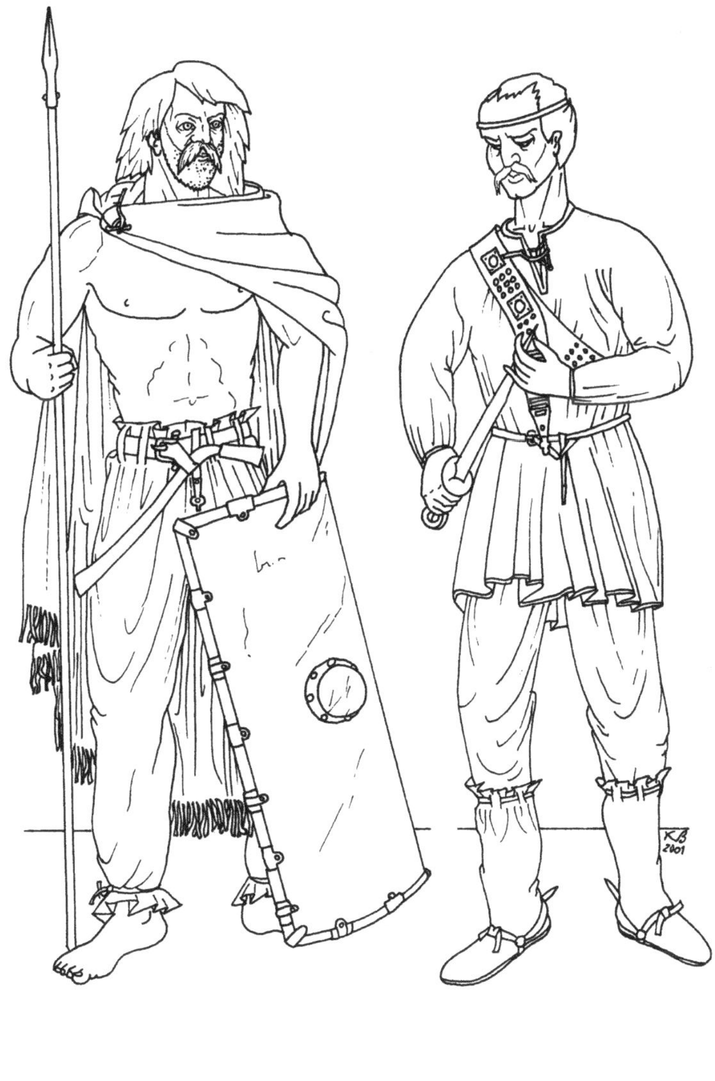

BAR International Series 1356
2005

Published in 2016 by
BAR Publishing, Oxford

BAR International Series 1356

The Amber Lands in the Time of the Roman Empire

ISBN 9781841718019 paperback
ISBN 9781407327860 e-format
DOI https://doi.org/10.30861/9781841718019
A catalogue record for this book is available from the British Library

BAR Publishing is the trading name of British Archaeological Reports (Oxford) Ltd. British Archaeological Reports was first incorporated in 1974 to publish the BAR Series, International and British. In 1992 Hadrian Books Ltd became part of the BAR group. This volume was originally published by Archaeopress in conjunction with British Archaeological Reports (Oxford) Ltd / Hadrian Books Ltd, the Series principal publisher, in 2005. This present volume is published by BAR Publishing, 2016.

BAR

PUBLISHING

BAR titles are available from:

BAR Publishing
122 Banbury Rd, Oxford, OX2 7BP, UK
EMAIL info@barpublishing.com
PHONE +44 (0)1865 310431
FAX +44 (0)1865 316916
www.barpublishing.com

Contents

I. Introduction

I.1. The Geography of the Amber Lands

The region of the Amber Lands may be equated to the ancient territory of the Old Prussians (in German "das Bernsteinland"). In the 20th century it was part of the Imperial province of Eastern Prussia (in German, "Provinz Ost-Preuβen"), known in 1772 as the eastern part of the Kingdom of Prussia. This territory was the last acquisition of the former USSR. In 1946 it became part of the Soviet Union, designated the "Kaliningrad Region". The area of the Amber Lands most densely scattered with archaeological sites is the Sambian Peninsular (Samland), its historical name being the "Amber Coast" (Bernsteinküste). This name corresponds entirely to the geological characteristic of a territory that is still the largest source of natural amber in the world.

Geographically, the Amber Lands include the western outlying district of the southeast sector of the Baltic Sea littoral, or "Southeast Baltic". This part of Central Europe, limited by the lower reaches of the River Vistula in the west, and the Neman in the east, has a complex landscape formed in the Late Glacial. Dating to this historical period are the amber-bearing layers of glauconit ("blue ground"), covered by a mass of the sandy soils carried by the glacier sweeping in from east Europe approximately 2 million years ago. Long before this cataclysm, the climate became intensely cold and the dense forests were eradicated by the thick ice arriving from the north – the territory of present Scandinavia. The retreat of the glacier northwards occurred at varying speeds and the receding mass of ice literally ironed the ground flat. When the temperatures began to rise again appreciably, the glacier also increased its retreat north. The result is the arc of lakes surrounding the coast of the southeast Baltic from the Polish borders to the Pskov region. The sandy landscape between the Vistula and the Neman (with its hundreds of lakes) is known as the Mazurian Lakes (in Polish, Pojezierze Mazurskie; the limits of the Warmińsko-Mazurskie wojewodztwo). One of the lakes, the Wyshtynetskoye, straggles the borders of Poland, Lithuania, and the Kaliningrad Region.

The Mazurian Lakes appeared in the final "Valdai" period, covering the territory of the modern Kaliningrad region from the south to the southeast. They separate the plains from the dominant dunes of the Lithuanian landscapes. The increasing speeds of the glacier heading towards the Baltic Sea eroded the coastal surface. High hills now only remain in the Ozersk and Nesterov districts bordering the Mazurian Lakes.

The ice flows deposited sedimentary rocks on the carboniferous layers, impregnated with veins of amber, which were deposited in the Cenozoic. Such was the origin of the amber ground-deposits in the area of the present settlements of Sinjavino (Kraxstepellen), Yantarnyi (Palmnicken), Grachevka (Craam), and Romanovo (Pobethen). Over time these amber deposits, and the development of trade in this much-cherished material, shaped the special character of the local history. (Dryakhlov V.N., 1988, s. 142, 143).

The line of the Baltic coast at the end of the glacial period gradually stabilized and began to assume the outline shown on today's maps. Only the sandy spits of the Baltic (Frische Nehrung and Kurische Nehrung) remained unstable (right into the Middle Ages), in turns forming small islands or disappearing under the waves entirely. As late as the 12th-14th centuries, the dune ridges of these spits were cut through by numerous channels, a section even being of artificial origin (Kulakov V.I., Tepljakov G.N., Puzakova G.S., 2001, s. 41).

At the beginning of the Holocene, in the post-glacial period, the broad-leaved forests (beech, hornbeam, oak, linden) began to cover the territory of the southeast Baltic, forming an undivided large forest ("grosse Wildnis"). This huge forest crossed the territories between the rivers Vistula (Weichsel) and Neman (Memel-Fluss) along a line running south-west/north-east, even in the early Middle Ages. The lands between the rivers Prokhladnaya (Frisching) and Lava (Alle), as well as the Kaliningrad peninsula (Kr. Fischhausen/ Samland), being the highest and plateau-shaped part of the southeast Baltic, were unforested. These lands, known in the early Middle Ages as Natangia (Natangen) and Sambia (Samland), were already inhabited in the Iron Age. The ethnic names of these tribes, the Aestians and Prussians, are known from numerous written sources from the 1st-13th centuries.

This work is devoted to the history of the Amber Lands in Roman times. It is based on archaeological data dating back to the beginning of the 2nd millennium AD, and studies the local material and spiritual culture of this territory.

I.2. The chronological framework of the research

Antique written sources since Tacitus (fig. 1), covering the 1st-4th centuries AD, refer to the Amber Lands under the names of Ozericta and Austeravia (Kulakov V.I., 2002a, s. 13), and the population as the Aestii. This ethnic name is of German origin and translates as "eastern people" (Laur W., 1954, S. 233). Maria Gimbutas (M. Gimbutienė), the well-known researcher of Balt antiquities, has questioned the identification of Tacitus' ethnic name "Aestians": "It is not established exactly whether the name "Aestians" appertains to all Balt tribes, or whether Tacitus was referring especially to the Balts or, more exactly, the Prussians, i.e. western

Balts, or only the amber gatherers on (the coasts of) Frischen Haff (the Vistula/Kaliningrad Lagoon - K.V.), which nowadays the Lithuanians call "Aistmarės" (Gimbutas M., 1983, S. 18). However M. Gimbutas and other archaeologists have no doubts in linking Balts to the Aestians; on the contrary, modern thinking is again leaning towards a Finno-Ugric origin for this population. This conclusion, which does not take into account archaeological data, is based on matching the ethnic names of "Aectians" and "Estonians" (Bajer H.-F., 2001, s. 367). European archaeologists consider the "Aestians", inhabiting in the time of Roman influence the lands between the rivers Vistula and Neman, to be descendants of the autochthons of the Bronze Age (middle of the 2nd millennium BC) (Kilian L., 1980, S. 131), and, in turn, are direct ancestors of the Old Prussians of the early Middle Ages (Jankuhn H., 1950, S. 62).

During the two-hundred-year history of Prussian archaeology, German, Polish and Russian researchers and antiquaries collected a great deal of material on the Roman period, mostly from gravesites. The presence of this unique corpus of European archaeology has provided a very sound basis for establishing a complete theory on the material and spiritual culture of the inhabitants of the Amber Lands from Nero (the time of the first contacts of the Aestians with classical antiquity) to the beginning of the Middle Ages, as marked by the end of the Hunnic wars.

I.3. The history of the study and a historiography of Aestian antiquities

On the map, the Amber Lands appear as a rather insignificant, north-eastern district of Central Europe, however they occupy a unique space in 20th century. Over the last few decades, European historians have concentrated on this period but the territory remains very much a blank on the archaeological map of Europe. Because of the difficult political fortunes of the former East Prussia since the war, Aestian antiquities were practically off-limits for archaeologists and the material is practically unknown by younger scientists, east or west.

The capital of the old Duchy of Prussia, formed in 1525, was the city of Königsberg (today's Kaliningrad), which, due to its Alberts-University (1544), was a centre of powerful educational and research activity. The history of the population in this region attracted scientists as early as the 16th century. J. Maletzki (1563) and M. Stryjkowski (1582) devoted their energies to the description of the spiritual culture of the Old Prussians. Their data on the relicts of local paganism was generalized by W. Mannhardt (Mannhardt W., 1936, S. 327-366). At that time the fixed sites of their material culture were of interest only for their potential wealth (Gaerte W. 1931a, S. 135-138). Prussian artefacts circulating between the antique dealers generated in the elite a desire to learn more about the distant past of their native land.

In 1790, "The Königsberg Physical and Economic Society", the oldest scientific association in Prussia, stimulated the growth of interest in Prussian antiquities (as a rule Bronze Age barrows) among local intellectuals. In 1844, under the initiative of Ernest August Hagen, the founder of Prussian art criticism, the Prussian Society dedicated to the study of local antiquities was formed in the capital. Its members were drawn from those who attended the "Physical and Economic Society." At first they regarded archaeological finds as curios. The curiosities received by the Prussian Society were studied out of context as works of art. In the middle of the 19th century, G. Berendt, K. Stadi, G. Bujak and J.Henshe, members of the Prussian Altertumsgesellschaft "Prussia", were the first to carry out excavations on the sites where ancient items were found. As a rule, the excavations were carried out at Roman cemeteries located on the Sambian Peninsula (Grebieten, Kirpehnen, Gaffken). Already by the middle of the 19th century, members of the Prussian Society began to pay attention to individual sites and the circumstances of the finds: from that time the Society prepared lists of finds with appropriate notes. From year to year the Society's own collection increased, finally providing the core of the East Prussian Provincial Museum which opened in 1878. Its first director was Otto Tischler, a founder of Prussian archaeology, who was a physicist and mathematician. The success that accompanied a display of the museum's star exhibits at the Berlin XI Congress of the German Anthropological Society (August 5-21, 1880) changed the course of the archaeological study of the Amber Lands. The magnificent publications prepared for this exhibition (Tischler O., 1879) and later ones (Günther C., Voss A., 1880; Tischler O., Kemke H., 1902), created exceptional interest in Europe in East Prussian antiquities.

The Prussian archaeological school was created in the middle of the 19th century. Its founders, Otto Tischler and Henry Kemke, not only gave an objective picture of Prussian antiquities, but also established the basis for a chronology of Europe from the first half of the 1st millennium AD. For this chronology (phase B-D), Tischler found approximately 2000 artefacts in 1879 while excavating the cemetery at Dollkeim, Kr. Fischhausen (the present village of Kovrovo, Zelenogradsk). Tischler ordered the materials in a typological series and his work was highly regarded among his contemporaries. Throughout the 20th century, Tischler's scale was supplemented by such recognized authorities of European archaeology, as Adalbert Bezzenberger, Hans Jurgen Eggers, and Kazimerz Godłowsky. Recently Jaroslav Tejral has added modifications to this chronological system, but Tischler's understanding of phases B-D remains unchallenged, and contemporary researchers need only specify certain nuances.

Today Ulla Lund Hansen's works on Prussian material are the most recent. They take into account Tischler's system (B_{1a} - 20-50 AD, B_{1b} - 50-70/80, B_2 - 70/80-150,

B_2/C_1 - 150-200, C_{1a} - 200-225, C_{1a}-C_{1b} - 225-250, C_{1b} - 250-275, C_2 - 275-325 - Lund Hansen U., 1987, S. 39) and that of Jaroslav Tejral. Tejral divided the materials of the Middle Danube stage D into phases D_1 (360/370 - 400/410 AD), D_2 (380/400 - 440/450 - stage Untersiebenbrunn), D_2/D_3 (430/450 - 470/480 - stage Smolin), D_3 (450 - 480/490) and D_3/E (470-500/510) (Tejral J., 1997a, S. 351). The hiatus between the chronological schemes of Lund Hansen and Tejral occupies the period between 325 - 360/370 AD; this corresponds, most likely, to phase C_2/D_1 (C_3?). In European archaeological science there was no common opinion concerning this phase. M. Martin calls it C_3 and dates it rather widely to 350-400 AD, or even a little later (Martin M., 1995, S. 680). However the temporal gap between Lund-Hansen and Tejral shows that this phase was probably shorter. The final phase ('the great migration of the population') is structured by J. Kowalski thus: E_1 - 490/510-525/550 AD, E_2 - 550-turn of the 6th-7th centuries, and E_3 - about 600-675. The last Amber Lands phase is linked to the so-called "Elbląg group" material (Kowalski J., 2000, s. 219-224). True, the typological aspects of Prussian finds allow us (rather conditionally) to move the border of the E_2/E_3 phase a little bit earlier in time to around 575 AD. Correspondingly the end of the E_3 phase shifts to about 625 AD. Today the chronological gradation of antiquities from northeast Barbaricum seems the most likely.

The monographs by Tischler and Kemke were the final works to be published in the 19th century on the Roman cemeteries of Prussia (Tischler O., Kemke H., 1902, S. 14-46). The material presented in this edition is, in effect, the bringing together of all the objects that give dates to the chronologies mentioned above.

In the second half of the 19th century, the members of the Prussian Society specialized in separate archaeological periods. Emil Hollack and Felix Peiser studied the cemeteries of the Latin and Roman epochs in Sambia and the area of the Mazurian Lakes.

The Society formed productive contacts with foreign colleagues. Oscar Montelius (in Tischler's time) (Hoffmann M., 1991, s. 89) and Oscar Almgren at the beginning of the 20th century (Hoffmann M., 1992, s. 100) became members of the Society. These major scholars worked with the Society's materials and actively included the Prussian finds. They used the antiquities of the Amber Lands to develop typological and chronological schemes which are still valid. This is particularly so of Almgren's typology of provincial Roman brooches, which is based in significant part on Prussian material.

A century of continuous archaeological investigations was summed up in Emil Hollack's works. In the textual explanation to the archaeological map (Hollack E., 1908a), the author detailed the brief characteristics of the stages of development of the material culture of the population of the Amber Lands in the antique (Aestian finds) and medieval (Prussian antiquities) periods. Hollack based his work on Tischler's chronological scale.

In researching Prussian burial rites from the 1st millennium AD, Hollack followed Tischler (Tischler O., 1889, S. 22) in concentrating on Sambian antiquities of the first centuries AD; they considered as highly characteristic those urn cremations overlapped by covering stones (Hollack E., 1908b, S. 146-155).

At the turn of the 19th century, trial trenches in cemeteries gave way to large-scale excavations lasting several seasons. The years before the First World War were especially productive for Prussian archaeologists in terms of fieldwork. At the end of the first decade of the 20th century, full archaeological reports began to appear in the Prussian Museum's archives, and they developed into a very important area of regional activity. They included full textual information and figures of finds excavated by the members of the Society. All the excavated material was at first handed over to the Prussia Museum. After high-quality restoration, some of the finds were transferred to the museums of Memel (Klaipėda), Ragnit (Neman), Insterburg (Chernyakhovsk), and Georgenburg (Mayovka). After the First World War, the materials accumulated in the storerooms of these museums began to feature intensively in scientific research. In particular some formed the basis for Nils Åberg's typology of Prussian assemblages, representing the era of the great migration (the great barbarian migration), which is still relevant to scholars today (Åberg N., 1919).

The next stage of development of the Prussian archaeological school began with the growth of the Weimar German Republic. It was characterized by advances in field methods, as well as by the scientific assimilation of the accumulated materials. Unfortunately, the Stalinist period affected the understanding of methods and results of Amber Lands archaeology between the wars (Kushner P.I., 1951, p. 113, 141, 142). As Nowakowski correctly notes (Nowakowski W., 1996, S. 8, 9), the strategy of archaeological research in Eastern Prussia between the wars was determined by two authoritative scientists, Herbert Jankuhn and Carl Engel. For the first time they carried out experiments on East Prussian material by modelling the ethnic and cultural factors of the first centuries AD (Jankuhn H., 1933a; Jankuhn H., 1933b). Unfortunately Engel's planned volume for the publishers Grafe & Unzer on Prussian archaeological sources dating from 1st-13th centuries AD (including a catalogue of sites) did not materialize. The whereabouts of Engel's manuscript remains unknown (Beran J., 1997, S. 138).

Excavations of Aestian cemeteries continued into the 1920s and 1940s, and there was a tendency to fully open entire sites. An example is the cemetery at Zophen (Suvorovo, Gvardeysk District), excavated in 1928, where 502 4th/14th-century graves were discovered (Prussian Museum general index VII, 11942) (Heym W., 1938).

Wilhelm Gaerte, who was Prussian Museum director until 1945, significantly improved the layout of the museum and popularised Prussian antiquities. His monograph "The Prehistory of East Prussia" remains a seminal work (Gaerte W., 1929).

The text published in 1937 by Engel (from 1929-1934 employed at the Prussian Museum) and Wolfgang La Baume (from 1938-1945 a state curator of the archaeological sites of East Prussia) is at first sight a popular summary of the history of the Amber Lands, well known to Köningsberg archaeologists by the beginning of the 1940s. In actual fact the book offers a detailed scientific programme of work for future generations of researchers into the material culture of the Western Balts. Thus, taking into account the current political situation, researchers into Aestian and early Prussian antiquities were attracted by the idea of isolating a German ethnic element (Goths and Vikings) on the outlying western district of the Balt world (Engel C., La Baume W., 1937). However western Balts, following Hollack, continued to be considered as Amber Lands autochthons (La Baume W., 1940a, S. 6).

This conclusion was maintained post-war by Soviet scientists, although their work on the Aestians was severely criticized by representatives of the old Prussian archaeological school (Šturms Ed., 1954, S. 84). The first Russian-speaking researcher of Amber Land antiquities was Frida Davidovna Gurevitch. She began in northeast Prussia in 1946, working there (with breaks) from 1949 to 1959. During this time Gurevich investigated 30 sites mainly of Prussian origin. She pointed to the originality of Prussian material culture all through the 1st millennium AD, thus taking up the thesis stated by Hollack in 1908 (Gurevich F.D., 1960).

In 1974 the Baltic expedition of the Academy of Sciences of the USSR Archaeological Institute was organized. It has been working since then in the Kaliningrad area under the leadership of the author of the present book. During the 1974-2002 seasons we investigated 336 archaeological sites from various periods in all districts of the region; 193 sites were located in the Zelenogragsk district, the historical centre of the western Balts. Of the monuments mentioned above, the following relate particularly to the theme of this present study:

Roman influenced cemeteries (1st-5th centuries AD) - 24 sites (Sambia - 15).

Roman influenced villages - 14 sites (Sambia - 8).

Cult stones and stone sculptures from pre-Order time - 3 sites (Sambia - 2).

Cemeteries dating to Roman times and the period of the migration of the "Hünenberg" ("Mountain of Giants") populations (Rantau - Neu Kuhren) were extensively excavated in 1985 and 1987-1993; Kovrovo (Dollkeim) was the subject of large-scale work in 1992 and 1994-2001. There were no correspondingly large excavations at settlements of 1st-5th centuries AD, except for the settlement of Klintsovka-Kamenka (Wiekiau-Michelau, 5th-6th century AD) where, in 1980, rescue work was carried out and remains of a pillared dwelling were excavated (Kulakov V.I., 1994, s. 96).

Between 1992-2002 the Natangian group (K.N. Skvortsov) of the Baltic expedition conducted several excavations, mainly at Roman cemeteries - Klein Heyde and Bolshoye Isakovo (Lauth) in eastern Sambia. The results have not yet been published (2002).

Since the war, most of the work undertaken by our Polish colleagues has been devoted to Aestian sites. The Sudawian expedition was organized in the eastern part of the Mazurian Lakes area soon after the end of World War Two; it concentrated on Roman cemeteries and the results were published by Jezi Antoniewitzc. In particular, he came to conclusions about the major boundary changes that occurred between the 4th-5th centuries AD in Balt societies between the rivers Nemans and Daugava, mostly connected with sharp changes in the economic situation of the region. Gimbutas came to the same conclusions based upon the significant volume of knowledge gathered in the first half of the 20th century about the Balts (Gimbutas M., 1963, p. 137). These changes were caused by tribal migrations leading, in particular, to the appearance in Szwajcaria (at the end of the 4th/middle of the 5th centuries AD) of individuals of a "yellow, pacific race" (Antoniewicz J., 1962, s. 7, 10).

The material, representing over 50 years of advances in Prussian archaeology, has allowed Jerzi Okulicz to publish a work summarizing the results of post-war Polish research (Okulicz J., 1973). His approach continues the tradition of Gaerte's work, issued half a century earlier. In Okulicz's opinion, separate Prussian tribal associations already lay claim to specific areas in the 5th century AD. The author lists Roman antiquities from the Amber Lands as being of western Balt culture, and divides them into Mazurian and Sabian-Natangian groups (sub-divided later in even more detail - Okulicz J., 1976, ryc. 1). The latter (Sambia, a part of the Vistula Lagoon coast and the Pregolya/Pregel River basin) did not change its area until the early Middle Ages (Okulicz J., 1973, s. 353, 367, 444). As the specified area coincides in plan with the outlines of the populated territory in Prussia from the 15th century (along a west/east axis, from Germau to Tammowischken, Mortensen H., Mortensen G., 1937, S. 19, Abb. 1-4), it is necessary to look for the reasons for such habitation first of all in the natural landscape characteristics of the region (i.e. an inconvenient landscape for agriculture, comprising sandy-argillaceous soils and bogs to the east of Sambia - Odoj R., 1970, s. 56)

In the middle of the 1970s, Jaskanis investigated the burial rites of inhabitants from the Amber Lands from Roman times. He came to the conclusion that from the very beginning of the 1st century AD, the rites of the western Balts (the "burial of cremation remains in the ground") were clearly influenced by Celtic traditions, and, to a lesser degree, by those of the Romans (Jaskanis J., 1974, s. 254, 268). There were no precise attributes of Amber Lands burial rites in this work, however Jaskanis' subsequent publication was a full catalogue of western Balt cemeteries from the 1st-5th centuries AD (Sambia, Natangia and the Mazurian Lakes area)(Jaskanis J., 1977).

The last 15 years of 20th-century Polish archaeology were marked by real advances in the study of Aestian antiquities. These advances were connected with the scientific activities of Woitsekh Nowakowski, a pupil of the well-known western Balt researcher Jezi Okulicz. The young Polish archaeologist undertook an initial analysis of antique imports originating from the southeast Baltic, basically along the Vistula "Great Amber Route" (Nowakowski W., 1985, s. 100). His conclusions correspond to the opinions of the Prussian school archaeologists (Engel C., 1942, S. 159). Nowakowski made an attempt to identify those Aestians (from the beginning of the 1st century AD) who inhabited the outlying district of Suebia (Germania Libera, Barbaricum) (fig. 1), within the circle of western Balt antiquities (Sambia and Mazurs) and the "wild Fenns", which, according to Tacitus, were outside Barbaricum (along with the people of the "shaded ceramics" culture)(Nowakowski W., 1990, s. 91-96). However, in the opinion of Shchukin, from the middle to second half of the 1st century AD, the Aestians/Balts might not have been the dominant Sambian tribe, having been pushed aside by groups of newcomers from the Danube area (the "wandering Veneti"), tempted by the prospect of amber and occupying, according to Pliny the Elder, the Eningia peninsula (Sambia) (Shchukin M.B., 1998, p. 206).

Nowakowsky indirectly confirms this obvious historical fact (Nowakowski W., 1995a, ryc. 1).

The tendency to "sovereignize" the Morąg, Węgorzew, Suwałki and Awgustów groups of western Balt culture of Roman times, as put forward by Antoniewicz and Okulicz (Okulicz J., 1976, ryc. 1), became current in Polish research of western-Balt antiquities. Nowakowsky has declared the first two as "Bogaczewo culture", existing, in his opinion, in the Mazurian Lakes area from phases A_3-D; the Awgustów group was shifted by him to the Suwalkian culture (Nowakowski W., 1986-1990, s. 32, 33). He compares the "Bogaczewo culture", whose indigenous western Balt traditions were strongly influenced by the Wielbark culture, with those Galindians mentioned by Ptolemy. The Sudowians were declared to be people of Suwalkian culture. Nowakowski identified the area of Galindians, following the designation of Peter von Dusburg, who, at the beginning of the 14th century, freely resurrected this ethnic name that was mentioned by Ptholemy in his "Chronicle of the Land of Prussia" (Nowakowski W., 1995a, s. 207). This does not necessarily corresponded to historical reality.

Nowakowski's monograph devoted to Sambian antiquities from the time of Roman influence (Nowakowski W., 1996), incorporating Alexander Humboldt's material (Germany), marked a considerable development for Polish archaeology. The basis of this research is data (unfortunately partly incomplete and with extremely brief correlation tables) from 502 graves from 15 cemeteries along the Amber Coast (the Zelenogradsk district of Kaliningrad and nearby micro-regions). In his book, Nowakowski publishes separate finds from 335 graves occurring from 42 flat grave cemeteries. During the preparation of his monograph, Nowakowski did not use material from the Kaliningrad museums, nor did he visit Sambia. The main merit of Nowakowski's work is the scientific use of the archive material of Martin Yan, and the fragmentary data published in the works of O. Almgren, N. Åberg, S. Bolin and H.-J. Eggers. However, his conclusions are extremely unreliable because of the absence of a catalogue of the burial assemblages used in the work, and a too generalized typology of grave goods (the types labelled as "brooch with 'eye'", "bent strap ends", "brooch with star-shaped foot", etc. correspond to European archaeological typology from the end of the 19th century), and the absence of ceramics, burial rites, and Sambian horse burials. The author, therefore, has rather mechanically divided the antiquities of the so-called "Dollkeim-Kovrovo culture" (the former Sambian-Natangian group of western-Balt culture) into six phases (1 - B_{2a}-B_2/C_1 (to the middle of the 2nd century AD); 2 - B_2/C_1-C_{1a} (to the beginning of the 3rd century AD); 3 - B_2/C_1-C_{1a} (to the beginning of the 3rd century AD); 4 - C_1-C_2 (to the third (?) quarter of the 4th century AD); 5 - from the end of the 4th/middle-third quarter of the 5th centuries AD); 6 - the end of the 5th/beginning of the 6th century - the end of Dollkeim-Kovrovo culture (Nowakowski W., 1996, S. 48-54). The inaccuracy of this system is obvious: for example, the material from the 7th/beginning of the 8th century (for example the brooches of the Sprossenfibel type) were wrongly referred to as phase 6.

Nowakowski's superficial acquaintance with the Sambian material means that he has noticed neither the sharp change in the local material culture of phase B_2/C_1 (Vityaz' S.P., Kulakov V.I., Medvedev A.M., 2000, s. 13), nor the formation of the Old Prussian culture at the end of "phase 5", on the basis of the Aestian antiquities (Kulakov V.I., 1994, s. 65). This last fact was already anticipated by Antonewicz (see above).

However the estimation of Nowakowski's work as a turning point is quite fair. Despite the errors and omissions, his monograph (the publication of his dissertation) marks the end of a "plot of silence" concerning Old Prussian antiquities, which existed in

European archaeology after the abolition in 1947 of the state of Prussia.

Okulicz briefly summarized the studies of Polish archaeologists into the problems of western-Balt archaeology from Roman times. He fairly criticized Nowakowski's shortcomings and urged for an analysis of the characteristics of Aestian historical ethnography. (He also repeated the thesis of "antagonistic relations" between the Aestians and the people of Wielbarsk (Okulicz J., 2000, s. 290, 291, 294).) Actually, material from the Kovrovo cemetery, which gave its name to Nowakowski's "Dollkeim-Kovrovo culture", shows the prevalence of the "Goth-Gepidae" in northern Sambia at least from the 3rd century AD (Kulakov V., 2000a, S. 596). This fact firmly removes the question of their "antagonistic relations" with the natives.

Information on the final phase of the development of the Aestians, and on the period of the great migration along the Amber Coast, was briefly submitted by this author in the catalogue of cemeteries, settlements, and finds (Kulakov V.I., 1990a). Antiquities from the end of the 4th-5th centuries AD are presented in the monograph that is devoted to the history and culture of Old Prussia from the early Middle Ages (Kulakov V.I., 1994).

For the initial analysis of Roman antiquities we collected material from 70 cemeteries (mainly from Sambia, but to a lesser degree from Natangia and the Pregel/Pregolya basin (fig. 2, 3)). From Hollack's time, the majority of researchers connect their attributes (characteristics of rites, stone coverings from the tops of graves, specific forms of ceramics and jewellery, etc.) with antiquities belonging to the inhabitants of the western part of the Amber Lands, and refer them to the Sambian-Natangian Group (SNG) of western-Balt culture, the centre of which was undoubtedly Sambia. The SNG area is separated from related antiquities of west and east Mazurs by the influential flood plains of the Pregolya (fig. 2). In the widest places (for example near modern Pravdinsk/ Friedland), during flood times, an area 1km wide is covered with water. 597 burial assemblages were collected from the cemeteries mentioned; they are submitted textually and graphically. Unfortunately, the possibilities of research are limited because there is no data of settlement archaeology, and no complex information on those graves from the cemeteries of the key Amber Land regions. Such information is submitted only for separate burials (Dollkeim-Kovrovo and "Gora Velikanov"/Hünenberg are exceptions), and thus it is not possible to illuminate all aspects of the history of the Amber Lands during the Roman period.

It is necessary to characterize briefly the attributes of Aestian antiquities from the first half of the 1st millennium AD and those of the neighbouring populations.

Chapter II. The outlying western district of the Baltic world in the 1st-5th centuries AD

II.1. Attributes of Aestian antiquities

The outlying western districts of the Balt world - the lands between the rivers Nogata/Nogat and Neman/Memel in the 5th-13th centuries were occupied by the Old Prussians, Scalovians, Lamatians and Sudins/Yatvigians. Reliable information on these tribes is to be found in the written sources from the early Middle Ages. The territories of these tribes as a whole are known by data extracted when researching burial sites (Tautavichus A.Z., 1980, s. 86). The unity of these tribes is understood on the basis of a shared language fixed by later sources (Mažiulis V., 1981, p. 8). The western-Balt culture from the first half of the 1st millennium AD (fig. 3, 4), dating back to the barrow culture of the 1st millennium BC (Sedovs V., 1992, p. 31, 44-46), and known basically by their burial sites, included all the tribal areas mentioned above. The territory occupied by this culture in Roman times included all the southeast Baltic from the Mazurian Lakes in the south to the present Lithuanian shores and the "Curonian Peninsula" (Kuržeme in contemporary Latvia) to the north. This culture is characterized by the prevalence of flat-graves with external cremations covered with stones. Bones were put into ceramic urns or in wooden receptacles (or bark, leather, etc.). The grave goods included brooches, buckles, bracelets, weapons, and everyday objects; there were no traces of burning. These objects, within the limits of the given culture, may be compared not only by nomenclature, but also by type (Jaskanis J. 1974, s. 211-244). The Aestian area corresponds to the Sambian-Natangian group of the western-Balt culture and covers Sambia (the Amber Coast proper), the Pregolya Basin, and the territory between the rivers Pasłęka/Passarge and Lava/Alle. The SNG is characterized by "snake-shaped" bracelets, bucket-shaped pendants, and ornamentation of the "double cross" type on the accessory vessels within some burial assemblages (Kulakov V.I., 1998, S. 629, ryc. 3).

The internal territorial gradation of western-Balt culture, which had already begun in the 1st century AD, is directly connected with the process of consolidation of those local tribes known later as Prussians, Scalovians, Lamatians, Sudins (Yatvigians), and Curians (Tautavičius A., 1987, p. 104-106). Nevertheless, as a whole, the material culture of the western Balts to the end of the 7th century AD preserves shared features. They are represented by archetypes of metal objects (mainly breast decorations: staff-shaped pins and other variants, plate brooches, crossbow-shaped and cross-beam examples (including Masurian forms), horseshoe-shaped clasps, etc.), and by styles of manufacture (copper-alloy, or, more rarely, iron base with silver overlay) (Vaitkunskienė L. 1981, p. 47-53). The designs of the local workshop are characterized by the conservative preservation of archetypal indicators extended over a maximum number of derivatives of alien forms (allochthons) and non-adoption of their variants. An important feature of the early medieval archaeology of western Balts is the inclusion of horse skeletons (or fragments of them), with items of horse harness, in the graves.

Certain events during the era of Roman influence, and especially pressure on the Aestians by their western and southern neighbours, resulted in the movement of a large part of the indigenous population of the Amber Lands to the east and northeast, as witnessed by the archaeological material found (Kulakov V.I., 2001b, s. 47, 48).

The group of coastal Aestians is the largest, in terms of the area occupied within the territory researched (fig. 4, S). This group inhabited an area from the Amber Coast (western Sambia) in the west, to the Upper Pregolya in the east. From the end of the 2nd century, there was among them a tradition established of cremations outside the graves and this continued until the 11th century AD. However, poor data concerning Aestians burial rites from Roman times only allows us to say that, for them, urn cremations took priority. Nuances of local ceremonials require special study. In the middle/second half of the 5th century, when the Old Prussians moved from the Sambian Peninsula to the mouth of the River Nogata and to the western part of the Mazurian Lakes, some features of the burial rites changed. Burnt bones were earlier deposited in urns of Grebieten type, or in pits around them, and horse burials were placed to the west of the graves of the horsemen, from the middle of the 5th century, the tradition of biconical accessory vessels in assemblages is preserved, but more often there are remains of a funeral pyre and horses are placed in the lower level of the grave, under the human remains.

Already by late Roman times at Suvorovo, Bolshoye Isakovo, and the former Grebieten cemeteries, there were pair burials with details of male and female costume (fig. 5). A similar tradition is also confirmed by medieval written sources (Gaerte W., 1931b, S. 126, 132), when in the upper levels of graves there were occasionally two or three local groups of burnt bones; one of these groups contained female jewellery. Ethnic indicators of western Balt grave goods for female assemblages from the 2nd century AD are cuff-shaped bracelets and cross-beam brooches; 3rd century AD artefacts include a pair of crossbow-shaped brooches.

II.2. The characteristics of the Aestian region

Aestian antiquities were known to Tacitus and to other antique historians as early as the 1st century AD (their descendants in the 13th century were the Sembs, and other enemies of the Teutonic Order, from the territory between the rivers Nogata and Neman). They had appeared on the Sambian Peninsula and neighbouring lands from the 1st century AD, and were soon distributed in the Pissa and Angrapa rivers basins and in various parts of the Mazurian Lakes. Except for Sambia, western-Balt antiquities from Roman times (ascending, according to their rites, to the traditions of the Sambian and western-Balt barrows) correspond with other groups of western-Balt culture (fig. 3, 4). A part of this territory (namely the southern and eastern coastal area of the Vistula Lagoon (Kaliningrad) occupied by the people of the Old Prussian culture around the 5th/beginning of the 6th century AD) already by the end of the 9th century shared the common place-name of 'Witland', mentioned in a text of Wulfstan. His message (included in the 'Chronicle of Orosius') also refers to Estland (the land occupied by the Balts) and the River Ilfing (on the western border of Estland) (Orosius, 1861, S. 732, 733). All these names are of German origin.

At the centre of Witland, precisely marked by the congestions of burials (fig. 2, 4), is the Sambian Peninsula (Samland). The antiquities of its cemeteries are typical of Germania Libera as a whole (fig. 5). All the territory between the Nogata and Neman Sambia is characterized by a moderate climate, an insignificant number of woods and bogs, and comprises mainly of sandy soils. There is no corpus of investigated settlements there from Roman times and it is difficult to trace the palaeogeographical and palaeodemographical situation in the southeast Baltic of the first centuries AD.

II.3. Areas of the western Baltic tribes beyond the Amber Lands

To the northeast of the area of the coastal Aestians, in the lower reaches of the Neman, there was a tribal territory inhabited by the common ancestors of the latest Scalovians and Lamatians, who, on the northeast coast of the Curonian Lagoon in the 2nd-6th centuries AD, constructed graves with stone tops (Tautavichus A.Z., 1980, s. 81) in accordance with the conservative traditions of western-Balt rites.

Ptolemy mentions a tribe of Sudins (their late ethnic name is Yatvigians) who were the southeast neighbours of the Sambian inhabitants of Roman times. Going back to the barrow-culture of the western Balts, the Sudawian culture of the 5th century appears to differ considerably from that of the united western-Balt culture, and at this time it is adopted by Awgustów and Suwalkian groups (fig. 4): at first the latter groups are territorially delimited. Their antiquities are united by the tradition of constructing stone and earth mounds above partial inhumations (in the majority of cases). The Sudawians renewed the custom of barrow and urn cremations during the phases C_1-C_2. In the East Mazurs, flat-grave cemeteries are gradually replaced by inhumations. True, it is difficult to call these "barrows": the barrow mound is often no more than 0.5 m high. The grave goods of both types of burials from Roman times are closely synchronous to Sambian antiquities (Pronin G.N., 1989, s. 68), however they stand out for their significant numbers of western-Balt objects (cross-beam brooches, cuff-shaped bracelets, and, less often, pins - fig. 6). During the Prudz phase (the name given by A. Bitner-Wróblewska for the "Sudawian" culture horizon that was synchronous to the period of the great migrations), the inhabitants of the Suwalkian Lakes area expanded their territory to the west, occupying the Goldapian Lakes (in the upper reaches of the Angrapa), and strengthening their contacts with the Sambian population (Bitner-Wróblewska A., 1998, s. 308, 309). All the area of the Sudawians/Yatvingians is characterized by the presence of a great number of lakes of glacial origin, surrounded by sandy dunes. The western part of the Mazurian Lakes has a similar landscape. From the 1st century BC to the beginning of the 5th century AD, there was a culture there called by Nowakowski "Bogaczewo". At that time, the "Bogaczewo" culture population had a material culture close to that of the Aestians of Sambia, and thus they were powerfully influenced by the Wielbark culture. Characteristic indicators of the "Bogaczewo" antiquities from the northwest part of Mazurs in Roman times are pins with triangular staff-shaped tops (Jankuhn H., 1950, S. 59, Abb. 10, 15). As the local culture developed (around the middle of the 6th century AD), the Mazurian cultural group appeared there with the direct participation of the allochtons of the Middle Danube area. This Mazurian group is first of all characterized by its cremations placed in perforated urns (Kulakov V.I., 1990b, s. 172).

The intertribal links between the Sudins (Yatvigians) and the Old Prussians have only been rather poorly investigated. It is possible to assert with confidence that a point of contact was situated at the open settlement of Gusev/Gumbinnen 1, which, on the evidence of items found while ploughing, was a Yatvigian trading centre, existing from the 2nd to 13th centuries AD, on the southeast border of Old Prussia. It is interesting that this site is located on a small sandy lens distinguished among clay and podsolic soils of other areas of the Gusev district of Kaliningrad. Thus the soil situation in this micro-region of central Nadravia shares close similarities with the Yatvigian territory of eastern Mazurs. It is possible that this aspect, which helped the Yatvigians conduct the traditional economy of their native land, involved them with the northern intertribal area, and provided a base for establishing Yatvigians-Sambian contacts.

Chapter III. Rites and ethnicity of the populations of the Amber Lands in the Roman epoch

III.1. The problem of the continuation of burial rites in the southeast Baltic area

Tracing the ethnic identity of the inhabitants of the southeast Baltic depends on the degree of connection between the late phase of Sambian barrow culture with the antiquities of those Aestians who inhabited the Amber Lands in the first centuries AD. This problem is considered by modern archaeologists of the Balts to have been solved. Already by 1974, Jan Jaskanis, analyzing the burial rites of the western Balts in Roman times, wrote: "In the early Roman period, barrows existed in the northwest region on the Sambia Peninsula, and ... in the northeast part of the Mazurian Lakes area" (Jaskanis J., 1974, s. 71). These barrow burials are believed to date not later than the 2nd century AD (Okulicz J., 1973, s. 367), and are linked to the Amber Coast region of the early Roman era by rudiments of the traditions of the early Iron Age. It is taken for granted that such vestiges of burial rites show the continuity of certain customs of Sambian inhabitants from the second half of the 1st millennium BC and link them to the Aestians. This population did not practise urn cremations within barrows, as before. Instead they adopted the inhumation rites of the inhabitants of eastern Pomorze and the Lithuanian coast (Kaczyński M., 1987, S. 31). This opinion was central to the thinking of Prussian archaeologists from the beginning of the 20th century (Gaerte W., 1929, S. 147), and is still current in European archaeology (Nowakowski W., 1996, S. 62). This axiomatic concept is not accepted by Łucia Okulicz-Kozarin, who is not convinced of the genetic connection between western-Balt barrow culture and the Sambian antiquities of early Roman times, although she agrees to a Baltic ethnic provenance for those local antiquities dating to the early Iron Age (Okulicz Ł., 1970, s. 155, 156). The evidence provided by the east Prussian excavations of the end of the 19th/beginning of the 20th century presents this problem in a new light(Kulakov V.I., 2000c, s. 372-389).

III.2. A new look at Aestian "barrows"

As the field research data shows, the burial mounds known from the Amber Coast in early Roman times are no more than 1.5m high and contain separate cremations, with or without urns, at bedrock level, or have corpses placed in grave pits that were excavated in antiquity from the upper level of the bedrock (Jaskanis J., 1974, s. 71). When plotting such objects maps of the Amber Lands (fig. 7, B) it is noticeable that, in contrast to the cemeteries of western-Balt barrow culture (fig. 7, A), they are only found in certain Sambian locations that are known as centres of a barrow tradition dating back to the Bronze Age. Are those burials noted on the map by Jaskanis (fig. 7, B) as grave mounds of the early Roman period really examples barrows? The catalogue of these grave finds, numbered according to Jaskanis' list (Jaskanis J., 1974, s. 277) and submitted here as Appendix 1, will help to answer this question.

III.3. Aestian "barrows" - an historiographic mirage

A review of the information collected on Aestian "barrows" of early Roman date allows us to draw the following conclusions:

A. There is no fundamental connection, in terms of burial rites, between the grave mounds of the western-Balt barrow culture and the Aestian "barrows" (Jaskanis) of the 1st century AD. In recent times, the difference between them becomes even more striking. As a result of erosion, a significant number of Sambian flat graves with multi-layered stone coverings have lost their former "barrow-shaped" appearance. For example, in 1872 Devitts described the "Gora Velikanov" cemetery in the Zelenogradsk district (wrongly marked in Jaskanis' list as Nr 350 - ZARECHJE") thus: "... stone wreath-shaped coverings 1m high, the loose surfaces of which were covered with grass. The burial urns (Grebieten-type) were decorated around the edges with finger impressions..." (Grunert W., 1944, S. 23). The excavations of the Baltic expedition (led by the Russian Academy of Science Archaeological Institute at Gora Velicanov) found burials from the late Roman era, overlaid in former times by substantial stone coverings that have more or less completely disappeared (Kulakov V.I., 1994, s. 63-65). Similarly styled "barrow-shaped" stone coverings, frequently with a matching concentric arrangement of stones, and with stelae in the centre of the coverings, are characteristic of the Wielbark culture of the Cecele phase, and are distributed (through stage B_2/C_1) from the Wielbark area over the Paslenka River to the east, as far as the Aestian territories (Bohnsack D., 1940, S. 23-25).

Such grave constructions are designated by Wołągiewicz, in terms of the Wielbark antiquities, as type 3, and he considers the initial area of their distribution to be only in Sweden and the island of Gotland, from where, in the 2nd century BC, the Goths introduced this style of burial construction to the territories around the mouth of the Vistula (Wołągiewicz R., 1986, s. 67). A Gothic origin for type 3 coverings in this territory is also feasible in terms of the 3rd century AD (Wołągiewicz R., 1986, s. 68). It is important to note that around 150 AD in the eastern part of Pomorze, barrows of a south Scandinavian shape began to appear (K. Walenta, type III - a stone mound 0.6 to 1.2m high, overlapping a circle of stones, in the centre of which the corpse was placed in a wooden coffin). This style indicates the next wave of Gothic migrants, moving into the territories bordering on the west with the Amber Lands (Walenta K., 1980-1981, 52, 94). This sort of new ethnic impulse might well have given rise to the migration of those tribes on the boundaries of Oikumene, and, as a result, to the creation of what is known as "Germanaric power". On the western boundaries of the Baltic, the new wave of Gothic migration was expressed first of all in the distribution of graves with weapons as discovered in the Wielbark antiquities, and, later, those of Chernyakhov; it is evidence that warriors from the Przeworsk area (at that time including the Masovia and Podlyasje territories) might have participated in the movement of the tribes of these cultures (Godłowski K., 1986, S. 146). All the above-mentioned nuances of the Wielbarsk culture from the boundary of the Lubovidz and Cecele stages (end of 2nd/beginning of 3rd century AD) are characteristic, as follows from the catalogue submitted above, as they are for Aestian "barrows". It is possible to assume that groups of the east German population, or newcomers from Scandinavia (most likely separate groups of warriors), crossing the Paslenka River, settled among the Aestians from the end of the 2nd century AD, and buried their dead in local cemeteries. Stone coverings over concentric structures, stelea, and the presence of weaponry and accessory vessels (with an S-shaped section) placed in the graves became distinctive features of the newcomers' burial practices. The accessory vessels show that part of the Przeworsk population (including the women, who were traditionally pottery-makers in clan societies) participated in these micro-migrations. In some cases bellicose newcomers buried their dead brothers-in-arms in earlier barrows as insert/secondary burials (for example, as at the cemetery of Chrustal'noye). By so doing, the participants in these burial rites could emphasize the closeness of their spiritual culture to the idea of 'under-barrow' burial. True, secondary burials in barrows from the Bronze Age, or the early Iron Age, were performed by the population of the Amber Coast at the end of the 1st millennium BC (Tischler O., 1886, S. 169), as well as directly before the arrival of the eastern Goths (Peiser F.E., 1919, S. 313). Instances of secondary phase B burials (inhumations and cremations without urns, under which, in five cases, horse burials were revealed - Engel C., 1931, S. 49, Abb. 3) are found in a barrow at the Lunino cemetery (Gvardeysk), attributed by Engel to western-Balt barrow culture, type VI mounds (Engel C., 1962, S. 39, 44).

B. There are no connections with the material from the barrows of the early Iron Age from those grave goods found in Sambian flat-grave cemeteries from the 1st to 3rd centuries AD. Nevertheless, J. Okulicz, in his review of the antiquities of the western Balts, wrote about the absence of cultural and ethnic changes in the southeast Baltic from the 1st century BC to the 1st century AD (Okulicz J., 1973, s. 353). Actually, significant differences in the antiquities associated with the barrow culture of the western Balts are already appreciable in the Amber Lands around 50 AD. Pottery vessels (including "Wiekau-type" mugs), weaponry, horse equipment (especially Vimose-type harnesses with copper-alloy plate nose-pieces and chain bridles), not to mention provincial Roman brooches, all demonstrate the close connection of the inhabitants of Sambia - at first with the population of the Przeworsk area, and, from the 2nd/beginning of the 3rd century AD, with representatives of the German tribes. Waves of their migrations substantially determined new ways in which the local material cultures development. However, modern scientific opinion has established that the presence of Wielbark innovations in B_2/C_1-stage Aestian territory was the result of a network of trading contacts along the Amber Route, the Vistula, and the Mazurian Lakes (Nowakowski W., 1996, S. 99). The material from the Chrustal'noye and Isobil'noye cemeteries, and other "barrows" of the Amber Coast, allows us to assume with confidence the presence there of German ("Gothic") population groups. The same eastern German sites on the ethnic map of the outlying districts of Barbaricum are explained within the framework of a new view of the Ojum problem. Toporov has fairly supposed that this legendary country, mentioned by Jordanes, reflected in the legends of the eastern Goths a range of separate micro-regions through which they were distributed among the non-German tribes along the route from the mouth of the Vistula to Tavria (Toporov V.N., 1984, s. 140). According to Jordanes (Gethica, I6 27: "This place ... is surrounded by an abyss, sealed by unstable bogs"), the description of "the desired country" of Ojum (in Gothic, Aujūm - "the country abounding with water" - Anfert'ev A.N., 1991, s. 116, 117) quite corresponds to the Sambia Peninsula surrounded by sea and lagoons and with its impassable high valleys of the rivers Pregolya and Deima. The significant number of traces of long-term residences of the Wielbark groups supports the assumption stated above that Sambia could have been the setting for the land of Oujum.

It was most likely that the Amber Coast's natural resources proved highly attractive to newcomers of different ethnic origins. In turn, the former inhabitants of Sambia – the western-Balt barrow culture - were obliged to resettle to the southeast and northeast of their ancestral homes, judging by the stone barrows found around the Mazurian Lakes and those cemeteries featuring stone wreaths on the Lithuanian shores (Tautavičius A., 1987, p. 104). Later in these regions, there are large tribal groupings of western Balts - the Sudins-Yatvigians and Curonians. Taking into account the facts mentioned above, and that there are no genetic links between the Sambian antiquities of the early Iron Age and early Roman periods, it is possible to assume that the western Balts from the 1st-2nd centuries AD did not constitute an ethnic majority on the Amber Coast. It would have been more favourable for the tribal consolidation of this new community if its territories were less populous than the Amber Coast, which from the time of Tacitus became an important crossroads for various populations. However, some of the indigenous population nevertheless remained on the native lands of their ancestors in the 1st century AD. One of the pottery burial forms of the Aestians (accessory vessels of biconical shape decorated with zigzag ornaments and fitted with pseudo-handles) confirms this fact. True, in the Bronze and early Iron Age such a form was characteristic not only of the culture of the western-Balt barrow population but also of many other tribes along the Baltic coast. In addition, it is necessary to regard the custom of placing several urn cremations, and even inhumations, under a single stone covering, as found sporadically at some Sambian cemeteries (Grebieten, Dollkeim, etc.), and in the Angrapa/Angerrap basin (Grunajki, Boczwynki), as a rudimentary feature of collective (family) under-barrow graves from the early Iron Age. Nevertheless, it is possible to doubt the validity of the axiomatic epithet, "the Golden Age of the Balts", when used for the antiquities of the Amber Coast from Roman times (Gimbutas M., 1983, S. 125). In effect, the 1st century AD became an era of astonishing development in the growth of Sambian culture, based on the phenomenal 'profits' of the amber trade. More tellingly, the light reflected from this "Golden Age" fell not only on the Balt populations, but even more so on the Germans.

III.4. Interpreting changes in rites of the Aestian area in the first centuries AD

Investigation of the real "barrow" situation in Sambia from Roman times, allows us to suppose that amber finds in the 1st century AD, in the period of intensive ethnic migrations in Barbaricum (Shchukin M.B., 1994, s. 190), involved the Sambian groups with eastern Germans (Wielbark culture). The Amber Coast became for newcomers "the desired country", Ojum. Ever since the classic works of the 'old school' of Prussian archaeology, the distribution of Sambian inhumation rites of the 1st century AD has been linked to a Gothic impulse (Gaerte W., 1932, Karte 1). This thesis is convincingly maintained by modern scientists (Nowakowski W., 1996, S. 62). Newcomers from the lower reaches of the Vistula, and, probably, some of the population of Gotland and the Danish islands, also brought to the Amber Lands the tradition of constructing coverings of concentric stone circles above their graves. Later this tradition was used by the poly-ethnic populations of the Amber Lands, including the Aestians. The Goths and Gepidae were trade mediators for the inhabitants of Sambia. This assumption is indirectly supported by the fact that in the Wielbarsk area there are no "barbarous" enamels of eastern European type (T.Stav'jarskaja), but they are found in bulk in the eastern-Balt area. In the opinion of our Polish colleagues, enamels were probably the favourite jewellery of eastern-Goth women (Bitner-Wróblewska A., 1991-1992, s. 129). As an aside, it is hardly necessary to add, as Bitner-Wroblewska does, that the "mutual hostility" of the Goths and Balts interfered with this mutual exchange. Most likely western-Balt enamels were too familiar to the inhabitants of the Wielbark area to arouse any interest as objects of direct use; they served as goods for trade with the outlying territories.

A special place in Aestian antiquities is occupied by the rich grave-goods of warriors such as those found at Chrustal'noye and Isobil'noye. Such assemblages combine different ethnic features of rites, and their splendour emphasizes the elite status of the deceased. They are characteristic of many tribal collectives of European Barbaricum, from early Roman times, and are known as "prince graves" (Gebühr H., 1996, S. 191): they are the graves of chiefs from the first "druzhinas". It is therefore necessary to recognize the important role of the poly-ethnic composition of the population of the Amber Coast in early Roman times. This specific feature, noted also in the archaeology of the later epoch, is just one of the special characteristics defining the rich history of the Amber Lands (Kulakov V.I., 2000, s. 380, 381).

IV. Typology and chronology of the antiquities of the Amber Lands in Roman times

IV.1. Dollkeim - the basis of Aestian chronology

The first chronology of Aestian antiquities was compiled by O. Tischler in 1880 on the basis of material from the cemetery at Dollkeim (today the village of Kovrovo in the Zelenogradsk district of Kaliningrad). This archaeological site is located on the northern coast of the Sambian Peninsula and is one of the most northerly points on the Vistula Amber Route in Roman times (fig. 8). Tischler proposed a Dollkeim burial chronology characterized by separate artefacts (Tischler O., 1880, S. 416). In his work the founder of Prussian archaeology constructed a series of typologies of artefacts in the process of development. Many decades later Tischler's chronological principles were constant not just for Dollkeim in particular, and Prussia as a whole, but also for all the European antiquities from the 1st millennium AD. Only in the third quarter of the 20th century were phases A-D thoroughly revised, and the Eggers-Godłowski system created; the Aestian chronology was also revised at the same time. Wojtech Nowakowski classified those Roman Sambian antiquities as being of the "Dollkeim-Kovrovo" culture (Nowakowski W., 1996, S. 48-54). Unfortunately, when creating his precise chronological scale of Aestian antiquities, by means of current computing methods (considering them as the Ouenedai of Ptolemy; Nowakowski W., 1995a, ryc. 2), Nowakowski was probably unable to sufficiently utilize the existing publications connected with the Berlin exhibition of 1880. As a result the Aestian cemetery burial assemblages (especially at Dollkeim) that were central to Nowakowski's cultural and chronological constructions were incomplete in his original 1996 publication. The results of his factorial analysis of the Aestian chronology were submitted above, and contained a number of obvious inaccuracies.

The dynamics of the development of the Dollkeim cemetery (which linked it to the Aestian culture) from the 1st to the beginning of the 6th century AD, as proposed by Nowakowski, was based mainly on the analysis of some of the artefacts found during the 1879 excavations. Unfortunately, some obvious cardinal changes in the composition of the grave goods, brought about by chronological, ethnic, and cultural factors, did not attract the attention of our Polish colleague. Nevertheless, Nowakowski considered it realistic to drop the name Dollkeim in favour of his new terminological concept, the "Dollkeim-Kovrovo" culture instead of the former "Sambian-Natangian group, Prussian culture" (the SNG) (Okulicz J., 1973, s. 366; Nowakowski W., 1996, S. 14). The abundance of the material, the limited possibilities of the methodology (above all limitations in the correlation tables), and the author's apparent lack of familiarity with post-war Sambian antiquities, prevented Nowakowski from taking into account those materials from Dollkeim and other Sambian cemeteries that contributed to the formation of Prussian culture from around 450 – 475 AD. This culture differs sharply in several respects (especially rites) from that associated with Aestian antiquities of Roman times (Kulakov V.I., 1994, s. 70). As a result of Nowakowski's work, the Dollkeim cemetery is considered nowadays as a model for Sambian antiquities from the Roman era and early phase of the great migration. The attempt by Nowakowski to revise Tischler's dating has already been mentioned above, and our Polish colleague has used a method of value correlation for certain indicators of the Dollkeim grave material (52 positions), selectively drawn from a plentiful corpus of finds from the cemetery. Correlating these indicators from 100 burials (Tischler excavated 250 graves from Dollkeimer-Berg, of which 104 have been published) the groups were plotted and interpreted by Nowakowski as a parameter of the stages of development of the cemetery from about 100 AD to the beginning of the 6th century (Nowakowski W., 1996, S. 21, 22, Tab. Ia, Ib). In all, Nowakowski considered the material from 42 Sambian cemeteries, the Pregolya River basin, and the coasts of the Vistula Lagoon, which fully characterizes the Aestian area in the Roman period. Despite the apparent completeness of Nowakowski's base tables (including the material from 335 graves), the representation of this material is doubtful. The author used either unpublished archival material from the old excavations (for example Martin Jan's collection housed in Warsaw), or the total textual information provided from the works of Almgren, Gaerte, Åberg, and other researchers; it is often impossible to attribute an item to the types and subtypes of these works.

The modern status of the material (by theme) only allows us with certainty to have a corpus of Dollkeim burial assemblages (128 graves), with figures, that can be validated by typological and chronological conclusions. In addition, material from 70 Aestian cemeteries has been incorporated in the present work. The territory included extends to the south (to the limits of the Wielbark and "Bogaczewo" cultures), and to the northeast (to the regions beyond the Neman cultures in the area of modern Lithuania). In total, 527 burial assemblages are analysed (including Dollkeim). It transpires that Nowakowski did not use data from all 138 burials from the 28 cemeteries. True, the majority of this material (except for the Dollkeim and "Gora Velikanov" cemeteries) is covered by individual assemblages or by their (not numerous) groups. This is, unfortunately, the result of various works published by the archaeologists of the Old Prussian school, which were rather short of information. All of this obliges us to base the typological and chronological analyses of the antiquities from the Amber Lands mainly on the material from the Dollkeim cemetery. The principles of revealing the chronology of burials from this site are based on the following:

1) the maximum information available on the burial rites, horse burials, and grave goods from Dollkeim, and the "complete" list of values of attributes for the assemblages excavated in 1879, 1992, 1994, 1998, and 1999;
2) to ensure the use (when creating a Dollkeim chronology) of chronological indicators – i.e. artefacts with an established conventional dating (so as to prevent forming illusory combinations (groups) of values of attributes -see Nowakowski's tables);
3) the arrangement of the assemblages with tables according to the chronological indicators of the closed assemblages, and to reveal the hierarchy of the importance of attributes for the formation of groups (from narrowly dated values up to those positions characteristic for longer intervals of time);
4) to confirm, using new information and modern excavation techniques, Tischler's practice of allocating burial assemblages. It is to be understood that in the formation of the chronological system, groups of artefacts (brooches in Tischler's system) and non-selective groups of values (Nowakowski's conceptualistic approach) will feature; but it does ensure that all possible information on Dollkeim burials will be taken into account.

IV.2. Burial rites and the Sambian-Natangian group

The burial rites of the Sambian-Natangian group have the following attributes (Jaskanis J., 1974, s. 268): a) the basic burial site is a bi-ritual, flat grave cemetery: from the end of the 2nd century the cremations completely supplanted inhumations; b) the majority of the last burials were orientated north or north-westwards; c) the cremations took place outside the burial site. They are located in simple graves, with or without the remains of the funeral pyre, and overlapped with substantial stone coverings: the quantity of urn burials far exceeds those without urns; d) the grave goods are classified by common categories and types of objects, among which the weapons (spears, battle-knives, shield-bosses) are widely distributed; e) at a number of cemeteries there are "prince graves" (warriors connected with the activities of the "Sambian ala", cf. Chapter VI) with large volumes of finds, including Roman imports.

By means of fractional analyses of the details of the rites, Jaskanis marked out the regional groups of the western-Balt culture appropriate to those areas of the later tribes, i.e. the Middle Ages: a – the Sambian-Natangian group is characterized by the distribution of burials where the horse (legs folded) was placed on its belly in a pit that would include a Grebieten-type urn containing a small amount of burnt bones; b – the east Mazurian group is represented by the Morąng and Suwałki subgroups. In the former there are no stone coverings above the cremations (Okulicz J. 1973, s. 447), while the latter is characterized by bi-ritual graves under stone and earth mounds (Okulicz J., 1973, s. 449) containing grave goods different in type from those of other groups in this culture. The Suval subgroup corresponds to the area of later Yatvigians (Kaczyński M., 1976, s. 253). The rites of the Aestians, briefly characterized in one of this author's previous works (Kulakov V.I., 1994b, s. 32) as bi-ritual flat graves, do not stand out among the corpus of other antiquities of the Germans and Balts from northern Barbaricum. From Roman times the inhumations (further - INH) within the limits of the southern part of the western-Balt area (including the SNG), are represented by 28 cemeteries (fig. 9), and a series of cremations (mainly in urns and with overlapping stone coverings and other constructions) from 85 cemeteries (Jaskanis J., 1974, s. 127, 137). The origin of inhumation rites in the Amber Lands is not clear. The appearance of this phenomenon is probably connected with Celtic and/or Roman influences (Schultze E., 1992, S. 205), as was the case in west Europe during the first centuries of our era. Unfortunately, before the publication of the material from the Isakovo and "Gora Velikanov" cemeteries, which were excavated to modern standards, our knowledge of the procedures of Aestian burial rites remains incomplete; it is based on the sketchy outlines of the publications of the archaeologists of the Old Prussian school. The elements of the external structure of Aestian burials (i.e. stone constructions) are a specific feature of the SNG. On the eponymous (according to Nowakowski) site of this group, at the flat-grave cemetery of Dollkeim, the various stone constructions are: Do-14[a], Do-28 - INH under a round-shaped stone covering on the perimeter, surrounded by a ring of stones; Do-75[a], Do-302 - cremations (further - CRM), in or on which the filling stone ("stele") remains (Malstein); and Do-310, Do-315[a], Do-316 – the "stele" is placed away from the grave, but connected with it in plan. Conventionally these features of Dollkeim burial rites have a southern Scandinavian origin and are connected with the later stages of the Roman era (see Chapter III). In the Aestian area (including at Dollkeim), stone rings (as a rule concentric ones) are found above flat graves at the cemeteries of Dubrovka (formerly Regehnen), Yaroslavskoye (formerly Schlakalken), "Gora Velikanov", Klevernoye (Gvardeysk district, formerly Drusker Forst-Espenheim, Kr. Wehlau), formerly Wengerin (Chernyakhovsk district, formerly Wengerin, Kr. Insterburg), Nova Bochvinka (Warmińsko-Mazurskoie wojewodztwo, Poland, formerly Neu-Bodschwingen, Kr. Goldap), and also on the northeast border of the Wielbarsk area in the former Pettelkau (Bagrationovsk District, formerly Pettelkau, Kr. Heiligenbeil). Stone stelae are found (as well as at Dollkeim) at the cemeteries of Kulikovo (formerly Elchdorf), Vetrovo (formerly Ekritten), Rovnoye (formerly Pollwitten), "Gora Velikanov" (Zelenogradsk district, formerly Kr. Samland), formerly Stuttehnen (Bagrationovsk district, Stuttehnen, Kr. Heiligenbeil), formerly Wengerin (Chernyakhovsk district). Stone rings are also found in the Curonian area, but they are not accompanied by finds of a non-Balt origin (Žulkus V., 1995, I pav.) and hardly serve as proof of the presence there of an alien ethnic element. But such finds are very frequent in the inhumations of the SNG and Dollkeim, especially in those graves overlapped by stone coverings and rings.

IV.3. Aestian grave goods

Aestian goods as a whole are not found within the contexts of Barbaricum material in Roman Empire times; they are presented in various typological schemes created recently by western colleagues. The schemes of O. Almgren, H. Jankuhn, and R. Madyda-Legutko, involved in the present work, are recognized by modern science and are quite current for Aestian material. They are concerned, first of all, with provincial Roman brooches, and their derivatives, as found within the burials of the SNG in the groups AI, II, II (fig. 10), group AIV (fig. 11), group AV (fig. 12), concave with tied clasps (fig. 13), and various types of crossbow-shaped brooches with cast (compact) pin-receivers (fig. 14). Brooches of the last version, going by two zones of their distribution, correspond to Sambia (the centre of the SNG) and the eastern part of the Mazurian Lakes (fig. 15) - to the area of the Sudins/Sudawians (from the time of the great migration). The Sudawian corpus of crossbow-shaped brooches (with a long foot), includes both prototypes of the Duratón-type clasps, and their late versions (especially with corrugated foot and ridge buttons). This feature of Sudawian brooches indirectly touches on the actuality of the Sambian "Balt reconquista" hypotheses, dated to the middle of the 5th century AD, and referring to the migration of a part of the Sudins/Sudawians to Sambia (see Chapter V, B). Earlier, in phases B_2/C_1-C_1, one of the few clues to the presence on the Amber Coast of separate groups of a western-Balt population are finds of cross-piece brooch derivatives of type-AIV (88 clasps). Thus, these derivatives in the area of the SNG differ appreciably from related types, but, far from being identical to the "Mazurian" brooches, are characteristic of examples from the southern zone of the western-Balt culture (Kulakov V.I., 2001a, s. 48, fig. 2).

Until now, the question of the derivation and development of brooches with "ring" elements remains unanswered (cf. fig. 16 (top)). Nowakowski, as well as Tischler and Kemke, differentiate only the large "crossbow-shaped brooches" (fig. 16, Do-70) from the corpus of clasps, dating them to 350-400 AD (Nowakowski W., 1996, S. 19). Analyses of these artefacts and related brooches monstruosa (fig. 17) are submitted in Chapter V.

Brooches with bulb-shaped tops (fig. 18), characteristic of Romanized German warriors in the early phase of the Hunnic era, are extremely rare finds in the Aestian region. On the contrary, one of the exclusive features of the historical ethnography of the Amber Coast population from Roman times are the plate disc brooches (fig. 19) and pins (fig. 20) found mainly in the upper reaches of the Pregolya River and (basically) in the eastern part of the Mazurian Lakes. Direct contacts with the Roman provincial culture of Julius-Claudius played a significant role in the formation of the corpus of plate disc brooches as an indispensable feature of female Aestian costume. The warriors of the XV Apollonian legion came to the Amber Coast from Pannonia between 51 and 63 AD. Their appearance stimulated the distribution of the phalers-type clasp in the vicinity (Koulakov V., 2000, s. 30, 33). This burst of activity along the Great Amber Route connecting Sambia to the Danube (fig. 8) was caused by this Roman expedition. It promoted the distribution among the Aestians of Pannonian belts (fig. 21) with characteristic Flavius-era ornaments of opus-interrasile style (Jankuhn H., 1933c, S. 30). In Sambia (more exactly its western part - fig. 22), as well as in Pannonia, these wide belts, which in the Empire were characteristic of legionary equipment, were worn exclusively by women and emphasized their waists. It is true that lock-hooks and buckles with fixed frames (fig. 21), with which these belts were supplied, reach back to Oksyv traditions. Buckles with moving frames appeared in the southeast Baltic as early as phase B_1, and, as a rule, do not fall outside the limits of the conventional typology of these artefacts of eastern Barbaricum (fig. 23). The same may be said of neck-rings, which, from the 1st century AD, became one of the ethnographic attributes of the costume of the inhabitants of the Amber Coast. Already by phase C_1, neck-rings (above all the types known as "mit dosenende" and "mit oesenende" - fig. 24) were common features of Balt antiquities found along the right bank of the Neman, and resulting mainly from the cultural influence from Sambia.

Some categories and types of Aestian grave goods, mostly dating to phases B_2-C_2-D_3/E, have until now remained outside the limits of conventional typologies. In particular this refers to the bracelets with semicircular heads, which our Polish colleagues call "snake-shaped". These are shown in fig. 25: their genetic development is traced horizontally; the vertical takes into account any connections with closely related types. The genesis of "snake-shaped" bracelets is linked to those with "flask-shaped" heads that are found in the lower Vistula area of the Wielbark culture (B_1, Lubovidz phase - Grabarczyk T., 1983, s. 16, 17). It is important to note these bracelets also from the Dollkeim material (Do-27b); in this cemetery these bracelets are presented in ring form (a temple or ear-ring perhaps) with a reduced diameter and widening at one end only (Do-11b). They seem strongly related to bracelets with "flask-shaped" tops from the Wielbark antiquities of the Lubovidz phase, and from Dollkeim graves, and to the "armilla-type" Roman military decoration bracelets. These bracelets were indispensable accessories for early Empire legionaries. The bulges on the bracelet ends helped fix them to a leather belt so that they could be worn as decorations on Legionary armour. It seems to indicate that only those warriors with Roman citizenship were individually decorated with bracelets. Federation warriors might receive such decorations only as a collective encouragement to their military division (Kolobov A.V., 1999, s. 67). The fact that such bracelets (and other types) belong only to female costumes from Aestian antiquities, and in the Wielbarsk area, is remarkable. In modern Polish archaeological science, there is an opinion that phase B_1 and later armillae from the Wielbarsk area influenced the formation of Przeworsk bracelets of the Kamieńczyk type, and early forms of the Wielbark "snake-shaped" bracelet (Andrzejowski J., 1994, s. 320). The latter is considered a Sambian phase B_2 derivative (and one of the indicators of the development of the Amber Route from Sambia to the Upper Vistula)(Andrzejowski J., 1994, s. 322). In Aestian antiquities (especially at Dollkeim), there would seem to be other influences at work in the evolution of "snake-shaped" bracelets. Examples of the Kamieńczyk type are really derivatives most closely related to armillae. Bracelets from Do-6 and Do-14^b differ from the Kamieńczyk type only in the doubling of the hemispherical ends to the rods. The other attributes (including the triangular rod sections) remain the same. However the bracelets that Andrzejowski counts as Sambian derivatives of "snake-shaped" bracelets are actually the next development of the Sambian variant of the Kamieńczyk-type bracelet. In recognizing this type it is necessary to understand the style of the Wielbarsk "snake-shaped" bracelets that continues (down to its degradation) in a line of Sambian variants of the Kamieńczyk type. It is impossible to explain in any other way the presence in all bracelet types from Pomorze and the lower Vistula of the double division of the "snake head" (Grabarczyk T., 1983, tabl. IX). This division is absent only from the bracelets of the Polab region (Andrzejowski J., 1994, ryc. 11). It is an indicative feature of the spontaneous occurrence of the Sambian variant of the Kamieńczyk-type bracelet, where versions are only encountered in full in the Dollkeim material (and coincide with small armilla rings).

Among Aestian weaponry, spears, battle knives (25-50 cm long; fig. 26) and socketed axes (fig. 27) are the most common finds. On the basis of such Aestian assemblages from Roman times, Nowakowski comes to the conclusion that the basic weapons of the Amber Coast inhabitants in the 1st century AD were axes with lugs, mentioned by Tacitus as "fustis" - "cudgels" (see Chapter V). From this Nowakowski affirms that there was a difference between the tactics of "non-German" Aestians and those of neighbouring tribes, and especially the Lugii living in the Przeworsk area. However, axes with lugs in the Roman period were mainly characteristic of those inhabitants of the western part of the Mazurian Lakes (Nowakowski W., 1994b, S. 391); in Sambia these axes are practically unknown in Roman times. In the cemeteries of western Sambia only socketed axes have been found (fig. 28) and they are characteristic of Aestian antiquities (Kulakov V.I., 2001, S. 50). As for burials, the artefacts from Ringgarnitur-Stil include battle knives with wide edges, some 30 cm in length.

Thus, differing significantly from the traditional Celtic battle principles of the Lugii horsemen and swordsmen, using the tactics of lightly armed spearmen, the Aestians corresponded to the German military traditions of phases C_1-C_2 (Raddatz K., 1976, S. 425-429), though with the added specificity of the use of the archaic socketed axe. It is necessary to note that, for Barbaricum antiquities from the 1st century AD, the densest congestion of burials with weapons is found in Sambia, and thus strikingly distinguishes this region from the lower reaches of the Vistula, i.e. the Wielbark area of the Lubovidz stage (fig. 28, 29). This phenomenon in Sambian antiquities shows the influence that the Przeworsk culture might have had on this part of the south-eastern Baltic at the time. This hypothesis is confirmed by the distribution among the Aestians of shield-bosses (fig. 30) and spurs (fig. 31) of types indicative of Przeworsk antiquities from the 2nd-4th centuries AD.

More rarely investigated is the Aestian pottery that is represented in graves by urns (fig. 32) and accessory vessels (fig. 33). In the earlier German archaeological literature, only large "bucket-shaped" urns (Eimerurnen) were known as vessels of Grebieten-type. In spite of the fact that, up to the present, the study of Aestian pottery of Roman times is difficult as it is represented mainly by figures, Nowakowski attempted to reveal separate types of artefacts of interest. He distinguished the earliest type of Aestian burial pottery – the biconical vessel with multi-piece handle that dated back to group IV urns of the late phase of the western-Balt barrow culture (Okulicz Ł., 1970, s. 40, tabl. XXI, 19,22). A generic connection with vessels of Do-31 type (fig. 32) with those "mugs" of "Wiekau Type" (with characteristic spherical body and symmetrical neck and foot) is clear. There is some connection between these "mugs" (actually accessory vessels) and the forms of Wielbarsk urns of type RW VIIIC (Nowakowski W., 1996, S. 59), and pots with handles of type RW XVA (Do-36). True, in contrast to Wielbarsk pottery, practically all similar Sambian vessels have handles with top edges that do not join the rims of the vessels, but fasten to the neck. In Aestian graves such accessory vessels accompany those urns with attributes of Wielbarsk pottery of types RW IVB, RW IC (Kumpf) and RW IB (fig. 32). The vessels of RW IVB type arouse special interest. In the Wielbarsk antiquities of the Lubovidz phases, the zigzag decoration on the "shoulder" separates the rough lower part of the vessel from the polished upper section. On Aestian vessels of the same type there are shaded figures instead of rough "triangles" (fig. 32, Do-31). This was interpreted by Gaerte as a representation of the magical German symbol for rain clouds (Gaerte W., 1939, S. 44).

In dealing with the cardinal changes of pottery types in the "Dollkeim-Kovrovo culture" between phases B_2/C_1, Nowakowski refers to the "Dollkeim-type" vessel as predominating in this era. Its body appears as a flat sphere with extended "neck" (fig. 33, Grei-19a). This form is known among the Wielbarsk antiquities of

"Cecele" phase as type RW XVII (differing from the "Wiekau-type mug" by the absence of foot and handle), on which the decoration is represented by bands of "double crosses". The occurrence of Amber Coast pottery with this decoration, which has no direct analogue in Balt antiquities (although such decoration on a vessel from Egliškiai is an exception among the Balt antiquities of the early Iron Age - Steponaitis V., 2000, p. 63), as well as examples from Wielbark and Przeworsk, is connected with the migration of some of the southwest Scandinavian inhabitants (most likely from the Jutland peninsula) to Sambia around 180-200 AD (Kulakov V.I., 1998a, s. 96). In particular, "Dollkeim-type vessels" are comparable to pottery of type 1c2 (phases C_{1b}-C_2) from the Hjemsted-2 cemetery in Southern Denmark (Ethelberg P., 1990, p. 78), which, in the 2nd century AD, was in the southern part of the Cimbri area, not far from the northern border of the tribal territory of the Saxons. Both in Scandinavia and at Dollkeim, such accessory vessels soon adopted a sharp rib on the lower part of the body (fig. 33, Do-28; analogue - Vedel E., 1886, fig. 305). They are known within the limits of the Baltic only in western Sambia (fig. 34). Simultaneously with vessels of RW XVII type, there are biconical urns with horizontal decoration fixed to the "shoulder", analogies of which are known in the same area of southwest Scandinavia from late Roman times (Vedel E., 1886, fig. 248). Urns with fixed decoration do not differ from other forms of urns of Grebieten type with poorly fired grey and ochre fabric; on the whole they have certain attributes of those receptacles that contained ashes of Scandinavians from the 1st century AD (Nerman B., 1934, Abb. 7,8). In the phase-C-D cemeteries of Sambia, and Cecele-phase cemeteries of Wielbark (Bezzenberger A., 1909, S. 109), and earlier in southern Sweden from the 1st century BC/1st century AD, urns were found as accessory vessels among the burnt bones (Oxenstierna E., 1945, S. 103, 104). Accessory vessels of subtype 2.3 (Kulakov V.I., 1994, fig. 22), and biconical vessels of subtypes 1.2 and 1.3, are the latest pottery forms in SNG cemeteries at Dollkeim (where cremations without urns are already known). These biconical vessels have convincing archetypes in the south Scandinavian material from the 3rd to 5th centuries.

So, except for the vessels with multi-piece handles, which are rare for Aestian antiquities, the B_1-B_2/C_1-phase pottery material of the Amber Lands is connected to the Wielbarsk traditions of the Lubovidz phase (including the spheroid-shaped classical vessels found among "Goths-Gepidae" burials - (fig. 33, Gru-VII)); and, later, up to the final phase of the great migration, it was connected with the pottery of the southwest and southern extremities of the Scandinavian peninsula. The principle of the arrangement of sites where analogues to the Aestian pottery from north European regions are found, is very interesting: from the centre of Jutland, across the islands Fyns, Zeland and Bornholm, and including the southern part of Scandinavia. These locations correspond to the most convenient waterway from the western coast of the Baltic along the eastern branch of the Gulf Stream (Golfstrom) to the Amber Coast.

As commented on above, one of the few cemeteries of the southeast Baltic, with complete information on excavated burials from Roman times to the early Middle Ages, is Dollkeim, situated in the centre of the historical ancestral land of the western Balts, i.e. Sambia. Analysis of as yet unstudied aspects of the typology of Dollkeim assemblages, as well as the conventional typological systems, enable the classification of the various descriptive attributes of the assemblages of the site investigated. The general number of these values is 153, and, for the convenience of presentation, the material is divided into chronological sequences.

IV.4. Descriptive attributes of Dollkeim burial assemblages

Value sequence for phases B_1-C_3:

Burial rites
1 - INH; 2 - INH, overlapping stones with a further ring of stones around the perimeter; 3. Subtype 1.1 (urn CRM); 4. Subtype 1.2 (CRM as group of bones); 5. Subtype 2.1 (CRM as bones, dispersed in "RFP" (remains of funeral pyre)); 6 - horse burial to the west of CRM; 7 - horse burial at the bottom of the grave, under CRM.
Grave goods (jewellery)
Brooches: 8 - AII, 36; 9 - AII, 42; 10 - AIII, 57; 11 - AIII, 58; 12 - AIII, 59; 13 - AIII, 60; 14 - AIII, 61; 15 - AIII, 62; 16 - AIII, 63; 17 - AIV, 72; 18 - AIV, 73, 19 - AIV, 74; 20 - AVI, 162; 21 - AVI, 164; 22 - AVI, 167; 23 - AVI, 169; 24 - AVI, 161 (MV), 25 - "large crossbow-shaped brooch" - var. Do-70 (derivative AVI, 168); 26 - var. Do-85; *buckles*: 27 - Madyda-Legutko (ML) D-17; 28 - ML D-20; 29 - ML D-22; 30 - ML D-29; 31 - ML E-14; 32 - ML F-2; 33 - ML G-40; 34 - ML H-18; *belt sets*: 35 - with hook clasp; 36 - "Pannonian"; 37 - cinculum; *strap-ends*: 38 - prototype Raddatz J, O-1; 39 - Raddatz J, O-1; 40 - Raddatz JII, 1 41 - *decoration (falera)*; 42 - *diadem*; 43 – *neck-rings* with funnel-shaped ends; *bracelets*: 44 - armilla; 45 - with hemispherical flat heads; 46 - iron *bucket-shaped pendant*; 47- *iron bell-pendant*; *beads*: 48 - amber Pauckenperlen; 49 - amber discoid; 50 - glass; 51 - glass of Tempelmann type 1.67; 52 - "enamel" (paste); 53 - mosaic; 54 - gold-glass;
Grave goods (weapons): 55 - battle-knife; 56 - a spearhead, of unclassifiable type; 57 - spearhead Skiaker 14; 58 - pilum; 59 - lancet-like javelin; 60 - socketed axe; 61 - iron "edge" with square section (arrow or awl?); *shield-bosses*: 62 - Vermand type; 63 - Zieling X type; 64 - Krgapol'vzew-Bazhan 10 type; 65 - Chorula type; *spurs*: 66 - Ginalski type B2; 67 - Szwajcaria type; 68 - Malaeshty type; 69 - Chorula type;
Horse equipment: 70 - curry-comb; 71 - a detail of Vimose headpiece (Proto-Vimose);
Tools and everyday objects: 72 - scythe; 73 - sickle; 74 - shears; 75 - "small knife"; 76 - knife with blade, decorated by sickle-shaped stamp; 77 - pincers; 78 -

spindle whorl; 79 - Roman bronze coin; 80 - razor; 81 - whetstone; 82 - piece of flint; 83 - piece of amber;
Pottery vessels: 84 - RW IB type; 85 - RW IC; 86 - RW IVB type; 87 - RW XVB type; 88 - RW XVIIIB type; 89 - Wiekau type; 90 - Do-102 type; 91 - Do-43 type; 92 - Do-46[a] type; 93 - Do-45 type; 94 - Do-306a = RW XVII type; 95 - Do-306b = subtype 2.3; 96 - Do-41 type.

Value sequence for phases C_2/D_1-D_3/E_1:

Burial rites (values correspond to the database for phases B1-C3)
Grave goods (jewellery) (hereinafter the values of the attributes already marked in the database for phases B_1-C_3 are accompanied by +)
Brooches: +22 - AVI, 167; +25 - "large crossbow-shaped brooch" - Do/70 (derivative AVI, 168); +26 - Do/85; 1 - AIV, 72; 2 - AV, 130; 3 - AVI, 163; 4 - AVI, 168; 5 - Schulze-Dörrlamm Iz Aa 6b; 6 - Schulze-Dörrlamm Iz Af 2c; 7 - Schulze-Dörrlamm Ix AF 1a; 8 - Schulze-Dörrlamm Ix Ca 4b; 9 - Kiew; 10 - Duratón; 11 - Schönwarling; 12 - Bitner-Wróblewska II; 13 - Breitenfurt, 14 - crossbow-shaped brooches of type 1; 15 - Vil'kantsy; *buckles*: 16 - ML H-3, 17 - ML H-14; 18 - ML H-16; 19 - ML H-29; 20 - ML H-38; 21 - ML H-43; copper alloy *bucket-shaped pendant*; 22 - pendant, definition of type is impossible; 23 - "long" pendant of E31 type; +47 - iron *bell pendant*; beads: +48 - amber of Pauckenperlen type; +49 - amber discoid; 24 - amber beads, definition of type is impossible; 25 - amber cylindrical beads; 26 - amber biconocal beads; 27 - amber washer-shaped beads, diameter is not less than 3 cm; +50 - glass; +53 - mosaic; 28 - glass beads of cube-octahedron shape; 29 - copper alloy beads; 30 – *neck-ring* from smooth or corrugated copper alloy rod with a clasp as a hook and a loop; *rings*: 31 - braided; 32 - with the interlaced ends; 33 - with the widened ends; 34 - spiral; strap-ends: 35 - tongue-shaped narrow; 36 - tongue-shaped wide;
Grave goods (*weapons*): +55 - battle-knife; 37 - knife-dagger; +56 - spearhead, determination of type is impossible; +58 - pilum (angon); +59 - lancet-shaped javelin; +61 - iron "edge" with square section (arrow?); *shield-bosses*: 38 - impossible to determine a type (frequently - fragments of boss); 39 - Dobrodzén; 40 - spur of var. 1.2.1;
Horse equipment: +70 – curry comb; 41 - bit with rings and two-piece snaffle;
Tools and everyday objects: +72 - scythe; +75 - "small knife"; +76 - knife with a blade decorated with sickle-shaped stamp; +77 - pincets; +78 - spindle whorl; +79 - Roman bronze coin; +80 - razor; +81 - whetstone;
Pottery vessels: +87 - RW XVB type; +90 - Do/102 type; +91 - Do/43 type; +92 - Do/46[a] type; +93 - Do/45 type; +94 - Do/306a = RW XVII type; +95 - Do/306b type = 2.3 subtype; +96 - Do/41 type; 42 - 2.3 subtype = RW XVIID type; 43 - RW XVIIIC; 44 - 1.1 subtype = Do/107; 45 - 1.2 subtype = Do/119; 46 - 3 type = RW XI.

Value sequence for phases D_3/E_1-F_1:

Burial rites (values correspond to the database for phases B_1-C_3)

Grave goods (jewellery) (hereinafter the values of the attributes already marked in databases for phases B_1-C_3 are accompanied by ++, in databases for phases C_2/D_1-D_3/E_1 - by +): *brooches*: +5 - Schulze-Dörrlamm Iz Aa 6b; 8 - Schulze-Dörrlamm Iz Aa 1a; 9 - Schulze-Dörrlamm Iz Aa 6c; +10 - Duratón; +13 - Breitenfurt, +14 - type 1; 10 - crossbow-shaped brooches of var. 2б; 11 - crossbow-shaped brooches of type 3; 12 - crossbow-shaped brooches of type 4; 13 - Neuwied; 14 - a derivative of animal-headed knob brooches of group 1; *buckles*: +16 - ML H-3, +17 - ML H-14; 15 - var. 1.2a; 16 - type 2; 16a - lancet-shaped strap-end; *amber beads*: +24 - determination of type is impossible; +25 - cylindrical; +26 - biconical; +27 - washer-shaped, diameter is not less than 2 cm; 16b - broken; +30 – neck-ring, as a rule, from twisted copper alloy rod with a clasp as a hook and a loop; 17 - a silver *capsule*;
Grave goods (*weapons*): ++ 55 - battle-knife; +37 - knife-dagger; ++ 56 - spearhead, determination is impossible; ++ 59 - lancet-shaped javelin; +40 - *spur* of var. 1.2.1;
Horse equipment: ++ 70 - carry-comb; +41 - bit with rings and two-piece snaffle;
Tools and everyday objects: ++ 75 - "small knife"; +78 - spindle whorl; 18 - piece of amber.
Pottery vessels: ++ 90 - Do-102 type; +42 - 2.3 subtype = RW XVIID; +44 - 1.1 subtype = Do-107; +45 - 1.2 subtype = Do-119; 19 - 1.3 subtype; 20 - 2.3 subtype.

IV.5. Chronology of the Sambian-Natangian group

The tables showing groups of associations are organized according to chronological indicators. They are represented in the typological tables by 84-105 recently dated objects-taxons, which are conventional within the framework of the stages determined in 1880 by Tischler from some material of this cemetery. The tables reveal three basic methods of burials that differ in their rites. Inhumations are characteristic for phase B_1 to the early part of phase B_2/C_1 (quite often the inhumation was overlapped with a stone covering and stone circle); for the late part of phase B_2/C_1, to phase D_2, urn cremations are indicative, and the material from the first half of phases D_3-F_1-F_2 reveal cremations in the form of charred bones. Already this general overview shows the legitimacy of the methods used in this present work. These methods show the dynamic character of the development of the Dollkeim monuments, accompanied by consecutive changes in what is the most conservative area of the archaeological culture - burial rites. Such striking shifts in the spiritual beliefs of the Amber Coast population, from the 1st,to the beginning of the 8th centuries, raise doubts on Nowakowski's theories of the existence, over eight centuries, in the Amber Lands of a united "Dollkeim-Kovrovo culture". As local antiquities of the great migration period (see Chapter VII) show, around the beginning of the second half of the 5th century AD, a Prussian culture appears within the Sambia peninsula and its environs. (The connection of this nomenclature with the most recent ethnicity of "Prussian" is conditional.) The Prussian culture is characterized first of all by the discontinuation of cremation urns (Kulakov V.I., 1994, s. 14).

IV.6. Rites as an indicator of Austeravia migrations

The earliest evidence for phase B_1 Dollkeim - an early part of phase B_2/C_1 (in effect, the beginning of the functioning of the SNG) - as a whole, is connected mainly with rites of inhumation. Jaskanis connects the appearance of this feature of rites on the western boundaries of the Balt world with the influence of the neighbours of the southern Balts – the Wielbark-culture population - although he considers inhumations in the Balt environment a territorially limited episode from the Roman epoch (Jaskanis J., 1974, s. 214, 215). It should be remembered that Wielbark antiquities are characterized not by a prevalence of inhumations, but by bi-rituality. A map of those cemeteries with Cecele-phase inhumations, which, according to Wołągiewicz, is one of the main indicators of Wielbarsk culture, shows an obvious "un-episodical" number of such sites in Sambia and its environs - numbering some 35 (Gaerte W., 1928, Karte). In addition, their quantity and density of arrangement along the Amber Coast is obviously higher than the parameters of the lower reaches of the Vistula. Inhumations marked for the southeast Baltic on the Gaerte-Jaskanis map (fig. 9) are not differentiated according to burial rites. From the sketchy data of the old excavations, stone coverings above the 1st-century inhumations AD are known in the Wielbark area (particularly in its "classical" western part - Kokowski A., 1995, Abb. 15), as well as in Sambia. Skeletons in both mentioned regions are orientated northwards (where such information is available). As Wołągiewicz points out, the Wielbarsk inhumations from the B_1 phase seem to have been influenced by Celtic and Roman customs that were introduced, during the era of Maroboduus, to the lower reaches of the Vistula by the inhabitants of Pannonia (Wołągiewicz R., 1981a, s. 151). Shchukin is of the same opinion, and already with relation to Sambia, and considers that there may well have been penetrations of groups of Celtic-Roman traders and craftsmen and women from the Middle Danube area, driven by the possibilities of commercial benefit, to the Amber Coast in Nero's time (Shchukin M.B., 1994, s. 226). These hypotheses deserve closer attention because neither the Scandinavians, who occupied the future Wielbarsk region until the middle of the 1st century AD (Wołągiewicz R., 1981b, s. 88), nor the people of the western-Balt barrow-culture in Sambia, employed inhumation as the foremost burial rite in pre-Roman times. These rites were spontaneously distributed in the first half/middle of the 1st century AD along two sections of the southeast Baltic coast, which were the most convenient for settling, the Sambian peninsula and Elbląg heights, called by Cassiodorus and Jordanes "the Gepedojos Island". In Sambia these rites were used by the population, which, during the Cecele phase, exceeded in number the population of the Vistula delta. To a considerable measure this depended on the geography of the highest section of the Sambian peninsula; at its western part, the region most suitable for settlement, it exceeds the area of "the Gepedojos Island" by a factor of six. Along the Amber Coast, distribution of flat graves with inhumations is not a logical development of the burial rites of the western-Balt barrow-culture. These antiquities, whose associations with the western-Balt culture of Łucia Okulicz were uncertain (Okulicz Ł., 1970, s. 155, 156), suddenly disappeared from Sambia in the 1st century AD, and are found in Roman times only in the outlying districts of the western-Balt territory - in the Mazurian Lakes and the Curonian areas. True, in the Lower Neman, Sambian inhumations were distributed simultaneously, however they did not feature stone constructions (Nowakowski W., 1997, p. 100). The only indication by Nowakowski of the link between the western-Balt barrow tradition and flat graves with inhumations, was the detection of a brooch of type AIV, 72 in Barrow 5 at the Klevernoye cemetery (former Drusker Forst-Espenheim, Kr. Wehlau) (Nowakowski W., 1996, S. 94): Actually, a phase-B_1 burial was placed into this late Bronze Age barrow (see Chapter III).

Urn cremations, which gradually spread into the Dollkeim and other Sambian cemeteries, from phase B_2 to the boundary of phases B_2/C_1 and C_1, at length replaced inhumations. From this time in the Wielbark area, the percentage of inhumations is also appreciably reduced. At sites created by the "Goths-Gepidae" in their movement southeast, started about 200 AD, inhumations were often completely absent (Kuharenko J.V., 1980, s. 24). Despite lacking complete data on Cecele-phase Wielbark rites, it is possible to speak confidently about the growth in number of urn cremations among the Aestians' southwest neighbours at the turn of the 2nd-3rd centuries.

Complex Do-7k is of special interest regarding Dollkeim burial rites of phase C_2. In effect this burial is a variant of the burial mound of type Walenta III. This type of southern Scandinavian burial construction appears about 150 AD in eastern Pomorze, marking the occurrence on the continent of a new wave of Scandinavian migrants (Walenta K. 1980/81, s. 52, 94). The arrival of these newcomers in the southeast Baltic regions, as already mentioned in Chapter III, brought about the replacement of the Lubovidz phase of the Wielbarsk culture by the Cecele phase. To this area of Aestian martial, the northerners brought new customs including urn cremations, horse burials (or rather sacrifices), and the placement of weapons into the graves. They also distributed the tradition of constructing concentric stone circles and stelae above the graves. As a result of the distribution of these innovations there are no traces of 1st century AD rites at Dollkeim until the middle of phase C_1.

A sharp change from urn cremations to burials without urns of subtype 1.2 in Sambian cemeteries marks the beginning of the formation (about 450-475 AD) of the 'Prussian' culture, the peoples of which began to be known as 'Prussians' not later than the 9th century. In phase E_1 the burial rites of the subtype 1.2 are gradually replaced by the strewing of burnt bones among the remains of the funeral pyre, a common feature of burial rites in Prussian cemeteries from the 6th century.

In contrast to other Sambian cemeteries, there was a non-standard parctice with horse burials at Dollkeim, which was generally unnoticed by Tischler's excavations with rare exceptions (in Do-15, Do-16, and Do-17 horse equipment was found), probably as a result of the field excavation methods used in the second half of the 19th century. Post-war excavations at Dollkeim have revealed that there were horse burials to the west of the horsemen's graves even in the 6th century, although in the field they are not traced without certain difficulties.

IV.7. The chronology of assemblages from cemeteries of the Amber Lands in the Roman epoch

There are three trends in the composition of burials at Dollkeim, marked by the distribution of the respective attributes of their burial rites. These trends progressively replace each other. The distribution of grave goods, reflected in the base table, forms a more diffuse system which is particularly visible in the table where the values of the attributes are naturally organized in groups with obvious chronological factors. Each of these groups is determined by combinations of values that are uniquely characteristic and do not require additional correlation. Similar work carried out with the burials of the "Gora Velikanov" cemetery (the southern suburb of Pionersky, north of Sambia - Kulakov V.I., 1997d, fig. 1) for phases D_1-E, confirms the actuality of the chronology offered for Aestian antiquities. The groups are as follows:

Group of burials for phases B_1-B_2 (fig. 35): mainly inhumations, especially for phase B_1. Male burials are characterized by the inclusion in the grave goods of a single brooch, knife, belt buckle, and less often the shield-boss. The grave goods of Do-7 and Do-30 attract attention (as do the artefacts from unnumbered burials (the former Ilischken, Gvardeysk District, former Kr. Wehlau - Nowakowski W., 1996, Taf. 92, 1-4)), as they are evidence of direct Aestian contacts with Roman legionaries in the 1st century AD (Kulakov V.I., 1997f, s. 362). A distinctive attribute of Sambian female graves of phases B_1-B_2 is the presence of numerous (up to six examples in one association) brooches of groups AIII and AIV. In burial assemblages these clasps are well associated with details of Noricum-Pannonian belt sets and with bracelets of the Sambian variant of Kameńczyk type. Thus, some of the brooches (types AIII, 62, AIII, 63, AIII, 46) serve as examples of the local manufacture of clasps of types AIII, 57 and AIII, 59. These "Prussian series" brooches (fig. 36) are characteristic not only of north Sambian B_{1b} and B_2 phases, but also of the Cecele-phase Wielbarsk culture; these brooches were later distributed in East Europe. The map of finds of these brooches in the southeast Baltic shows they had two areas of concentration - in Sambia, and in the eastern part of the Mazurian Lakes region (the Upper Angrapa/Angerrap - Nowakowski W., 1996, Karte 3). Except for those details of Noricum-Pannonian belt sets characteristic of western Sambia (fig. 36), other phase B_1-B_2 Dollkeim grave goods of nonferrous metals do not differ widely from the Wielbarsk sets of the Lubovidz phase. In particular, the number of brooches corresponds to the principles of female costume distributed in the time of Julius Claudius, across wide areas of Barbaricum, over the territory between the Elbe and the Vistula, as well as in southern Scandinavia (Tempelmann-Mączyńska M., 1989, S. 222). The Absence in the Lubovidz-phase Wielbark graves of weapons and tools suggests the "female character of the grave goods" (Wołągiewicz R., 1981a, s. 151). This conclusion is also credible for Sambian burials of the 1st century AD. Thus Sambian antiquities from the 1st century AD that can be classified as grave goods and their types are quite comparable and close, but not the same, as the attributes of the Wielbark finds.

The overwhelming majority of 1st-century AD Sambian pottery conforms in typology to Wielbarsk burial vessels. The oldest Dollkeim burials (Do-4, Do-17), on the basis of brooches of types AIII, 42, AIII, 62, AIII, 63, may be allocated to the second quarter of the 1st century AD. This dating is older than the earliest date of the Wielbark culture determined for the lower reaches of the Vistula (the second half of the 1st century AD)(Wołągiewicz R., 1981b, s. 88). Indirectly, the seniority of Sambian antiquities as a whole, and Dollkeim in particular, over those materials from "the Gepedojos Island", is confirmed by the development in the Wielbark area of a Sambian variant of Kameńczyk-type bracelets (see above), distributed first of all in phase B_1 along the Amber Coast. Nowakowski treats this phenomenon wrongly as a result of the amber trade (Nowakowski W., 1989b, s. 145, 149). The Dollkeim material shows a reverse course of events: from the middle to third quarter of the 1st century AD, the craftspeople of Sambia created a style of female jewellery (bracelets with double spheres on the ends of the rod, and "Prussian series" brooches) that was claimed later by people of the Wielbark culture. Both this culture, and the Sambians', of phases B_1-B_2 only very rarely put weapons into their burials (fig. 38).

The group of burials of phase B_2/C_1 (fig. 35) corresponds to a transitive stage of Dollkeim development. During this rather short-term phase (150/170-200), inhumations are replaced by urn cremations; they are characterized only by male burials, in which were found about 150-180 socketed axes and details of horse headpieces (predecessors or the earliest forms of the Vimose type). In the male burials of this group there were further examples of "iron spikes of square section", which appeared earlier in phase B_2 and which Gurevich considered as arrows. There are no figures of these artefacts, and the rather brief descriptions suggest they were instruments for piercing (awls, javelin-heads), or other items of the toolkit. As is known, the ancient Germans used pin-like tools from nonferrous metals for making "concept marks" (Begriffzeichnen) and symbols of the early Runic alphabet ('futhark')(Torsson Ed., 2002, s. 72). It is probable that the Germans along the Amber Coast, in their use of the above marks for magical and other purposes (Kulakov V.I., 2002b, s. 440-446), also used iron awl-shaped "styluses". Wielbark-type burial pottery gradually went out of use in phase-B2/C1 Sambian cemeteries. Only urns of RW IB type and "mugs of Wiekau-type" are retained from the wealth of former shapes. The tendency towards a change in the attributes of the local culture, as shown in the burials of phase B_2/C_1, is finally realized in the following group.

Group of burials of phases C_1-C_2 (fig. 35). Urn cremations are the sole form of rites for this group. As for grave goods, the impression is given that, for this historical period (from 200-325 AD) at Dollkeim, men-warriors exclusively were buried. In their graves, spearheads, battle (and every-day) knives, Szwajcaria-type spurs (male associations), and clay spindle whorls (female associations - fig. 42) predominate. The presence of these categories of grave goods in the later antiquities of the Amber Coast, up to the late Viking phase, is remarkable. One innovation, found in a group of phase-C_1/C_2 burials, is the presence of Roman bronze coins among the warrior graves. They obviously had a magical significance as a protecting object (Nowakowski W., 1996, S. 75, 76), and are symbols of the leading roles played by soldiers in the control of the amber trade during the boom of "the golden century of the Balts" around 175-250 AD (Kulakov V.I., 1997e, s. 59). In the C_1-C_2 phases of the Aestian cemeteries, provincial Roman brooches and their derivatives replaced the style that featured a "tied foot" (fig. 39). Clasps of type AVI, 167, decorated with ringed elements (Ringgarnitur-Stil, in effect a style of wire decoration), are the main attributes of both C_1-C_2 phases of Sambian antiquities, and synchronous with the Cecele phase of the Wielbark culture. Details of female costume dating from this phase sharply differ from the principles of Lubovidz-phase costume. They consist of pairs of brooches and an insignificant number of glass, copper alloy, or amber beads. Bead necklaces were supplemented with iron bell-pendants, which, during the C_2 phase, were replaced with bucket-shaped pendants. It is interesting to note that a special jewellery set, convincingly connected with the Gothic tradition, develops later and far from "the Gepedojos Island" - in the territory of the Chernyakhovsk culture (Tempelmann-Mączyńska M., 1989, S. 223). The above-mentioned brooches of AVI, 167 type are represented in the Dollkeim burials by individual examples, and have male associations. Pairs of derivatives of these clasps - "large crossbow-shaped brooches" (Grosse Armbrustfibeln - "Grosse ABF"), dated to phase C_2, correspond to female associations. The ceramic material from the described group of burials also breaks with the previous tradition, and is characterized, as a rule, by urns of Do-43 type (Grebieten-Urnen), and accessory vessels close in style to type RW XVII. These vessels, referred to by Nowakowski as "Dollkeim type", are decorated with strips of a number of polished figures as "double crosses". One of the few indicators of the presence in Sambia of C_1-C_2 western Balt material is the clasp of individual brooch derivatives of types AIV, 72 and AV, 98 (fig. 40), related to "Mazurian brooches". These derivatives, which appeared in the western-Balt environment during phases B_2/C_1 (fig. 41), together with plate bracelets and pins, are conventional indicators of the traditions of southeast-Baltic autochthons. Such artefacts are absent from the Dollkeim cemetery, and this clearly indicates Nowakowski's mistake: he has used the name of this site as representative of the western Balt as a whole - in his opinion the SNG. But all the mentioned finds are characteristic of the Aestians of the Mazurian Lakes, and of the territory between the rivers Instrucha/Inster and Lava/Alle in the 2nd to 3rd centuries AD: "Mazurian brooches" are mostly found there in pairs in female burials. In the Dollkeim material, early forms of crossbow-shaped brooches serve as paired clasps of female dresses (fig. 42).

The group of burials of phases C_2/D_1 - D_2 (fig. 43). In its rites, this group continues the tradition of the previous stage of the development of Sambian antiquities. An innovation, however, is the distribution of urns with horizontally applied decoration (type Do-41), known before in some island monuments of the western Baltic. The number of female burials grows appreciably (8 from 21 assemblages of this group), and site Do-162, from its grave goods, contains the remains both of a woman and a man. From phase C_2/D_1 (= C_3, or about 325-360/370 AD), pairs of "large crossbow-shaped brooches" (fig. 44, 45) become a leading attribute of female Aestian association; in the beginning, the brooches feature backs bent at 90 degrees under one corner, and later (phases D_1 and D_2) there are examples with arched backs. During the whole period described, only these clasps, from the entire heritage of late Roman times, are maintained in the repertory of the finished products of the local jewellers. (In the antiquities of the eastern neighbours of the Amber Coast, up to the 7th century, their size increased.) This process has a functional sense: the wide spring of the brooch promoted the closer fitting of the fabric to a woman's body; this might well answer the aesthetic ideals of the inhabitants of the southeast Baltic. From phase C_2/D_1 the male burials, together with their spears (fig. 46), battle-knives, socketed axes, and shield-bosses, which were all by now becoming traditional grave goods, also began to contain individual brooches of type Do-85 ("Grosse ABF"), and the still rare clasps with cast pin-receivers. On the contrary, in the late burials of phases C_2/D_1 - D_2, brooches with 'star-shaped' feet appear. They are categorized by Bitner-Wróblewska as group II. In association with these clasps, there are finds of spearheads, knives/daggers, and plate spurs. The assemblages with numerous warrior grave goods described above (Do-109, Do-146, Do-151, Do-161, Do-163), in part dating from the time of the campaign of the multiethnic hordes of Radagaisius into Italy, might be considered as "optimates". According to the text of Olympiodorus, this was the name given to the commanders of groups of 10-15 soldiers in Radagaisius' army (Skrzhinskaja E.Ch., 1999 s. 128, 219). If, in phase D_2, the Dollkeim graves contained the remains of both "optimates" and ordinary warriors (a terminology offered first by Olympiodorus - "foederati"), the situation radically changes from the end of phase D_2.

The material included in the *group of burials of phases* D_2-D_3, is, in the early part (the period between 400 and 450 AD), characterized by the next change in form of the burial rites: urn cremations now being replaced by groups of burnt bones (var. 1.2). These new rites are linked. In a variant of cremation 1.2, instead of urns, other receptacles are used for the bones. These were made of organic materials (wood, leather, fabric) that have not survived. In phase D_2 the grave goods included many crossbow-shaped brooches, in effect derivatives of clasps of "Grosse ABF" type. The 'Duratón' type is a common clasp of the time (perhaps the third quarter of the 5th century)(fig. 47,2). The connection of the population who used these brooches to the formation of the Prussian culture has been proved (see. Chapter VII). However, according to the Dollkeim data, it is necessary to attribute the genesis of this culture, referred before to the middle of the 5th century, a little bit earlier, to the second quarter of the 5th century: the time of the height of Attila's power. In the Dollkeim male burials of phases D_2-D_2/D_3 containing crossbow-shaped brooches of Duratón type, only spearheads are found. This set is much more modest than the weapons complex met with in phase D_2. Although in this phase, and with phase D_3-D_3/E_1 graves, prestigious objects of the time, those decorated with Sösdala-style elements, are found, the burials of the "optimates" (at least in comparison with the assemblages of the previous era) are not noticeable in Sambia. Female assemblages of phases D_2-D_3, in which, as before, the rites of urn cremation are seen, no longer contain the large "crossbow-shaped brooches" that were earlier the important feature of local female costume. The fashion for these female accessories switches to the area along the right bank of the Neman. In the third quarter of the 5th century they were replaced by brooches with an "animal-headed" roundel. Both in the Halibo area and Sambia, these younger contemporaries of brooches with "star-shaped" feet are an ethnic attribute of the Vidivarii.

From the evidence of Dollkeim the table of finds most typical for the SNG is drawn from the times of the Roman Empire (fig. 48). The relative chronology of these artefacts is supported by the coming together in graves of their separate types and subtypes. The table of chronology of the Dollkeim brooches is based on the same principles (fig. 49). This corpus is distinguished from similar finds from other cemeteries in the Amber Lands by the absence of "Mazurian brooches" and the other clasps allocated in the bottom right-hand corner of the table in the group, shown by a continuous broken line. Indirect confirmation of the legitimacy of the offered chronology of Aestian antiquities is shown by the dynamics of the consecutive accommodation of graves of different periods in the territory of the Dollkeim cemetery (fig. 50).

In the base table the final *group of burials of phases* D_3-F_2 (fig. 51), deriving from the framework of the chronological period researched, shows the initial stage of Prussian culture. For burials of the second half of the 5th century, as well as for the assemblages of the 6th to the turn of the 7th/8th centuries, cremation in the form of a group of burnt bones is the norm. By the end of the period it is gradually replaced by a ceremony involving the strewing of bones, carried to the gravesite from the pyre in a "temporary urn" (vessels of subtypes 1.2 and 1.3, throwbacks to various south Scandinavian pottery forms of post-Hunnic time). In the beginning of the 4th century, the custom appears for a man to wear a single brooch, and a woman a pair of clasps. This custom was maintained at Dollkeim and all over Sambia up until the turn of the 7th/8th centuries. It is interesting that women used the simplified forms of brooches of Neuwied type while the men preferred the crossbow-shaped brooches of subtype K2 - rather simplified derivatives of clasps of the Duratón type. The widened parts of their feet are often decorated with a "slanting cross". These widened parts are a development of square figures with the same cross, known on earlier brooches of the Duratón type; these were distributed in western Europe in late Hunnic times and are indicators of Sambian and late Wielbark cultural and ethnic influences (Kazanski M., 1999, fig. 1, 11-13). The location on the map of the southeast Baltic of the above-mentioned brooches shows the process of some unification of the features of the culture of western-Balt tribes from the 6th century (Åberg N., 1919, Karte VII). From the beginning of the 6th century, weapons are very rarely found in Aestian male burials. At the late stage of the great migration of the people, these burials are necessarily accompanied by horse burials. Female assemblages, besides an accessory vessel and a pair of brooches, quite often include copper-alloy neck-rings (at the beginning of the period these are fashioned from smooth rods, from the second quarter of the 6th century from twisted rods), and occasionally spindle whorls already appear in female assemblages from the 3rd century. With the burial rites (and in the grave goods of the Old Prussians) from the beginning of the 6th century, there is evidence of all the attributes of rigid regulation, characteristic of the local population up until the end of the 10th century.

Unfortunately, there is no data on those settlements of Roman Empire times excavated over the wider area, so it is impossible to create a multi-plane representation of the history of the Amber Lands in Roman times. However, the information contained in the burial sites of this period of Baltic history illuminates the separate aspects of the material and spiritual cultures of the Veneti, the Goths, and the Aestians settled in the Amber Lands in the first half of the 1st millennium AD.

V. Genesis of the tied Baltic brooch

As shown by Nowakowski's chronological research, one of the most important groups of artefacts from the local Aestian archaeological material is the tied brooch with folded leg (Fibeln mit umgschlagene Fuß - Nowakowski W., 1996, S. 21). Although these finds occupy a key position in the chronology of the Amber Lands in Roman times, they have not received the study they merit. This chapter aims to fill this gap and 40 sufficiently well documented assemblages of SNG grave finds of tied brooches are investigated here.

The brooches comprise the category of clasps with "folded leg", subdivided, according to Ambroz, into categories: the "one-piece" type (the "body" of the brooch and its pin are made from one piece of metal rod, and the "bowstring", as a rule, is fixed at the top = obere Sehne, pc. 13, A-67), and the "two-piece" type (the "body" of the brooch and its pin are made from various pieces of rod or wire and the bowstring is fixed at the bottom = untere Sehne, fig. 13, Grei-73) (Ambroz A.K., 1966, p. 10).

One-piece brooches of the Ambroz 1 variant, directly continuing the La Tène tradition of east Europe in the 1st century AD (Ambroz A.K., 1966, p. 48), are absent from the archaeological sites of the southeast Baltic in Roman times. This form of clasp has a remote similarity to the brooch of type AVII, 204 (fig. 13, Kr-F). In this area are found the late variant 2 clasps of type AVI, 158 of the central European series (fig. 52. They are referred to by Ambroz as dating to the 3rd-4th centuries AD (Ambroz A.K., 1966, s. 59). Grave goods containing such brooches, from the burial assemblages of the southeast Baltic, allow us to extend their usage from around 250 AD, up as far as around 400 AD (Kulakov V.I., 1998c, s. 42). The insignificant number (no more than 5 examples) of such clasps found in the documented sets of grave goods, apparently raises doubts as regards the classic thesis of brooch study in Russian archaeology and especially concerning the genesis of the two-piece concave tied brooch (according to Ambroz, the basic corpus of tied brooches appertains to this subgroup in eastern Europe during Roman times). This thesis is as follows: "one-piece concave brooches" [i.e. artefacts from the 1st century AD - K.V.] are the closest to them [i.e. to tied brooches of series I "with elbow-shaped bent back" - K.V.]. Areas containing these brooches and others coincided only in the territory of the Old Prussians... where the appearance of two-piece concave tied brooches with narrow foot might be supposed" (Ambroz A.K., 1966, s. 62). Unfortunately, the period of the 1st century AD in the southeast Baltic is a part of the temporal hiatus between the late La Tène and late Roman Empire phases as regards the distribution of one-piece brooches. Modern western European colleagues do not see in Aestian regions the origin of the two-piece tied brooch (or, come to that, any other region of Europe in the time of the Roman Empire - Godłowski K., 1994, S. 484). However Godłowski's caution, which seems obviously deserved, and puts Ambroz's conclusion in doubt, loses its appeal in view of the finds of crossbow brooches in the historical region of Altmark (Land Mecklenburg-Vorpommern, Ost-Deutschland). These finds represent "barbarous" versions of provincial-Roman brooches of groups AV; such imitations are designated by Almgren as type AVI, 181. There is also a version shaped as a bow-construction brooch - type AVI, 157. The main difference of these brooches from their Roman prototypes is the replacement of the soldered plate pin-receiver with a tied one, and representing the bottom part of the foot, bent at acute angles and attached to its middle section with wire at the end of the bent part of the foot (originally soldered to the pin-receiver). This is not an evolution of the tied brooch in a "barbarous" environment, as Ambroz, Godłowsky and others have thought, but an original break from the technology used by German craftsmen in the early phase of late Roman times. Occurrences of similar forms of brooches in central Babaricum, according to finds in Altmark as well as Thuringia and Jutland, date from the end of phase C_{1a} (by U. Lund Hansen C_{1a} = 200-225), i.e. shortly before 250 AD (Schuster J., 1995, S. 108). Similar brooches with a wide foot are known from Aestian areas only at the cemetery of former Grebieten (Sambia) (Nowakowski W., 1996, Taf. 37,3), but they are rather more distributed in central and southern Lithuania in the future tribal area of the Zhemaits. This brooch is one of the most popular styles with the Balts from the wide repertoire of tied forms. True, the Lithuanian material is not dated earlier than the 5th century AD (Tautavičius A., 1996, p. 198), and is not linked chronologically to the finds from Altmark, Thuringia, and Jutland. Thus, the question of the direct "import" of tied brooches of type AVI, 181 is not relevant for the Baltic. However the fact of the presence in this part of Europe in Roman times of a great number of two-piece tied brooches, as well as the origination of this style of clasp only in the Amber Lands (as Ambroz suggests), demand explanations.

The first task is to locate the earliest Baltic variant of the two-piece tied brooch. Extremely numerous in late Roman times, such brooches are the norm in both male and in female assemblages (fig. 52). By their general outlines they copy provincial-Roman brooches of groups V (fig. 12, 92,2,2a), referred to as a whole by Lund Hansen as from phase B_2 (Lund Hansen U., 1987, S. 39). Even the twist in the backs of these brooches (especially well defined in types AV, 136-137) is imitated by crossbow examples found in the Amber Lands. It is logical to arrive at the conclusion that the craftspeople of this region, as well as their colleagues in Jutland and certain other sites in Germania Libera, copied the most popular "barbarian" brooches of phase B_2. Of course, if the inhabitants of Germany as a result of this action acquired AVI, 181 tied brooches, the inhabitants of the Amber Lands, as a result of their creativity, had examples of type AVI, 162. Such iron clasps appear in phase C_{1a} in the environment of the Przeworsk culture (Nowakowski W., 1995, s. 30). This very area, instead of the Amber Lands, is the basic territory of distribution of the one-piece brooch in European Barbaricum, as the distribution map of tied brooches shows (fig. 54,1). Two-piece brooches are most numerous in the Lower Vistula and Sambia (i.e. in the area of the Wielbark culture and the SNG, fig. 54,2).

The main difference of the western Balt and Przeworsk brooch from the type AVI, 181 is the rod back of elongated oval shape, or even rectangular section, instead of the plate and wide back. Such brooches, with a lengthy development, exist up to the beginning of the 8th century in the form of clasps of subtype K4 (Kulakov V.I., 1990b, s. 151, 152). At the threshold of the Middle Ages, the ancestors of the last Zhemaits imitated Mazurian crossbow-shaped brooches of subtype K3 by manufacturing brooches with a tied foot (Vaitkunskienė L., 1995, 171 pav. 1). As experience shows, the production of a crossbow-shaped brooch with a tied foot does not demand special professional skills and requires only copper alloy rods and wire. These very clasps, by the end of the 4th century AD, became (in the Baltic) prototypes for crossbow-shaped brooches with cast pin-receivers, which were distributed all over Barbaricum. The transitional form of clasps was found by Åberg in the material of the cemetery at Warengen, Kr. Samland (Zelenogradsk region) (Åberg N., 1919, Abb. 50, S. 54, 55).

The problem of the genesis and development of tied brooches with ring elements (Fibeln mit Ringgarnitur = RG, fig. 16) is much more difficult. Their major attributes are rings of corrugated wire decorating the brooch foot, and a decorative "button" at the point of connection of foot and spring. These attributes unite such artefacts with brooches of type AVII, 211 (fig. 55). Ambroz, following Åberg, counts these clasps as belonging basically to the Baltic, where they, in his opinion, appear in the 3rd century AD (Ambroz A.K., 1966, p. 67). Godłowski dates the appearance in the western Balt and Wielbarsk areas of such brooches of type AVII, 211 to phase C_{1b} (about 250 AD). These clasps have the distinctive feature of a high, soldered plate pin-receiver, and were made in the above-mentioned areas, as well as in the Przeworsk territory, up to phase C_2 (Godłowski K., 1994, S. 485). These brooches are the simplified derivatives with strongly curved clasps of type AIV, 74 (phase B_2 - Lund Hansen U., 1987, S. 39). True, if the "body" of these brooches, according to the traditions of manufacturers in the Roman provinces, was made from separate components united at the centre of the foot by a graceful discoid band (fig. 56); the AVII, 211 brooches only imitate this complex design. Their feet are wound with one or several rings from corrugated wire. Probably, it was enough for the wearer just to give the impression of having a "real" provincial-Roman brooch. It is only possible to guess at the places of manufacture of these "fakes", but the connection of these objects with the Roman *limes* is obvious from their means of manufacture. As well as those examples shown in figures 13, 57, 58 ('RG' brooches with a bent foot), such finds are known from Eis-24, Palwe-1k, and Detlevsruh-12.

The appearance of brooches of type AVII, 211 coincided with the expansion of trade with Rome in the "barbarian" environment, and, according to market forces, and the general fall in quality (Rostovtsev M.I., 2000, s. 168). The beginning of the Gothic wars in 251 AD and the complicated military situation in the southern areas of the Great Amber Route interrupted the Germans' established flow of provincial-Roman goods from the eastern part of the Empire. This market had taken root for a long time in Germania Libera; and consequently, in the second half of the 3rd century AD there was the real problem of substitute production to contend with. This problem of 'fake' Roman consumer goods had already arisen by the end of the 2nd century, after the trading restrictions implemented by the Empire following the Marcomannian Wars (Kolosovskaja J.K., 2000, s. 116). This radical development in the nature of the material culture of Barbaricum (between about 180 to 260 AD) can be nicely demonstrated in the example of those brooches of type AVII, 211.

Returning from the Balkan theatre of the Gothic wars, migrating groups of Przeworsk-culture people might have brought to the borders of the Amber Lands a variety of methods for manufacturing "Sarmatian" (according to the terminology established in Poland) brooches with wire elements, harking back to the brooches of the "Lebyazhje series" (A.K.Ambroz). Vagrant groups of craftspeople from Panonnia and Dacia distributed their prototypes northwards (Mączyńska M., 1999, s. 90), downstream of the Vistula (i.e. along the Great Amber Route); as a result manufacture began on the Amber Coast of their derivatives, AVI, 167 with bent foot. The intermediate form, between the provincial-Roman and two-piece brooch, has been found at the cemetery of Pruszcz Gdański (Andrzejowski J., 1992, p. 163, 164; pl. 5, g). These are wire brooches of type AIV, 81, made, probably, even before the Gothic wars, in phase C_{1a}. Such artefacts were the productions of local "barbarous" handicraftsmen, and appeared as a consequence of the "famine" of goods after the above-mentioned trade restrictions by Rome after the Marcomannic Wars. This "famine" might have pushed some of the Goths-Gepidae to search for new contacts with the antique world, which led to the Gothic wars in the Balkans: the fact of the manufacture by provincial-Roman masters of such substitute products (in the Amber Lands type AV, 123 - fig. 4) is indicative. Such artefacts are rare in the Amber Lands, as are, however, proper brooches of type AVII, 211 (Sc-29) and their derivatives. It cannot be said that their derivatives (with bent foot) became a real trademark of the foundries within the Amber Lands from late Roman times, as is visible from the number of finds of these brooches in the southeast Baltic (fig. 59). For their manufacture a method of "tying" the foot to fashion a pin-receiver was applied; this method was already mastered along the Amber Coast for brooches of type AVI, 162.

Being a "barbarous" modification of the brooches of types AIV, 81 and AVII, 211, the mentioned clasps, united by Åberg with the type A2 (fig. 57), which he dated to the second half of the 3rd century AD (Åberg N., 1919, S. 13), are far from the high quality products of Sambian craftspeople of phases B_1-B_2. Clasps of type AVI, 167, decorated with ring elements (a style of a wire decor) are a leading attribute both for Sambian antiquities of phases C_1-C_2, and other areas of the SNG, and for the synchronous Cecele phase of the Wielbark culture. In the Sambian region, as well as along the right-bank of the Neman, 'RG' brooches came from Sambia proper (Åberg N., 1919, S. 13). The above-mentioned brooches of type AVI, 167 are represented in Sambian burials as individual finds and are part of the male assemblage. The Roman coins from Sambian cemeteries serve as a 'terminus post quem' for such finds: Adrian (117-138 AD); grave 46a Dollkeim (Bolin, 1926, S. 211); Faustina Younger (till 175 AD); grave 128 Greibau (Bolin St., 1926, S. 213). In many cases the Roman coins are not readable, but, being for the SNG a steady chrono-indicator, their presence dates the assemblages to phases C_1-C_2 (Vityaz' S.P., Kulakov V.I., Medvedev A.M., 2000, s. 14). Besides this, brooches of type AVI, 167 correlate with graves with "cross-beam" brooches (Do-48 - Tischler O., Kemke H., 1902, Taf. III, 17; II, 14,15; Grei-21 - Tischler O., Kemke H., 1902, S. 26). Such cross-beam examples (including "Mazurian" ones) were used by people of the western-Balt culture in phases B_2/C_1-C_1 (Kulakov V.I., 2001, fig. 2). Besides the above-mentioned brooches, those of type E3 are found in Do-254, Do-258, Do-299, Do-260, Do-303, and Do-306.

In conclusion, it is necessary to note that the clasps of type AVI, 168 are found in the material of the Amber Coast (fig. 60, in Gru-II with a Trajan coin). Their ring (it is more exact to say spiral) element is only in the centre of the foot and in effect continues the wire end fastening pin-receiver to the foot. These brooches make up approximately one third of that corpus of 40 tied clasps, which are known up to the present in the adequately documented burial assemblages of the southeast Baltic. In Aestian antiquities, together with brooches of type AVI, 168, there are the so-called "large crossbow-shaped brooches" of type Å2 (fig. 58) with ring elements (= grosse ABF), approximately 1.2 – 1.5 times longer than the average brooch of type AVI, 168 (compare the brooches in fig. 59).

The question of the genesis and development of the simplified 'RG' brooches (types AVI, 168 and Å3) is still open. Nowakowski, as well as Tischler and Kemke, from the corpus of clasps, only identify the "large crossbow-shaped brooches" (grosse ABF - type Å3), dating them around 350-400 AD (Nowakowski W., 1996, S. 19). As to their origin, Nowakowski (and some other modern archaeologists) only informs (following Almgren) that they belong to a "Prussian variant" of the crossbow-shaped brooch of groups A, VI and AV, II. A.Z. Tautavičius deals with the above-mentioned brooches of type AVI, 168 (Å2) and divides the Lithuanian material into groups I-III, dating them to the 4th/beginning of the 5th centuries (Tautavičius A., 1996, p. 195). The leading stylistic attribute of "large crossbow-shaped brooches" is their corrugated wire rings around the "body" of the brooch. This feature is connected to early Wielbarsk stylistics such as the "S-shaped" necklaces, certain decorative elements of "snake-shaped" bracelets, and by other jewellery (W. Nowakowski: Lubowidz-Stil, or, more precisely, Ringgarnitur-Stil) from early Roman times. Ostensibly the feature appears in Sambia "... from the east Pomorze and area of the Vistula delta, as well as from remoter Sarmatian areas" (Nowakowski W., 1996, S. 57). Actually, the material from Sambia and other Aestian cemeteries show that 'RG' brooches are a local continuation of a line of brooches of types AIV, 74, AIV, 81, and AVII, 211 (type Å2). The mapping in the southeast Baltic of those sites of brooch finds "with ring elements" (fig. 59) confidently points to Sambia as the initial micro-region for the distribution of these artefacts (Åberg N., 1919, Karte I, Fibel Typus Abb. 1-2). Their simplified retorts started to appear simultaneously with brooches of type AVII, 211 in the southeast Baltic: these brooches are correlated in burials. For example, this phenomenon is noted in:

Do-46^b - urn CRM under a stone covering. Grave goods: copper alloy brooch of "large ABF" type, brooch with ring elements of type AVI, 167, iron plate spur of type Swajcaria, shield-boss of type Chorula, a shield grip, a broad knife (Tischler O. = "konvex"), urn, scythe-head, iron hook and other finds (Tischler O., Kemke H., 1902, S. 20). Do-46^b contains, among other objects, the remains of a man/warrior and is dated by the brooch finds to around 400 AD. The shield-boss from Do-46^b appertains to phases B_2-C_2 and allows us to date Do-46_b not later than 290 AD (Kazanski M., 1994, p. 437).

As well as the simplified RG brooches mentioned above, such artefacts have also been found in Do-59, Do-251, Do-254, and Re-28(a).

It should therefore be concluded that the formation of not only the stable attributes of the brooches of type AVI, 167, but also the large ABF clasps, was complete in the Amber Lands by the end of the 3rd century AD. The given examples show the rapid and almost spontaneous occurrence of these prestigious types of cloak fasteners in just a few decades during the second half of the 3rd century AD. During this time the simply designed two-piece brooch (that fastened the shirt collar) continued in use. In the Curonian part of the western-Balt area, according to the assemblages from the cemetery at Žwiliai (Klaipėdos raj., Lietuva), the large ABF clasps, and even the two-piece brooches with bent back, were in use up to the 4th/middle of the 5th centuries AD (Astrauskas A., Gleiznienė G., Šimėnas V., 1999, 9, 10 pav., p. 137). In the 5th century, late modifications of the large ABF brooches began to appear, moving along the trading routes to the northeast of the Amber Lands and Lithuania, i.e. to the Balts of the Daugava river basin and to the south Estonians tribes (Aun M., 1985, s. 78).

The question of those brooches of types AVII, 212 and AVII, 216,217 (fibulae monstruosi), derived from clasps of type AVII, 211 (fig. 16, the bottom line, 17), occupies a special place in the research of Baltic crossbow-shaped brooches. Godłowski, following J. Werner, ascertained their northern origin (and more exactly, Jutland). He dated their occurrence to phase C_{1b} and connected their distribution in Barbaricum with the Moldavian sector of the Chernyakhovsk area (Godłowski K., 1994, S. 485). Actually, the type's high soldered pin-receiver is characteristic of the style of antiquities of the Baltic western coast from late Roman times. In some cases this pin-receiver was decorated with a copper alloy mount, with a silver overlay, in the form of an extended oval. It is represented by two "monstruosa brooches" from the Amber Lands (Gru-D, G(Nord)-177, a find from the Lauth cemetery in 2000). The above special mount is present on an example from the Kaunas museum, and is considered a Scandinavian import dated to phase C_2 (Mihelbertas M., 2000, S. 65). It seems that, until now, no mention has been made of the fact that that genesis of this oval mount occurs from those brooch prototypes with type-A3 clasps, on which the face sheet of the pin-receiver has, at first, a vertical functional rib: soon this rib assumes a purely decorative function. Such artefacts are found in the Kirpehnen cemeteries (Nowakowski W., 1996, Taf. 84,1), Eisliethen-194 (Nowakowski W., 1996, Taf. 53,1), Eisliethen-261 (Kulakov V.I., 1998, fig. 5), and Greibau-150 (Jentzsch A., 1896, Taf. II, 10).

This phenomenon is hardly accidental. On the contrary, the connection of those attributes belonging to the brooches of the Amber Lands and Jutland, at the beginning of phase C, is an independent confirmation of the important role of the separate Aestian area (a key point in the barbarian line of communications) that developed on the eastern outlying districts of Barbaricum after the Marcomannian Wars. As well as the other, elite, "trophies", found in the lake sanctuaries of Jutland, the monstruosa brooches were the gifts of a great number of tribes living far from this complex of sanctuaries that was sacred to all Germans (Kulakov V.I., 2001, S. 53). There can be no gainsaying that some of these magnificent clasps were made by master craftspeople who were very familiar with certain details of the material culture of the Amber Lands. Along this part of the Baltic Coast, one find (unfortunately only one) of a brooch is known that may be associated with some confidence as a local variant of a monstruosa ornament (Åberg N., 1919, Abb. 10).

At the end of the Roman era, the wire fastening of the tied foot in the centre section of the "body" of the large ABF brooch was modified by the western Balts into a silver mount with a facetted pattern; frequently an example will have a cast pin-receiver. It is true that this mutation, which resulted in the appearance of crossbow-shaped brooches of variant 3в1 (type Å4) (Kulakov V.I., 1990, s. 152, 153) at the last stage of the great migration phase, is characteristic not of the Prussian culture but of the Mazurian cultural group and antiquities of the Neman delta. These are the stages of the genesis and development of tied brooches in the southeast Baltic from the 2nd to the 4th centuries AD. It is possible to summarize the conclusions reached from the analysis of the given category of finds:

1. One-piece brooches of variant Ambroz 1, directly continuing in East Europe in the 1st century AD the La Tène traditions, are absent in the archaeological sites of the southeast Baltic in Roman times. Finds of artefacts of late variant 2 of clasps AVI, 158 (a Central European series), referred to by Ambroz as 3rd/4th century AD, do not support the hypothesis of this outstanding researcher on east European antiquities concerning the formation, in Aestian territory of the 1st/2nd century AD, of two-piece brooches from one-piece examples.

2. Two-piece brooches occur in the southeast Baltic, according to a find from Ostruda, in phases B_{1b}-B_2. By their general outlines they copy provincial-Roman brooches of group V (phase B_2). The masters of this region, as well as their colleague in Jutland and some other points of Germania Libera, copied the most popular "barbarian" brooches from the phase B_2 group V. It is possible, however, that, taking into account the active connections of the southern Scandinavians and Aestians in the 2nd and 3rd centuries AD, there was a direct transfer of the production methods of tied brooches from the western coast of the Baltic, or from the German areas of Pomorze, to the Amber Coast. There is nothing surprising about a "community of customs" between the Germans and the Aestians, based on the closeness of their cultures and languages, which Tacitus remarked at the very beginning of the 1st century AD (Dini P., 2002, s. 60). It is thought that the Scandinavian ethno-cultural impulse is reflected in the existence in the western-Balt area of stone circles and "stelae". They are characteristic of the Wielbark culture of the Cecele phase, and are distributed in phase B_2/C_1 from the Wielbarsk area, over the Pasłęk (Passarge-Fluß), to the east, and into Aestian territory (Bohnsack D., 1940, S. 23-25). Earlier direct indications of an opportunity for the eastern-Gothic "arrival" along the Amber Coast were limited only to this fact. Now, probably, support for this version of events may be found in the tendency towards a distribution of tied brooches from west to east. These finds occur simultaneously on sites in western Sambia and in the lower reaches of the Vistula, around 170-200 AD, and pointing (for the latter region) to the formation of the Cecele phase of the Wielbark culture, when "...the major expansion of the Wielbark culture in a southeasterly direction begins" (Kuharenko J.V., 1980, s. 64).

3. The distribution in northeast Barbaricum, between 180 and around 260 AD, of "substitutes" of provincial-Roman production, brought about, in particular, brooches of type AVII, 211. Vagrant groups of craftspeople from Pannonia and Dacia distributed these products, and also other artefacts with wire elements, northwards along the Great Amber Route. As a result, manufacture began on the Amber Coast of derivatives of brooch type AVII, 211 - clasps of type AVI, 167 with a bent foot (fig. 60).

4. Clasps of type AVI, 167, decorated with ring elements, are features of Sambian antiquities of phases C_1-C_2, as well as other areas of the SNG and synchronous Cecele phase of the Wielbark culture. In some burials they coincide with "ordinary" crossbow-shaped brooches and in some assemblages with ABF brooches. Irrespective of the social treatment of such assemblages, this feature shows that the period of the appearance and establishment of all three above-mentioned related styles of brooch was rather brief (within the limits of one or two generations during the 3rd century AD). Such specificity is characteristic for typological mutations of goods from outlying districts of Barbaricum, and influenced not only by the provincial-Roman workshops, but also by the unstable traditions of various tribes and social groups. Hence the impression arises of some "barbarous" tribes with their material culture exhibiting an inherent "scrappiness" (for example, the simultaneous use of several types of brooches, buckles, and so forth). In comparison with this ambiguous picture, the antiquities of the Vikings appear uniform, with the borders of closely related groups of material poorly distinguished.

Details of female costume during the Cecele phase of the Wielbark culture sharply differ from the costume elements of the Lubovidz phase; and far from the Baltic, in the Chernyakhov area, they link up with the Gothic tradition. Probably, the occurrence of brooches of type AVI, 167 among female costume along the Amber Coast is also not too far away from the Gothic tradition (although obviously with a regional bias). It cannot be excluded that this archaeological reality reflects the result of a process of distribution, up to 366 AD, of the influences of "Gothic power" on the inhabitants of the Amber Lands, which is "attributed not so much to the victories as to the prudence of Germanaric" (Gibbon E., 1998, s. 91). Certainly nobody is able to confirm so radical a definition of the forms of contact between the Goths (whose ethnic unity is not proven with certainty) and the Aestians. However, the ethno-cultural affinity existing between them (Kolosovskaya J.K., 2000, s. 31) allows us to suppose an easiness of cultural contacts and exchanges between the subjects of the Balts, Amals and Aestians, who might also have participated in the creation of "Gothic power". This assumption is in harmony with the distribution plot of monstruosa brooches.

VI. The beginning of the amber trade with Rome in the time of Nero

The Amber Coast in its narrowest geographical sense corresponds to the western part of the Sambia Peninsula. Here exists the largest deposit in the world of high-quality amber-succinum, resting on the sea shelf in a coastal zone 1km wide (Koulakov V., 2000, p. 29). This unique feature had made the region well known to the many peoples of Europe since olden times. The attraction of amber for the representatives of ancient civilizations is quite easily explained: the petrified resin of prehistoric trees was a fine material for manufacturing ornaments and decorations (pendants and necklaces), as well as serving as incense and as an ingredient of early medicines (Moshkov N.N., 2002, s. 5-11). For its peculiar and particular physical qualities, amber was considered sacred and associated with the cult of the Sun. In antique myth, amber is seen as the fragments of Phaeton's chariot itself; and among the western Balts the legend of an entire, and storm-destroyed, amber palace belonging to an underwater god lives to this day.

Already in the early Bronze Age (the end of the 2nd millennium BC) the archaeological material from Sambia contains a significant share of objects from alien cultures; in themselves indicators of the development of the amber trade. The prospect of participating in the exploitation of amber, and the direct control of sources along the Amber Route, involved Sambia directly with its western neighbours. The people of the Unětice culture, who had arrived from the west, contributed to the formation of the culture that has since been found in Sambian barrows (Engel C., 1933a, S. 190). The development, 900-700 BC, south of Sambia, of the Mazurian-Varminian sub-group of the Lusatian culture was the result of this process (Kulakov V.I., 2001b, s. 125, 126).

In the early Iron Age, the trade routes connected the Halstatt area and the antique world with the Amber Coast along the Elbas and Oder rivers. The intermediaries in these contacts evolved into the settlements of the Pomorze culture (Okulicz Ł., 1976, s. 107) that occupied the southern coast of the Baltic, to the west of the mouth of the Vistula, in the middle of the 1st millennium BC. At this time Sambia was the centre of an area of western-Balt barrow culture, the monuments of which were left by the very earliest representatives of the western-Balt ethno-linguistic community (Kulakov V.I., 2001b, s. 126). The disintegration (1st century BC to 1st century AD) of this barrow culture coincided with the beginning of the formation of the tribal structure of the western Balts: the trading of amber between Sambia and the Mediterranean dates from this time. Information on this period, which for centuries determined the special character of the antiquities of the Amber Coast, is extremely important, and modern methods of studying local antiquities allows us to reconstruct the historical events along the Amber Coast at the beginning of the 1st century AD.

The German origin of the ethnic name "Aestian" (see Chapter I), points to the informants of Tacitus, who first mentioned these people as speaking a Balt (by origin) language, and being the ancestors of later Prussians. The first detailed information on the contacts between the Aestians and Romans is a message from Pliny the Elder, written after 59 AD, in "Natural History" (Pliny Starshiy, 1937, s. 109). In Chapter 42 the author informs of the actions of the fleet of the Roman commander Germanicus, in 14-15 AD, off northern coast of Europe, and mentions the centre of the amber trade - "the Island of Austeravia". Taking into account the affinity of this name to the ethnic name "Aestian", it is possible to assume that "Austeravia" was a name for Sambia, surrounded as it was by sea, lagoons, and its rivers: it was also later considered as an island by the German chronicler Adam of Bremen.

Germanicus' ships opened up the Amber Coast for Rome. In Chapter 45, Pliny informs us of a trip made there by Roman horsemen in 54-68 AD (Kolendo Z., 1981, s. 28). Travelling from the Roman province of the Danube, Pannonia (there is a mention in the text of 889kms being the distance between Carnuntum, situated to the east of modern Vienna, and the Amber Coast), the Roman troopers visited "local trading points and coastal areas". These words indicate that there were locations for the extraction and processing of amber in Sambia at the middle of the 1st century. In 1994, M.B. Shchukin hypothesized about the specific goal of the above-mentioned expedition. In his opinion, the purposes of this trip were to undertake "... negotiations with leaders of local tribes about terms of trade and the movements of merchants, etc; the mission was more diplomatic than commercial" (Shchukin M.B., 1994, s. 225). Moreover, the discovery of a gravestone in one of the churches of southern Slovakia, provides Shchukin with an actual name of one of the horsemen mentioned by Pliny - Qvintus Atilius Primus - the composition of the expedition (soldiers of the XV Legion based in Carnuntum), and date of trip - 62 AD. By agreeing with Shchukin's assumption, the absence of a corpus of Roman "imports" in Sambia in the early 1st century AD becomes easier to understand.

Until now, the archaeology of Sambia during Roman times was studied on the basis of the analysis of separate finds of provincial-Roman origin. The new departure in this present work is the study of the closed burial associations. The most suitable for this purpose is the material from the cemetery at Dollkeim-Kovrovo, located in northern Sambia. It is only from this Aestian and Old Prussian archaeological site that we have 20 brooches made in the Roman province, dated to the first half of the

1st century AD (Günther C., Voss A., 1880, Taf. 7, 8). Although there were no Roman coins of this date at Dollkeim, there are other Roman goods besides these brooches.

In grave Do-7 (fig. 61), a complete legionary belt-set of type cinculum has been found and dated from Roman military history to 15-70 AD (Connoly P., 1981, p. 51; Deschler-Erb E. 1991, S. 58); the item is the major feature of soldiers' equipment in Julius-Claudius' Empire. It has been considered that the absence of such finds in Wielbark antiquities confirms the absence of any military contacts between the early Empire and the population of the southeast Baltic (Madyda-Legutko R., 1983, s. 119). This conclusion is disproved by (in addition to the cinculum find mentioned above) the detection in grave Do-30 of a faler (mount; fig. 62,4), of early Roman style and being a detail of cavalry equipment. By its size and ornament (concentric circles), this faler dates from the middle of the 1st century AD, when these objects turn from functional details of horse equipment into military decorations (Maxfield V.A., 1981, p. 92). Similar finds, with no rivets on their reverse for harness fastening but rather clasps for wearing on clothes, are known from four graves at Dollkeim. Furthermore, grave Do-30 revealed iron, rectangular mounts, which, from their descriptions, relate to details of Roman body armour made of iron plates. The picture of the originality of early Roman "imports" to Dollkeim is added to by finds of 6 belts, from female burials, with pierced ornamentation in the style of opus interrasile, and linked to the decoration found, in the time of Nero, on the belts of Roman legionaries and the harnesses of their mounts (Werner J., 1970, S. 40, 48). Prototypes of the openwork belts found at Dollkeim are also known in Pannonia.

It is, therefore, possible to state that several Dollkeim graves contained members of the Sambian elite from the beginning of the 1st century AD, and that their grave goods included items that echoed the Roman legionary equipment in the time of Nero. The form and ornament of these objects deeply influenced the further development of the local material culture from the 1st century AD.

This phenomenon is unique, and was obviously connected to a certain event with important significance for the local history. With a sufficient measure of probability it is possible to link this event with the expedition of Primus. As noted by Shchukin, the orders of this envoy were to make initial contacts with local clan leaders and sketch concrete mechanisms for the development of the amber trade and guarantees of its security. This undertaking would certainly be marked by the exchange of appropriate gifts, for example Roman military pallium cloaks with brooch fastenings (i.e. type AIV, 72), which might later have found their way into Dollkeim's graves.

The absolute situation of these brooches, which are provincial-Roman by manufacture, is not totally clear. Today a Jutland origin for their prototypes (AIV, 71), and their occurrence in the southeast Baltic (especially in the Wielbarsk area), with an individual impulse from phase B_1, is taken for granted (Nowakowski W., 1999, s. 281, 287). It is possible that our Polish colleague is mistaken, and that such brooches with "tympana" on their reverses do not appear in Jutland, but are only reproduced there by "barbarous" artisans and based on the imperial designs. Moreover, on one of these brooches, from the collection in the Prussia-Museum (fig. 41, AIV, 77), there is Roman figure engraved on the pin-receiver, perhaps part of an inscription "LEGIO XV". This artefact was probably one of "the guest gifts" (donum hostiae) provided by Atilius Primus' companions. It is difficult to relate the details of all the legionary equipment found in the Dollkeim graves to all the gifts mentioned above. These "barbarian" objects, according to the inhumation rites known, should be associated not with the Aestians-Balts, but, probably, with the Veneti. It is also improbable that these finds were trophies or the products of transit trade.

Apparently, the soldiers of the XV Legion who formed the guard of Qvintus Atilius Primus found it hard resist the opportunity while on the Amber Coast to take personal advantage of the situation. Struck with the abundance of amber, the legionaries, having nothing except their own equipment to exchange, gave to the Aestians their items of costume, decorations (falerae), and even military belts. According to Pliny, the largest piece of amber brought by them back to Rome weighed 4.26 kg, and the legionaries of Carnuntum appeared in the grip of "amber fever", and had to immune themselves to the temptation of seemingly infinite riches, similar to that experienced by Columbus's Spaniards, 1500 years later, when exposed to the dazzling gold of the New World. As with the Aztecs of Mexico, Sambia's Aestians were apparently rather blasé towards the treasure their land abounded in. As Tacitus wrote, "They (the Aestians) also search the sea (for amber) and, alone among amber gatherers, also the shallows and along the coast for the material known as glaesum. But the question of its nature and why it appears, they, being barbarians, no nothing of; you see, for so long has amber just been lying around with whatever else the seas throw up, and no lust for luxury has even provided it with a name. They do not use it in any way; they collect it in its natural state and deliver it raw to our merchants, who, to their amazement, reward them for so doing" (Tacit, 1969, s. 372). So, the Roman legionaries, to their astonishment, acquired bargain-rate amber along the coast, exchanging their equipment in return for valueless "electron" (the ancient Greek mythic name for amber). Being few and honouring their status, the Roman party did not apply to the Aestians the same ruthless and dishonourable measures, which, 1500 years later, were used against the Aztecs by the gold-blind conquistadors of Hernan Cortes.

It should be concluded that the professionally run expedition of Primus, which suddenly enriched many of the ordinary members of his team, opened the Great Amber Route from Sambia along the Vistula and Danube to the limits of Empire. The arrangements achieved by the Romans with local tribal leaders provided the stability of trade, and, on the Roman side, the negotiating process was supported by gifts. These gifts, and also those objects of legionary equipment that served initially (and perhaps spontaneously) as means of exchange during the undertaking of the embassy mission, were imitated by local craftspeople some time afterwards. The fact that so many features of Roman legionary equipment were adopted as fashion accessories by Aestian women is remarkable. Probably the gifts of the newcomers were totally alien to the equipment traditions of warriors along the Amber Coast: the fierce men from the north seemed eager to decorate their women with exotic Mediterranean gifts.

Indirect evidence of the development of trade with Rome began to appear from the end of the 1st century AD with the deposition of Roman "imports" among the burials and hoards of the Amber Coast; thus initiating a "Golden Age" for the Balts, in which, in all periods, the business partners of the Aestians and Old Prussians preferred to visit the Amber coast themselves. This determined the special character of the antiquities of the southeast Baltic, including the appearance of elements of various cultures from antiquity and the Middle Ages.

VII. Archaic religious rites and practices of South-Eastern Baltic populations

The extremely complex ethnic situation that arose in the Amber Lands in Roman times from the incessant movement of various population groups, makes it almost impossible to generate an integral picture of the local rites and religious practices observed in the first half of the 1st millennium AD. However, by means of certain archaeological research and on the basis of the data from Prussian folklore, we may recreate elements of the basic archaic views of the Aestians and other inhabitants of the nearby lands to the northeast of the outlying districts of Barbaricum.

"Each domestic holiday of the present Ndravians, Scalovians and other (inhabitants of Prussian Lithuania), was celebrated with its own major ceremony. The fête or holiday began with a drink. They prepared a jug or a vessel of beer, inside which a wooden scoop or spoon (samtis) was placed. Several kauszelen, i.e. drinking bowls, were called for and they were stood on a table or on half a quater (?). The Kauszele... might have no handles, or perhaps one for taking a better hold. The former had a (ritualistic) purpose and was thrown over the head, and was called szwencziama (i.e. "Sacred Kauszele"), and was preserved... by the invocation of prayers. [...] Kauszelen with handles were also considered sacred". In such a way, in 1681, the well-known collector of western-Balt folklore, Matteus Pretorius, described sacrificial bowls - indispensable objects for the inhabitants of Prussian Lithuania and a feature of pagan rituals in their final stages (Mannhardt W., 1936, S. 560). A "vurskhait" (or priest chosen by the community to carry out sacrifices), represented in the book of H. Maletius (1561-1562), holds in his hands a kauszele-bowl in the shape of flat sphere; it has no handle and is held ready to accept the ritual splashing blood of a sacrificial goat (Kulakov V.I., 1994, fig. 79). A similar function for a sacrificial bowl is mentioned in the Icelandic Landnamabuk when describing a temple created on Cape Tor by "Torolf the Bearded" from Mostr (Pennic N., Johns P., 2000, s. 220). This detail allows us to suggest a significant affinity between the rituals of the Old Prussians and Scandinavians in Viking times, always bearing in mind that the "import" of ideological views is not possible. According to Pretorius, kauszelen were also used by "wedding priests" – Szwalgones - splashing beer from a bowl in the eyes of the newly married (Mannhardt W., 1936, S. 553), as well as by house-owners when making various domestic prayers. The sense of these prayers is found in the 12th "commandment of Videvut", "... whoever kills the respected servant of a god, *through whom all friends of the dead receive* (from the god) *power and force*, they too shall kill him..." (Kulakov V.I., 1994, s. 150). The authenticity of Pretorius' data is proved by V.N. Toporov, however this most competent researcher has not arrived at a final opinion concerning the ethnicity of the people who carried out the religious practices described by Pretorius (Toporov B.N., 2000, s. 317, 319, 323). He calls them "Prussians" or "Lithuanians" living in "East Prussia", but this name for a part of the southeast Baltic was only established in 1772). Nevertheless, based on the concrete argument of the correctness of Pretorius' data, as submitted by Toporov, it is possible to try to reveal some archaeological correspondence of kauszele from the archaeological material of the Baltic, and, at the same time and whenever possible, to determine a tribal origination for those Balts who used the ritual utensils described by Pretorius in their ceremonies: these problems have not attracted the earlier attention of archaeologists and ethnologists.

In the burial antiquities of the Curonians, Lamatians and Scalovians, left by them on the Lithuanian shores and the right bank of the River Neman in its lower reaches from the 5th-9th centuries, there are tiny, pot-shaped vessels with heights ranging from 3 to 11cm (Taitavičius A., 1996, p. 268). In eastern Lithuania there is only one site of similar pottery from the time of the great migration; this is the Obel'aj cemetery (Urbanavičius V., Urbanavičienė S., 1988, p. 19, 19 pav.) that was, perhaps, under appreciable western-Balt influence. Similar vessels in form and size, were associated with the northern part of the Curonian area - the coastal zone of present southwest Latvia (Ozere I.A., 1986, s. 48-57).

The above-mentioned tiny vessels, from the territories to the east of the River Neman, are connected exclusively to the antiquities of the western Balts, and their religious nature cannot be in doubt. It is true, in view of their extremely small size, that it is hardly possible to compare them with Pretorius' kauszelen: the miniscule volumes of the handleless Curonian ritual vessels could not have contained the amount of liquid ("bier", "zemynelaudams", "palabindams", "papildams" - Mannhardt W., 1936, S. 577-579) described as being involved during sacrificial libations. The form of the Curonian vessels was not adapted at all for such functions: the officiant had to hold the vessel in his teeth and throw it over his head during the rites without the help of his hands (Mannhardt W., 1936, S. 564). However, even the tiny Curonian vessels, which do not correspond to the attributes of kauszele, are exceptional among the antiquities of Lithuania. As a whole it is necessary to recognize that among the ancient populations of the main part of present Lithuania and Latvia, there is no tradition of depositing cult pottery vessels in graves - although this does not exclude the hypothetical use by the eastern Balts of ritual vessels (not necessarily clay objects) in religious ceremonies. However there is no proof of this in the archaeological records of Lithuania and Latvia.

However, the burial antiquities of the Prussian territory do provide plentiful material for the study of cult accessory vessels of small forms and within the limits of the relevant chronological period. First of all, it is necessary to note the wide circulation of similar vessels (as a rule flat and hemispherical) over the immense spaces of Eurasia that were occupied by the Indo-Europeans in the Bronze Age. In the antiquities of the Celts, the high sacral character of such vessels is emphasized by the find, in Stettung, of the sacred chariot used in the ritual processions of the people of the Halstatt culture (fig. 63,1), and there is a figure of the supreme female deity (Mother Goddess), who holds on her head a hemispherical flat vessel. This artefact allows us to assume, with a certain share of caution, that such vessels might be viewed as symbols of the heavenly sphere of consciousness by the Celts (and other Indo-European tribes?). In Sambia and Mazurs at the end of the 4th phase of the Bronze Age, at the turn of the 6th-5th centuries BC, the formation of the western-Balt barrow culture begins, and one of the main attributes of the burial pottery of this culture is the use of urn covers, mainly bowls of flat spherical form (fig. 63,5-8). That they were made especially to cover urns may be seen by the presence of a special lip around the edge of the internal surface of the bowl (Okulicz Ł., 1976, ryc. 106, 9). Okulich-Kozaryn has attributed these vessels to type VI ("bowls"), referring back to the traditions of the Lusatian and Pomorze cultures (Okulicz Ł., 1970, s. 30). Antiquities of the southern and northern Bronze/Iron Age coastlines of the Baltic also include this above tradition of covering urns with various vessels, including those with hemispherical outlines. In southern Scandinavia, the lower sections of broken vessels were quite often used for covering urns; in the areas of the Pomorze culture (northern Poland), urns containing the ashes of women had flat covers (fig. 63,3), and those containing the remains of men used goblets of conical/spherical form (fig. 63,2) as coverings (not found in settlements). Quite often the lids had "star-shaped" ornamentation (an argument in favour of the bowl being a symbol of the heavenly sphere), later used by the population of the Amber Coast in Roman times on accessory vessels and spindle whorls (Kulakov V.I., 1998a, s. 94, 95, fig. 19). However, on the lids of urns from western-Balt barrows, which, in the beginning of the Iron Age, connected the united traditions of the Lusatian and Pomorze cultures, the ornamentation no longer resembles a "star" (even in view of the possible degradation of the astral composition), but a representation of the seems of the bone plates of the skull (commissures)(fig. 63,5,8). During the final development phase of the western-Balt barrow culture, biconical urns (for burnt bones) were gradually replaced by wide-mouthed pottery receptacles. Small vessels did not disappear from grave goods; they acquired the status of one-handled accessory vessels (fig. 63,9). Such vessels are not found in the synchronous material from settlements and this accentuates their special religious character.

In spite of the significant weakening of the Balt ethnic presence along the Amber Coast in the 1st/beginning of the 5th century AD (Kulakov V.I., 2000b, s. 380), accessory vessels, the lower part of which retained their hemispherical outlines, provide the majority of finds from the graves of the "Veneti" (by Ptolemy), Aestians (by Tacitus), and Vidivarii (by Jordanes), who comprised the multi-ethnic population of the left bank area of the Neman in its lower reaches. These small vessels are divided into several types, among which the most popular in the 1st century AD was the "Krug des Typs Wiekau" (fig. 63,9): the parallel to this vessel is known from the ceramic material of Gotland around 175-300 (Almgren O., Nerman B., 1923, S. 142, Nr 474). Later, in phases C_1-D_2, "vessels of Dollkeim type" became popular (fig. 63,10). During all the periods of Roman influence and the early phase of the great migration, these accessory vessels, frequently parallel with shield-bosses (fig. 63,10), covering urns, are decorated (including astral designs), underlining the high sacred status of these pottery objects (Kulakov V.I., 1998a, s. 92-118). As a rule, there are no finds in these vessels. It is possible to assume that, in the 1st/middle of the 5th century AD, they served as receptacles for organic products (first of all sacrificial food or drink), intended for the grave (for cleansing and fragrances associated with the cremated remains?), the traces of which did not survive.

Not only pottery vessels served the Aestians as receptacles for the above-mentioned contents. At the flat-grave cemetery of "Gora Velikanov", a sacrificial object (H-224) was found in 1992. It was located in an oval pit (1.36 x 1.00m size and O.61m deep) in the bedrock. In the southern sector the pit was covered by two stones, under which was found a broken vessel with modified rim (removed after firing). To the southeast of the vessel was a fragment of a decorated accessory vessel. To the north of these pottery finds, in the lowest part of the grave, was the sawn down top part of a human cranium without traces of fire; the saw line passed approximately through the middle of a forehead. The cut part of the skull could contain approximately 250 grams of liquid, corresponding to the volumes found in a range of artefacts: the H-224-type accessory vessels; vessels from the burials of the western-Balt barrow culture; the majority of "Kruge des Typs Wiekau"; and "Dollkeim-type" vessels. The custom of cutting off the tops of skulls was linked to the Old Prussian territories by finds at settlements such as Zedmar (Kr. Darkehmen), Plossen (Kr. Rössel), and Bauditten (Kr. Mohrungen). Gaerte dates these finds to the Neolithic and early Iron Age (Gaerte W., 1929, S. 17). In his opinion, the bottom parts of these skulls were used as cult bowls, and the author also mentions a similar custom among the southern Germans from the time of the early phase of the great migration. There is a fragment of a decorated vessel (from a sacrificial object H-224) in the archive of the Russian Academy of Science Archaeological Institute dated to this very time - about 450 BC. Kulakov V.I., 1992, Nr 17428). An almost complete parallel of a vessel and fragmented human skull was uncovered in a 1st-century AD sanctuary in Oberdorla (Türingen, Ost-Deutschland), and the crushed skull was found directly inside the vessel (Jankuhn H., 1968, S. 68).

The formation of the Prussian culture in the 5th century AD in Sambia and on the coast of the Vistula Lagoon, is marked by the transition of the use of ceramic urns to the use of receptacles, for burnt bones, made of organic materials (leather, wood, bark, etc.). The urn tradition did not disappear from the western region, populated by the Galindians (?) of the Mazurian Lakes. Here are found accessory vessels in the form of hemispherical bowls on ring feet, known here, and on southeast outlying districts of Sambia (Suvorovo/Zohpen cemetery), from the middle of the 5th to the end of the 6th centuries AD (Nowakowski W., 1989a, s. 115).

At the end of the 5th century AD, the Old Prussians established the rite of strewing burnt bones in graves. From this time the hemispherical accessory begins to disappear from burials and the role of the cult vessel in burial rites changes. During the late phase of the great migration, and in the early Middle Ages (6th-9th centuries), the Old Prussians used "temporary urns" - vessels employed for carrying the burnt bones from the funeral pyre to the grave. Thus the vessel, which, by its use, now belonged, with the bones it contained, to the next world, was thrown with force into the grave, its fragments strewn there among the bones and trampled down into the earth. Originally these "temporary urns" were biconical, handmade vessels (end of the 5th to the beginning of the 8th centuries), but by the turn of the 10th/11th century were wheel-made vessels of western Slavic types, as well as copper alloy dishes of the Hansaschüßel type. In the latter case, the Old Prussians used the trophies plundered from Christian temples in the Lower Vistula areas; these trophies, already from the beginning of the 11th century, began to serve the western Balts from Sambia and its environs as prototypes for the manufacture of small, copper alloy bowls (Kulakov V.I., 1994, s. 43). A similar tradition is encountered in southern Scandinavia (Trotzig G., 1991, p. 121-130). However, although the Vikings of northern Europe regarded bronze bowls/dishes as drinking vessels in their graves (including cult practices), in the Old Prussian lands these metal receptacles employed in their burial rites acted as temporary urns. By their forms, these bowls/dishes met the requirements of the local, archaic religious practices, reflected in the archaeological record of the southeast Baltic from the 5th century BC to the 6th century AD. This, of course, does not exclude at all the bowl/dish functioning as receptacles for sacred drinks at non-cemetery sacrifices.

The similarity of the style and dimensions of the Sambian copper alloy "dish" (bowl) of the late phase of the Viking epoch, to the cult vessel represented in the hands of the priest/Vurskhait in Maletius' book (fig. 64) is amazing. Such vessels were probably called kauszelen by Pretorius. From the 17th to 19th centuries, they were made not only of clay but also of zinc/bronze, and wood - including the sacred oak (Mannhardt W., 1936, S. 556). The above-marked similarity illustrates the continuous tradition of the manufacture and use in religious (sacrificial) ceremonies of bowls/kauszelen, found in the southeast Baltic from the 11th, through to the 17th century. Although there are no finds of sacrificial bowls in the archaeological material of the Prussian territory of the "Order" period, the fact that native religious traditions were preserved in this time is convincingly proved by the burial antiquities of eastern Sambia and Nadravia/ Nadrawien (Kulakov V.I., Valuev A.A., 1999, s. 80-85). The populations of these parts of the former Prussian area in the Order period consisted of autochthonic inhabitants and groups of western and eastern Balts who migrated, not always voluntarily, to the Amber Lands in the Order period. These peoples became the direct ancestors of the inhabitants of Prussian Lithuania, whose ceremonies, at the end of the 17th century, were described by Matteus Pretorius. There are clear grounds for associating the rites described by Pretorius with the religious practice of the Prussians. Evidence of the use of sacrificial bowls, from the turn of the Bronze/Iron Age, of standardized forms (in the 17th century they were called kauszele), are known only from the archaeological material of the Prussian territory. Finds of sacrificial bowls and goblets are found in this territory from the middle of the 1st millennium BC. There are standardized cult bowls in the material of the southeast Baltic from several periods dating from the 5th century BC to the 11th century AD, therefore the absence of such finds in the 6th-10th centuries should be not be interpreted as the interruption in the tradition of the use of such vessels, but, most likely, as their being then manufactured from organic materials (essentially wood) and not deposited in the earth.

At the present phase of investigation, the interpretation of the main sacrifices, and the role of cult bowls in the rites of the population of the historical German-Balt border regions (the Amber Lands) may be viewed as follows:

1. For the Indo-European tribes of the Bronze and early Iron Age, the hemispherical-shaped cult vessel was the most common. Hemispherical lids for the urns, imitating the tops of human heads (and even copying in several cases the design of cranial bones themselves), are comparable with Neolithic sanctuary finds and early Roman skull "cups" that were used as sacrificial vessels. It is reasonable to assume some sort of mystical conformity at the moment of sacrifice between the human skull and the clay sacrificial bowl. The association of an individual to his or her progenitors, especially during the burial ceremony, is most deeply reflected in the Rig Veda and associated with the basic truth of the parallel between the microcosm and the macrocosm (Gamkrelidze T.V., Ivanov V.Vs., 1984, s. 828). This Indo-European association (and also other groups: see Meletinsky E.M., 1980, s. 510) is also reflected in ancient German mythology: "The flesh of Imir/became land, / bones became mountains, / the *skull became sky* /and his blood became sea" (Rechi Waftrudnira, 1975, s. 206). Special attention to the human head was paid at the repeated, and obviously ritual, openings of graves for further religious purposes at the cemeteries of Chernyakov, as well as Wielbark, cultures (particularly in the southeast regions of the latter. Mączyńska M., 1998, S. 299). A similar phenomenon is marked in the Prussian Order period from the 13th to 15th centuries at the Alt-Wehlau cemetery, and in the Prussian folklore tradition of the 18th century. (Kulakov V.I., Valuev A.A., 1996, S. 494). During almost all of the 1st millennium BC, the inhabitants of the Amber Lands, in contrast to their eastern neighbours, adhered precisely to the tradition of the use of hemispherical cult vessels - symbols of the magic connection between the human skull and heavenly sphere.

2. After the weakening of the Balt ethno-cultural presence in Sambia and its environs, the sacrificial receptacles used in burials gradually lost their hemispherical form; this was probably because of specific changes to their ritual function. Many of these vessels, however, still have handles, or their rudiments, allowing these bowls to be held with maximum care in two hands, as related by Pretorius in his description of the later rituals (Mannhardt W., 1936, S. 562). In many other Indo-European areas at the end of the 1st millennium BC, and in the first half of the 1st millennium AD, the hemispherical bowl was used for transporting victims to the gods. The vessels are known in antique Europe as phials and paterae (from here we get the Christian "potir"), in India as gkhats. Such forms (some large) were characteristic of the outlying districts of the western-Balt area (the Mazurian Lakes).

3. The Balt reconquest (recoquista) took place in the southeast Baltic from the second half of the 5th century AD, and resulted, finally, in the revival of the custom of the local use of hemispherical sacrificial bowls. Although from the 11th century they were made from copper alloy, they nevertheless, both by their form and volume, were fairly similar to other ritual vessels of the early Iron Age. This once again shows the very deep Indo-European roots of the religious practices maintained by the western Balts, and that their archaic beliefs lasted up to the time of the creation of the Kingdom of Prussia (Königtum Preussen).

4. The ancient inhabitants of the Amber Lands and their descendants, the Old Prussians, probably did not restrict the use of the kauszele-bowl simply to the offering of libations to the gods. The idea of the flow of life-giving liquids (in the ancient German perception - "the fiery bridge") connected souls (alive and/or dead) with the world of the gods, which was a home for ancestors too. This symbolism represented the main essence of the religious practice of the representatives of the Indo-European tribes - a declaration of the mystical affinity (if not the participation) between a single person (the microcosm) and the world of the gods (macrocosm). In the material from Pomorze and the western-Balt barrow cultures, the bowl is placed rim down over the urn. This signifies one of the participants of the sacrifice - a person who departed for the other world and for whom this last drink of sacrificial liquid might be considered "the fiery bridge" to heaven. (Compare with the practice of ritual ablution and the application of fragrances to burnt remains in Hindu ritual.) Probably, therefore, the relatives perceived the dead person as an envoy of a family/clan/tribe to an assembly of gods and to the uncountable generations of their ancestors. This hypothesis (or at least the idea of a mystical connection between the inhabitants of the earthly world and their ancestors) is supported with a general opinion that the "... burial rites are a form of dialogue between human generations" (Tokarev S.A., 1985, s. 87). The belief that dead envoys could bear offerings from living tribesmen to the inhabitants of the next world was probably already held in the Stone Age (Vandermeersch B., 1970, p. 299). The ancient Germans recognized in the shape of the ritual bowl a representation of the curve of the skull of the god Imir (the sky). To them the shape echoes the final dwelling place of the sacrificial victims, and a reminder of the prayers connected with it. The inhabitants of Prussian Lithuania in the 17th century might well have symbolized in the same way (i.e. a "message" from clan members) the sacrificial flow from the kauszele to the participants in the ceremony and onto the sacred stones or earth; the same libation that was in the end drunk by the officiants in a ritual of prayer and song. The importance of these rituals, probably the basis of later Prussian religious practices, is underlined by the preservation for more than two millennia of the standard shape and dimensions of the particular sacrificial bowl – the 'kauszele'.

VIII. Mounted escorts along the Great Amber Route

As already mentioned in Chapter II, among the synchronous antiquities of Europe, the burial assemblages of the Aestians are distinguished by the presence of ceremonial horse burials. This is confirmed by the numerous finds of items of horse equipment in the burial grounds of the south-eastern Baltic. Jan Jaskanis has designated horse burials as specific features of the burial rites of the western Balts throughout the 1st millennium AD (Jaskanis J., 1974, s. 171). It was earlier supposed that that idea for constructing horse graves in central Europe (including the Baltic) in the 3rd to 5th centuries AD came to the Danube region from Sarmatia. However the presence of horse burials and/or sacrifices among the Germans already by the 1st/2nd century AD predates the start of a wider Sarmatian influence (Müller-Wille M., 1970/1971, S. 188).

In northern Europe, the distribution from the 5th century AD of horse burials was determined by a cultural impulse from the nomadic (Hunnic) area of eastern Eurasia (Klindt-Jensen O., 1957, p. 247). Jaskanis has not excluded the problem of the role of eastern influence on the Balts. In addition, he has interpreted horse burials among the populations of the Amber Coast from the 1st-4th centuries AD and later as a reflection of the important role of the horse in the religious ideas of the local society. The horse was a symbol of many beliefs - a symbol of fertility, a representative of chthonic forces (an agent for the transport of a dead soul to the next world), an indicator that the animal's owner participated in hunting and shared a religious belief - an organic feature of an autochthonic spiritual culture. The burial of a horse with the necessary equipment for riding was an essential part of the burial assemblage of an adult Baltic freeman from the middle Iron Age to the early Middle Age periods. In the southeast Baltic the graves of horsemen and mount shared a common plan (the horse to the west of the human remains, or, by phase B, as a last variant, at a lower level than its rider - Jaskanis J., 1974, s. 169). There are some exceptions to these layouts, as with the separate horse graves in the cemetery at Mojtyny, woj. Warmińsko-Mazurskie (ehem. Moythienen, Kr. Sensburg, and falling within the contact zone between the western Balts and the German ethno-cultural areas) and in some regions of Lithuania. In these areas the horse skeletons are not connected at all with burials of people. In the German areas it is recognized that Balt horse burials have a chronological priority over similar assemblages (Jaskanis J., 1974, s. 250, 251). Significant archaeological successes in recent years in the accumulation of data on horse equipment finds in Barbaricum allow us to return again to the question of horses and horsemen and their burial assemblages in Roman times along the Amber Coast.

In 1994, Susanna Wilbers-Rost, a graduate of Jan Jaskanis from Goettingen University, and former pupil of Klaus Raddatz, the well-known representative of the Prussian Archaeological School, published her dissertation on the problems of the origination, development, and distribution of horse headpieces with bronze-chain reins (Wilbers-Rost S., 1994), categorized also as Vimose and Thorsberg types. Unfortunately, this capital work has remained unnoticed by modern researchers into the antiquities of the Baltic and wider Barbaricum.

The major part of this German researcher's work is a catalogue that takes into account practically all the find spots of horse equipment from Roman times in European Barbaricum (Wilbers-Rost S., 1994, S. 164-213). A primary goal of this catalogue was the maximum gathering of data on headpieces, and their details, for further classification and dating. Wilbers-Rost's predecessor was the Polish archaeologist, Tadeusz Baranowski, who studied separate details of bronze horse headpieces and came to the conclusion that these unique artefacts - of Celtic and Roman influence – originated on the Jutland and Sambian peninsulas in phase B_2, and spread from there into Middle Europe (Baranowski T., 1973, s. 467).

Wilbers-Rost's catalogue provides data on 140 finds connected with horse headpieces in Barbaricum. The basic areas of their distribution in the 1st-4th centuries AD are southern Scandinavia (in all, 35 find spots from Jutland and Sweden - 8 in burials, 19 from places of marsh/lake sacrifices, and 8 of uncertain provenance) and the southeast Baltic (the Sambia peninsula and the Mazurian Lakes; in all 49 find spots). In Sambia and its nearest districts, according to Wilbers-Rost's catalogue, 31 find spots of artefacts of interest to us are known. Of them, 10 provided no information on the contexts in which the headpiece details were found. The other 21 sites were burial grounds in which 38 burials contained the remains of horse equipment from the 1st-4th centuries AD. Thus, in contrast to northern Europe, there, in the conventional centre of the western Balt area, the details of the horse headpieces and the entire complex of horse equipment finds were not located in places of sacrifices but in burials. Therefore, the thesis mentioned above by Jaskanis in relation to horse burials being a main feature of western Balt burial rites seems to be true. Certainly, this conclusion would be valid in case of Sambian burials of men buried with their horses, intended to transport their souls to transcendental heights. Such a picture of Old Prussian beliefs is evident from the Christburg contract of 1249, which reflected the burial rites of the last pagans of the Amber Lands (Pashuto V.T., 1959, s. 501). The Old Prussian culture, dating back to the middle of the 5th century AD, is characterized by the presence in the lowers sections of male graves (less often, and only in the 6th century, located to the west of the horseman) of the whole or part of the skeleton (and/or skin) of a horse with full riding equipment.

On the Amber Coast it appears that there is no similar situation with horse burials within the limits of the later Old Prussian area in the 1st-4th centuries. Of the 38 burials containing headpieces, or their details, collected by Wilbers-Rost, only in 6 graves are there one or two horse skeletons. One of the burials (not catalogued by Wilbers-Rost) is from a cemetery at Yaroslavskoje in the Zelenogradsk region (ehem. Schlakalken, Kr. Samland). This burial assemblage, resembling a flat barrow (grave 1), is in the form of a circle of stones, 5m in diameter and covered by two or three layers of stones. In the northern part of this circle there is an inhumation with a northerly orientation, and the edges of the graves are marked out with lines of stones. Among this warrior's grave goods was a copper-alloy brooch of type AV, 125, and dating grave 1 to the phase B_2/C_1 - 150-200 AD. To the west of the grave two buckles and some horse teeth were detected (Jankuhn H., 1939, S. 246, 247). This assemblage, which has no analogies in terms of its rites in the local material of previous date, is considered one of the most ancient horse burials in the southeast Baltic (Nowakowski W., 1996, S. 63).

Similar "barrow-like" stone coverings are characteristic of the Wielbark culture of the Cecele phase, and are distributed in the B_2/C_1 phases from the Wielbarsk area over to the River Pasłęk to the east, in the land of the Aestians (see Chapter III). Thus, neither in its grave goods, nor in its burial rites, does grave 1 at Yaroslavskoye have anything that would connect it to local traditions, not even echoes of the earlier western-Balt barrow culture. As we have seen, the succession between the rites of this culture and those of the antiquities of the Amber Coast from Roman times, and, assumed Jaskanis, is not supported by the archaeological data (Kulakov V.I., 2000b, s. 372-389). It is necessary to elucidate the validity of this conclusion in the case of horse burials, and more exactly with the minority of them known from the Amber Coast in the beginning of the 1st century AD. This will allow us to discover the origin and specificity of the horse equipment that is regarded as being an exclusive feature for Aestian antiquities. For this purpose a catalogue (Appendix 2) has been compiled of the most widely distributed elements of headpieces from the 1st-3rd centuries AD in Barbaricum. The catalogue is divided into two parts according to the dating of the artefacts. The datings were based both on the thoroughly researched conclusions of Wilbers-Rost (Wilbers-Rost S., 1994, S. 43-68), and on the chronology of the accompanying grave goods from the burial assemblages.

The second part of the catalogue details the materials connected with the Vimose-type horse headpiece (phase C_1), characterized, in particular, by the presence of copper-alloy parts of type Z4. The later type (Thorsberg) of headpieces (phase C_2) is not found in Sambia; this form differs by having an internal reinforcing rib made of chain rings. Prototypes of a similar processing of headpiece details are found in grave goods at the Soldatovo/Friedrichsthal cemetery, as well as harness buckles with a thickened internal frame. According to finds of U-type buckles (usually connected with Marcomannic antiquities in the territory of modern Czechia - Raddatz C., 1957, S. 25, Taf. 1,1), the burial assemblage from Soldatovo dates to phase B_2. The distinctive status of the horse owner is emphasized by the find of a bronze Roman hand-bell (tintinnabula), used by German "barbarians" on their horse headpieces (Nowakowski W., 1994a, S. 138).

The information on the Prussian material in the above-mentioned catalogue strongly suggests that the earliest burials with horse equipment known so far in the southeast Baltic are Grave 21 at Lugovskoye, and Grave III at Povarovka (fig. 65). It is obviously premature to classify them, and the assemblage with a horse skull at Yaroslavskoye, as horse burials. Using Lund Hansen's chronology (Lund Hansen U., 1987, S. 39), the brooches found in the assemblages may be attributed to phase B_{1b} - around 60-80 AD (B_2 by S. Wilbers-Rost). They are therefore earlier than the Grave 1 finds at Yaroslavskoye.

The spatial distribution within Barbaricum of the most popular items of horse harness in phases B_1-B_2 (fig. 66), shows their concentration in three regions: the territory between the rivers Oder and Vistula (NrNr 64, 61 and 71), the southeast Baltic (only in Sambia - 7 find spots with associations of 2 or 3 types of various headpiece details), and the western part of the Baltic coast (in Mecklenburg - Nr 45, and on the island of Fyn - Nr 22). The earliest phase-B_1 artefacts (of headpiece Nr 52) are in direct proximity with the *limes* of the Middle Danube, in a zone of intensive Celtic-Roman contact during the 1st century AD. Both Baranowski and Wilbers-Rost illustrate the Roman and Celtic sources of Vimose-type headpieces, whose prototypes are shown as phases B_1-B_2 (fig. 66). It is indicative that already in the era of Hellenism, and following Scythian and Persian tradition, magnificent harnesses were a distinctive mark of the elite cavalries of the states of the "Diadoches".

Celtic material from the 1st century BC to 1st century AD shows that the Celts favoured headpieces with bronze chain-shaped rein-fixings. Following the decline of the La Tène period bronze chain-shaped fixings, forerunners of the Vimose-type headpiece began to appear in all Celtic regions. These, as well as bits with straight and Π-shaped snaffles (Baranoswki T., 1973, s. 461, 462), were preserved in western and central Europe up to the period of Augustus. Up to the 1st century AD, the Celtic principle of bit arrangement was taken into account by the manufacturers of Roman cavalry equipment. It should not be forgotten that already from the time of Gaius Marius' reforms (around 105 BC), and up to the conquest of Dacia by Trajan (101-107 AD), the allied Roman cavalry was recruited from the "barbarians" - the Celts and Germans (Taratorin V.V., 1999, s. 126, 132). Among the finds made in the legionary camps of the 1st century AD, there are still found bronze harness details approximating to Rh1-type strap fixings (Connoly P., 1981, p. 235, 236, fig. 4), and the predecessors of Vimose-type headpieces (conditionally the Proto-Vimose type). In this connection it is possible to believe with some confidence that the early form of Vimose-type headpieces, based on Celtic design, was standard issue for Roman horsemen and auxiliaries.

In particular, details of Proto-Vimose headpieces were used on the sword-belts of the "aucsillarians-quadi", and found in the burials of southwest Slovakia. These contain analogues of Rh1 strap fixings among the assemblages categorized by finds of AIII, 57 brooches, and shield-bosses of type Jahn 7 (Kolnik T., 1980, Taf. CXLII, 9; CLIV, e), dating to phase B_{2a} (end of the 1st/beginning of the 2nd century AD - Godłowski K., 1992, s. 72). The direct familiarity of the inhabitants of the Amber Coast with Roman soldiers and their equipment took place at the beginning of phase B_{1b} (more precisely between 51 and 63 AD) as a result of the start of the Carnuntum diplomatic mission in Sambia by Atilius Primus, the Roman eques (Koulakov V., 2000, p. 33, 34). In the course of the spontaneous exchanges, the Roman soldiers willingly parted with elements of their equipment, receiving precious (for them) amber in return. In particular, the remains of obvious Roman headpieces with bronze details are found in graves at the Kovrovo cemetery (15 - phase B_2/C_1; 16 - phase B_1; 17 - phase B_{1a}) (Tischler O., Kemke H., 1902, S. 17). The headpiece mounts from grave 15 still retain their enamel details. It is from the time of Atilius Primus' mission that Grave 21 at Lugovskoje and Grave III at Povarovka date (see above) were constructed. This Roman horseman was accompanied by group of soldiers of the XV Legion (LEGIO XV APPOLLINARIS PIA FIDELIS) who earlier, in the time of Emperor Claudius, were billeted in Savaria (Kolosovskaya J.K., 1973, s. 76)[1]. This legion could have been detached to escort Atilius Primus between 51 and 63 AD. After 63 AD the XV Legion was transferred to Judea where it took part in the victorious storming of the mutinous city of Jerusalem.

Synchronous to the earliest Sambian finds, examples of Celtic-Roman headpieces are known from Cekanow (Nr 61) and Jezow (Nr 64), far to the south of the Amber Coast, in the territory between the rivers Oder and Vistula. These locations are situated in regions under the western influence of the Przeworsk culture, for which similar artefacts are not uncommon, along the most difficult section of the Amber Route (the pass over the watershed in a region of warlike tribes, and shown on fig. 66 as a dotted line - by Urbańczyk P., 1998, fig. 2, with the additions of the present author). Although information on the objects found accompanying the headpieces and their details - Proto-Vimose (phase B_1) and Vimose (phase B_2) types - in the Polish territory is absent, it is relatively complete for the Prussian material. An example of such an assemblage is the catalogue of goods from Grave 16 at Kovrovo (an inhumation under a stone covering): copper-alloy brooches AII,42 and AIII,61 with iron details; the skull of a horse (?) with bit; iron socketed crow-head (plough-head?); two spearheads; a pair of bronze (?) spurs of type Ginalski B_2 with a decorative ring at the edge; a accessory vessel (Typ Wiekau); knife; curry-comb; pincers; whetstone; piece of amber; two "wooden handles" and other objects (Tischler O., Kemke H., 1902, S. 17). This set of iron weapons and tools is standard for other Sambian warrior graves from the period including phases B_1-D_2 (Vityaz' S.P., Kulakov V.I., Medvedev A.M., 2000, s. 14, 15); the early stage of this period includes copper-alloy headpieces. According to the data of H. Jankuhn, the crow-head and horse skull with bit found in grave 16 at Kovrovo were located some way from the stone covering (Wilbers-Rost S., 1994, S. 202). This burial contained the remains of a man/warrior, dated by the brooches and spurs to stage B_1 (0-80 AD). Kovrovo 16 (and other assemblages marked in both parts of the catalogue) contained no artefact that could correlate to western Balt antiquities of Roman times, known, for example, from the territories of western Lithuania or the Mazurian Lakes.

Of special interest are the rites of those Amber Lands burials that contain headpieces of types Proto-Vimose and Vimose. Three decades ago, Jan Jaskanis, in his researches into the burial rites of the Aestians in Roman times, ascertained *a priori* a western-Balt origin for the bi-ritual tradition of the 1st-4th centuries AD in the southeast Baltic, and the uniqueness of local cremations in urns of type Grebieten (Jaskanis J., 1974, s. 212, 213, 226).

Actually there is no basis for linking the western Balts with the bi-ritual tradition of the Amber Lands of Roman times. In the Przeworsk area, which is located to the south, the appearance and gradual distribution in phase B_1 of urn cremations and a large number of weapons, coincides fairly well with the movements of the German military groups from the southern Baltic coast. In turn, the distribution in the above-mentioned area of inhumations and urn cremations without weapons, but with provincial-Roman jewellery (as a rule in female burials) is interpreted as the northerly migration of some of the Marcomanni and Quadi (Niewęgłowski A., 1981, s. 73). Most likely, this migration should be interpreted as an adjustment of matrimonial connections between the ethnically related peoples. It is likely that similar ethnic processes also initiated similar rites in the southeast Baltic.

The similarity of rites and nomenclature of grave goods in the male cremations within the Przeworsk area and Sambia attracts attention. They both feature the placing of burnt bones in S-shaped urns, or egg-shaped section (urns

[1] This information was provided by O.A. Tinjajev.

of type Grebieten) with rough external surfaces and rims impressed with fingerprints. Conversely, some of the Przeworsk grave goods (two or more spearheads, shield-boss, knife, the razor, trapezoidal fire-steel) were not found inside urns (Niewęgłowski A., 1981, s. 84), while in Sambia they were. The second distinction between these groups of burials has to do with their chronology: in the Przeworsk area, such urn cremations are distributed in phase B_2 and do not continue up to the final phase of development of the Przeworsk culture, whereas in the Amber Lands they continued into phases B_2/C_1-D_2.

It is important to note that, in Sambia, some specific details of the grave goods (plough-head and bridle-bit) were not connected directly with the weapons or the costume details of the deceased, and were placed outside the grave. The elements of horse equipment (without actual horse skeletons) in the assemblages, which have been taken into account in the catalogue, were arranged in this way (so far as the information is provided); also the head of the horse (entangled in its harness remains) was placed in the same way in Grave 1 at Yaroslavskoye.

In the Celtic material from the late La tène and early Roman epochs, there is a wide spectrum of analogies - not only for the headpieces of type Proto-Vimose, but also for the material that accompanied such headpieces in Sambia. The obligatory Celtic warrior kit of weapons and tools consisted of spearheads, shield-bosses, knives, socketed axes and axes with lugs, scythes (type "Lithuanian"), shears and razor. This set was identical for the late La Tène settlements in Gallia and for those late Celtic antiquities in the territory of modern Czechia (Filip J., 1951, s. 326, ryc. 72). Celtic masters continued to work in the provinces of Pannonia, and also in the first centuries AD, evolving the traditions of their culture, in particular the features of horse harnesses (Fits J., 1986, s. 271). It is true that Celtic burial rites did not include the putting into graves of the above-mentioned tools and details of horse equipment. In any event, it is necessary to search among the prototypes of Sambian and Jutland headpieces (of phases B_2-C_1) for artefacts accompanying the headpieces of the Amber Coast graves in those late Celtic antiquities to the south of the Carpathian Mountains. This is convincingly confirmed by a significant number of finds of headpieces along the Amber Route that crossed the imperial *limes* a little way into former Celtic territory. There is no doubt that the output of Celtic and Roman craftspeople was also found in the Przeworsk area, where warrior graves of phase B_1 and B_2 contain two-edged swords, spearheads and javelin-heads (framei, for Celts - grosf), shield-bosses, spurs, and, much less often, axes, scythes, shears, and fire-steels (Godłowski K., 1981a, s. 81-87). True, the groups of Przeworsk weapons 1-3 (phases B_1 and B_{2a}) and 4-5 (phases B_{2b} and C_{1a}), allocated by Godłowski (Godłowski K., 1992, s. 72-75), coincide with the synchronous material of Amber Coast warrior graves only by the nomenclature of spearheads and details of shields. By the number of items in this rather standardized kit, the material (especially from phase C) is much more representative than the Przeworsk one, and much closer to the synchronous German warrior kit of stage Adler 3 (Adler W., 1993, S. 124, 125; Beilage 2).

Studying the events that happened in Barbaricum from the 1st century BC to the 1st century AD, Shchukin has proposed that the Celtic-German cultural diffusion begins after the appearance of the Marcomanni in the territory of modern Czechia in 6-9 AD (Shchukin M.B., 1994, p. 191). Taking into account the debatable issues in his hypotheses concerning the destinies of the German and other tribes in the times of Octavian Augustus, there is a persuasive logic in his conclusion that, within the limits of Maroboduus' "power", the Celtic masters might have aspired to distribute the "Romanized" products far to the north, along the Amber Route that connected Sambia and Aquileia (Shchukin M.B., 1994, p. 194, 195). The caravans might only move with their assistance along the river valleys, supplying the inhabitants of the Baltic coast, in exchange for their amber, products in the opus interrasile style, as well as coins and provincial-Roman brooches. In Sambia and the Mazurian Lakes area, and obviously not without the influence of the newcomers, centres of amber processing appeared by the middle of the 1st century AD, distributing their precious products not only to southern Europe, but also to the adjoining (from the northeast) Balt region (Bliujienė A., 2001, p. 173, 182). In the Aestian burials of phases B_1, the Celtic traditions of jewellery directly specify the ethno-cultural roots of their authors. Wołągiewicz adheres to the same point of view, but already with reference to early Wielbarsk antiquities (Wołągiewicz R., 1981a, s. 151).

The changes in the ethnic situation, resulting from the spreading Celtic influence along the Amber Route to the region of interest to us, were fixed by written sources. Tacitus places the Aestians, whose western Balt origin does not so far give rise to doubt, in Sambia and to the east of it during the first half of the 1st century AD (Nowakowski W., 1990, ryc. 1). For the period between 50 – 170 AD, according to Ptolemy, the lands between the rivers Ουιστυλασ (Vistula), Κρονωσ (Prohlad-Naja/ Frisching or Pregolya), and Ρουδων and Τουρουντοσ (Gilge and Neman?), were occupied by the Veneti and Velti (Velites or the Celts?), and the Aestians were moved northwards from Sambia (Ellegård E., 1987, p. 14, 15). Thus the stabilization of the Amber Route that resulted from Atilius Primus' mission exposed Sambian amber to ethno-cultural impulses from the south, including former Celtic areas. Probable indicators of this impulse are finds along the Amber Route and in Sambia of copper-alloy horse headpieces and accompanying Celt-like grave goods; although the principles of their arrangement in burials (both inhumations and cremations) are far from Celtic ritual norms.

This phenomenon, until now looked on by archaeologists as evidence of the historical ethnography of the Aestians, may be explained by events that took place on the western boundaries of Barbaricum. The strengthening of the Rhine *limes* in the 1st century AD compelled the Roman command to involve more widely the Celts and Germans as foot soldiers (foederati), lightly armed velites and auxillaries. As a result, numerous burials appeared in the local cemeteries of the left bank of the Lower Rhein. These were carried out according to local rites, and were, moreover, filled with Roman weapons and equipment. It is supposed that these burials contained the remains of "barbarous" warriors who appeared after Augustus' military reforms – semi-regular (iuventus) and professional (alae, turmae, cohortes) troops recruited from among the local men. Overshadowed by the great Roman eagles and fully conscious of the honour of serving under the leadership of the Emperor, these "barbarians" considered it their duty to give the weapons and equipment supplied to them by the Romans to their lost brother-in-arms as they set off for the next world (Waurick G., 1994, S. 23-25).

The same situation probably applied to the Amber Coast at the beginning of the 1st century AD. Actually, the grave goods of those burials that contained elements of horse equipment included the full equipment of cavalryman from "barbarous" horse "ala" (ala - about 250 individuals). From Caesar's time, such horsemen were called desultores (Latin: "jumping off", an early analogue of dragoons), accompanied by lightly armed soldier-velites. If necessary these warriors made up the number of the Roman foot-soldiers auxiliaries. It is unreasonable to considering all the buried soldiers-horsemen in Sambian cemeteries as veterans who had earlier served on the distant *limes* and returned to native penates. Most likely the owners of horses with magnificent bronze headpieces were members of one or more legions of cavalry divisions billeted on the Lower Danube *limes*. Its "barbarous" contingent (conditionally the "Sambian ala" or SA) might have been billeted after Atilius Primus' mission to Sambia, and, as a convoy division, was called upon to protect the amber mines and the movements of caravans along the Amber Route. Only by such a resolution of this problem (of the initial phase of the existence of Proto-Vimose-type horse headpieces along the Amber Coast) is it possible to explain not only their Celtic-Roman origin, but also their scarcity along the Amber Route and in Sambia and the grave goods finds in the burials of horsemen, which is unprecedented for Barbaricum. The list of artefacts characterizes all the needs of the horse warrior: a thrusting spear adapted for battles on horse and foot, a throwing javelin, for hand-to-hand combat a socketed axe, axe with lug, and heavy battle knife. There was a whetstone for his weapons as well as a currycomb and shears for his horse. Nourishment was very occasionally provided for by cone-shaped plough-heads and/or portable scythes of "Lithuanian" type. (In some instances the plough-head was replaced by a socketed axe of "wide Celt" type; the late examples of these that have been found in some Byelorussian assemblages are categorized as early La Tène "imports". These are known in the area of the "shaded ceramic" culture from the 1st century BC.) The horsemen of the Amber Coast practised a system of "slash and burn" agriculture, as demonstrated by the presence in their burials of fire-steels, socketed, narrow-edged axes and "humpbacked" knives (Krummesser). The axe was used for cutting, and thus might serve as a head for a plough, the knife for hacking tree branches, and a fire-steel for the burning needed to clear a plot for future ploughing. (It should be said that the finds of fire-steels are more characteristic of phase C_1). The majority of the listed finds was connected with the horse - a warrior's principal companion - therefore the presence of the headpiece (in some cases with the other grave goods in the urn) is to be expected among the artefacts of a soldier from the "Sambian ala". Because these items (including the headpieces) are found in complete sets, it is impossible to conclude that they were the products of Aestian trade with Rome: military equipment was not delivered to "barbarians", particularly as full sets. This conclusion has been recently confirmed by Jaroslav Tejral, who connected numerous finds of Roman militaria (2nd century AD) from the north of Carnuntum with the long-term presence of Empire military divisions on the southern extremity of the Amber Route (Tejral J., 1997b, p. 117). In contrast to the amber deposits of Sambia, far from the imperial borders, which were guarded by the "barbarous" soldiers of the "Sambian ala", regular Roman troops supervised the approaches to the *limes* and the crucial southern section of the Great Amber Route in the disturbed years during the Marcomannic Wars.

The grave goods of Sambian horsemen of phases B_1 and B_2 help us to reconstruct the main aspects of their responsibilities. Their main function was the protection of the Amber Route and its amber deposits. From the 1st-4th centuries AD, the journey along the Amber Route took about 7 weeks from Carnuntum on the Danube up to Sambia (Wielowiejski J., 1980, s. 129). As well as providing escorts, the soldiers of the "Sambian ala" were trained to repulse sudden assaults made on the amber caravans as they followed the rivers, valleys and watersheds. As the burial items, including headpieces, show, the fighters held their enemies at bay with spears and javelins, thus fully corresponding to the tactics of Scandinavian horsemen in Roman times (Engström J., 1992, p. 48). Sambian warriors did not use long swords (at the beginning of the 1st century AD such swords were used in Europe only by Celtic cavalrymen) in close battle, preferring to fight with axes and battle knives at close quarter: thus desultores-auxillaries became infantrymen. A warrior tied his horse to an arch-shaped, bronze fitting (fig. 65,7) secured under the horse's lower jaw: such objects are so far only known in Sambia and Ireland. The antipathy of Sambian horsemen to long blades is nicely illustrated in Grave 34 at Chrustal'noye by a spatha (fig. 67, a), which was shortened to the length of a gladius, the weapon of the foot-legionary. This assemblage was accompanied by a headpiece of type Vimose (fig. 68).

This is a variant of the solution to the problem relating to the composition of Sambian weapons in the 1st-4th centuries - a problem that taxed Nowakowski. Our Polish colleague arrived at an incorrect conclusion about an axe find and its role in a set of Aestian equipment (Nowakowski W., 1994b, S. 387-389). In his opinion, Tacitus ostensibly referred to an axe by the term fustis (Latin "cudgel", a favourite German and late-Prussian throwing weapon). Actually, among the soldiers of SA the axe was intended first of all as an everyday tool; the main weapon was the spear.

Details of prestigious horse equipment, as well as other items of military kit, were used to signify to the next world that the deceased was of a high social status. The sacral sense of grave goods as indicators to the gods, and messages from the living to their ancestors, carried by the departed "envoy" (see. Chapter VII), is quite comparable by its religious significance to the phenomenon of sacrifice, i.e. the same message to both gods and ancestors. True, in the latter case the demonstration by "objects" (sacrifices) is not transmitted by a deceased "envoy", but is "sent" directly to a deity via a special gateway connected with water (a lake or river setting) reserved especially for sacrifices.

As our experience of the Barbaricum "prince burials" shows, leaders of clans, tribes, or in other ways "supporters" of the Empire, were buried with grave goods that indicated the high social status of the deceased. This often included horses with full equipment, as they were considered (as they were in the later Old Prussian culture) as a means of transport to the next world (Steuer H., 1998a, S. 170, 171). This is a feature of the chieftain burials at Lübsow and Hassleben-Leuna that is characteristic of northern continental Germany in the 1st-2nd centuries AD. For warriors of the German "followers", obliged to accompany the leader into battle and to sit together with him at the feasting table (as reflected in the traditional hero's way of life – "einkherians" in Valhalla having fallen on the battle-field, as in "Older Edde"), the presence of a horse within the burial context was not presupposed (Steuer H., 1998b, S. 550). The social status of those individuals, in whose burials details of horse equipment were found, should be interpreted according to the above tradition. As the catalogue shows, from the 8 burials in Scandinavia with headpieces of phases C_1-C_2 (a time of major changes to headpieces in the areas of Barbaricum), only 3 contain horse remains (in one case there are two horses) and these may correlate to "princely" burials. There is a similar proportion in Sambian graves; from 38 graves with headpieces, according to the catalogue, only 6 revealed horse bones. This also allows us to conclude that the majority of the investigated assemblages belonged to professional soldiers, not to "leaders" or ordinary members of society (as later in the Old Prussian culture). In "prince burials" of the Aestian area (see Chapter III), horse remains of are not found, with the exception of Grave I at Izobil'noye.

Finds of the most common elements of horse headpieces, from the 1st-3rd centuries AD, as well as from the beginning of the 1st millennium AD, are presented in three groups (fig. 69), more amorphous on the outlines than from phases B_1-B_2. This applies particularly to the territory of modern Hungary, where the provenance of headpiece finds (NrNr 01, 138-140), and even if they have come from an archaeological site at all, are unknown. With some caution, it is possible to believe that these finds belong to the Upper Teiss basin, which, at the turn of the 2nd-3rd centuries, divided the tribal areas of the Sarmatians and Quadi. It is probable that the headpiece finds that are connected with the activities of the horse troops testify to the unstable situation after the Marcomannic Wars. In this period there are no finds along the Amber Route, but individual headpieces are known to the east and west of its southernmost line which was broken by the thrusts of the Marcomanni (166/167 and 170) towards Carnuntum and further along the imperial road to Aquilea. Probably, and linked with the legends spread by the "barbarians" of the great riches of the southern sections of the Amber Route, the former natives of Scandinavia decided to acquire all these riches in one fell swoop and without exchange procedures. The movement of the German and Roman military troops during the Marcomannic Wars, which were unsuccessful for the "barbarians", stimulated the distribution in central and northern Europe of a good many artefacts that were characteristic even for phase B_2 (Godłowski K., 1994a, S. 118). This conclusion is also applicable for headpieces of type Vimose.

The impact of the Marcomanni, and the reciprocal movements of the Roman armies to the north along the valley of the River Morava as far as Přibice, resulted in a period of instability for the Amber Route. As a result, the activities of the Sambian ala became less relevant. (Beforehand, from the middle of the 1st to the second third of the 2nd century AD, they had been deployed on the Amber Coast to undertake convoy operations and accompany trading caravans on the line of the Vistula - the Upper Oder and the Morava.) This is confirmed by the absence of finds of Vimose-type horse headpieces along this line. The Hungarian finds marked on fig. 69, and rather inadequately dated, may be traces of the activities of Sambian ala, who returned to the site of their former deployment, as well as other cavalry divisions of the foederati on Roman service. In Sambia proper, and only in its western part, and originating from the amber mines to the east, four headpiece find sites are known (fig. 69, map). Among them is an early headpiece ("a") from Grave III at Povarovka (fig. 65,6-9), a "trophy" which has found its way into a grave of period C_1. In addition, the assemblage at Kulikovo (fig. 61) contains a direct indication (rare among antique bronze coins from Lucilla) of the Marcomannic Wars, as a terminus post quem. Information on the presence of horse bones in these Sambian assemblages is lacking.

There are also no horse burials on the sites where Vimose-type headpieces of phases C_1-C_2 were found. Over the same period in eastern Denmark they appear in incomparably greater numbers than for phases B_1-B_2 (fig. 66). This reverse interrelation between Sambian and Danish finds of headpieces is to be expected.

Based on convincing typological and chronological evidence, Wilbers-Rost has proved that the Vimose-type headpieces originated in Sambia. As with the earlier brooches of type AIII, 57-61, these headpieces appear at the end of the 2nd century AD in great numbers on the east coast of Denmark, mainly in marsh or lake settings used for sacrifices (Wilbers-Rost S., 1994, S. 101, 102, 107). Wilbers-Rost, and many Scandinavian researchers, looks on this phenomenon as a result of the widespread trade that developed between the inhabitants of the opposite coasts of the Baltic. However, it is incontestably proved that the many weapons and objects of horse equipment that have turned up in sacrificial sites in the geomantic centre of the German area (Podosinov A.V., 1999, s. 344) - Jutland, and on the islands of Fyne and Seland - are the spoils of German "followers" that they received on distant campaigns. Especially large sacrifices of military spoils were made at the lake sanctuaries of Denmark in the 2nd-3rd centuries AD, i.e. just after the Marcomannic Wars. These religious rites were undertaken to emphasize the military superiority and identity of the early German "followers" (Müller-Wille M., 1999, S. 63, Abb. 72). It is also quite apparent that, for the northerners, the Vimose-type headpieces were also military spoils received in Sambia.

The German incursion might have been a consequence of the distinct weakening of Roman positions on the Amber Coast as a result of the Marcomannic Wars; the Sambian ala was withdrawn, or cut off, from the Danube *limes*. Like Marcomanni, at the southern section of the Amber Route, the German troops (perhaps from east Jutland) attacked Sambia with the intention of robbery and the control of amber deposits. This was not reflected at all in the supply of amber to the Empire, where, in the middle of the 2nd century AD, the demand for amber soared (Tempelmann M., 1985, S. 109). However, after the end of the Marcomannic Wars, in the Wielbark and Przeworsk areas and the region on the right bank of the Neman, the scale of amber "imports" sharply increases because of the elimination of the SA from Sambia. Among the "imports" in phases C_{1b}-C_3, mushroom-shaped beads (type Tempelmann 471h) predominate exclusively, although these beads are extremely rare in the Amber Lands proper (in Sambia, at Grebieten and Bol'shoye Isakovo) and are considered to be connected with Wielbarsk and Przeworsk antiquities (Kazansky M.M., Mastykova A.V., 1998, s. 103). Even the most superficial analysis of these finds shows that the makers of these mushroom-shaped beads did not understand the nature of amber and its potential for the manufacture of artefacts. These workers wasted a great deal of the amber raw material when sawing the beads, and could not possibly belong to the Balt populations, who, over many centuries, were fully acquainted with amber and its properties. Further proof of the distinctiveness of the German newcomers who appeared in Sambia during phases B_2/C_1-C_1 is shown by the extremely inept way they tried to mimic Roman drop-shaped amber beads or the amber heads of Roman spindles (Ganzelewski M., Slotta R., 1996, S. 424, 425). Analogues of such beads are known from regions of the Roman *limes* in phase C_{1b} in the form of glass, and even bronze, pendants (Tempelmann M., 1985, S. 112).

According to numerous finds on the islands of the western Baltic of grave goods that were previously characteristic of Sambia (not only headpieces, but socketed axes as well - Thomsen P.O., 1993, p. 26, fig. 5), this military operation was highly successful. Moreover, it is possible to assume that significant numbers of inhabitants from the Danish islands moved to Sambia, a fact reflected in the material from the local cemeteries (Vityaz' S.P., Kulakov V.I., Medvedev A.M., 2000, s. 12). They brought with them the traditions of urn cremations (urns with rough surfaces), as well as cremations without urns, accompanied by articles such as accessory vessels (Typ Dollkeim), weapons (as at Godłowski 6 - Godłowski K., 1992, S. 74), and horse equipment (including simplified variants of Vimose-type headpieces, already without copper-alloy details), and the custom of constructing "prince burials", stone circles, and "stelae". In the Przeworsk area, the distribution of similar features of rites (true, without horse equipment) took place earlier, in phase B_2, and is connected with the movement from the north and west of new groups of Germans (Nieweęgłowski A., 1981, s. 91). However if these traditions were not established for a long time among the peoples of the Przeworsk culture, the customs of the German newcomers to Sambia were maintained down to the first half of the 5th century AD.

The distribution in the 3rd century AD of burials to the south of Sambia with similar attributes (in the Mazurian Lakes area and also rich in amber deposits) shows the tendency of the newcomers to expand a zone of control in the southeast Baltic. It is true that, almost without exception in this area of the Sudins/Sudawians, riding equipment (including local forms of headpieces) correlates in the cemeteries with horse burials (Jaskanis J., 1968, s. 90). There are rudiments of local rites in this part of the Baltic (above all the barrow tradition) and there can be little doubt about the preservation here of the western-Balt ethno-cultural identity.

The study of the origin and distribution of Vimose-type headpieces in the southeast Baltic during Roman times allows us to draw the following conclusions:

A. From phases B_1-B_2 burials with details, or with a full set, of horse equipment (only from assemblages with sufficient information on the grave goods and burial rites) are known in the territory of the southeast Baltic only within the limits of Sambia (fig. 71); there are 21 in number, in part included in the catalogue submitted above. Of these, 6 burials contain a horse skeleton, and in three cases the remains of the horse's skull are revealed. A horse skeleton with Vimose-type headpieces was only found at the Zarechie cemetery. Concentrations of burials of this group in Sambia are usually found near the amber deposits and/or the natural borders of the peninsula (fig. 71). This confirms the above assumption regarding the disposition along the Amber Coast (c. 63-67 - 167-180) of a certain cavalry division (the Sambian ala) that looked after Roman interests in and around the world's most extensive amber deposits. Concentrations of remains also convincingly mark the entry points to the SA of small groups of horsemen (on the west side of the peninsula at the basic deposits; on the north side of the peninsula at the harbour that was most convenient for trade ships to the east of the modern cape of Gvardeysky/Rantauer Spitze). It is difficult to determine definitely the ethno-cultural composition of the SA that obviously included the "barbarians"-auxiliaries. It is possible that a certain (and significant?) part of the "ala" was made by Romanised Celts who earlier lived on the left bank of the Danube (modern Hungary). Constantly billeted in Sambia, the horsemen-desultores, and those infantrymen accompanying them in case of need, provided the escorts for the amber caravans. Probably, during their travels from the Amber Lands to the south in the second half of the 1st century AD, they would incidentally capture groups of the Wielbark population (above all women), who left details of their costume in the Sarmatian area on the Danube (Mesterhazy K., 1989, S. 187-190). After their termination of service, the SA veterans that were recruited (quite logically) from among the natives came back to their native settlements and were eventually buried in local cemeteries. Special features of their burial rites (full sets of military equipment in the graves) closely resemble the rituals that were carried out at this time by the Celts-foederati of Rome on the Rhine *limes*. In both cases the participants of the burial ceremony, when constructing a burial the general function of which was everywhere and always the "maintenance of posthumous existence" (Smirnov J.A., 1997, s. 39), testified to the high rank and honorary title of the buried man as a soldier of Great Rome by the careful composition of his grave goods.

B. The events of the Marcomannic Wars (166-180) interrupted the activities of the Amber Route and made the further stay of the Sambian ala on the Amber Coast impossible. Involved with the riches of the Amber Route, the German "followers" broke through not only its southern extremity, but also, probably, carried out successful operations at its source - in Sambia. It is not reflected in the number of horsemen burials known here (28 burials, not all of them included in the Wilbers-Rost catalogue. Of them, 11 graves contained a horse skeleton, and 3 graves had horse skulls - fig. 72). It is true that on the Baltic sites the number of Vimose-type headpieces decreases sharply, but they appear in bulk at the lake sanctuaries of the islands of Fyne and Seland (Kulakov V.I., 2001, S. 55), and are dropped as sacrifices into the rivers of the north of continental Germany. Probably, they were spoils received by the Germans on the Amber Coast when crushing the remnants of the Sambian ala. The newcomers to the lands of the Veneti and Velts accepted the customs of their recent enemies and continued to protect the amber mines, not for Rome, but for their own benefit. The magnificent headpieces soon disappear from the burials of the allochthons, and during phase C_1 the former wide spectrum of grave goods is reduced to spearhead, battle knife, socketed axe, scythe, and fire-steel - the minimal military and economic kit. Thus, trying to develop the lands to the east of Sambia (in search of commodity markets of amber or its new deposits) the Germans came to the confluence of the rivers Instruch, Angrapa and Pissa, leaving horse burials there too (fig. NrNr 72, 181). Here these prestigious rites were adopted by the local western Balt population. This corresponds to the norm, widespread in Barbaricum, of borrowing the customs of those tribes who were famous for their achievements. "The Germans most of all appreciated their own originality and their primordial customs: if they imitated somebody, then it was only those who had the greater glory." (Budanova V.P., 2000, s. 111). In their environment, the eastern part of the Mazurian Lakes (fig. 72), horse burials (their rites and grave goods being rather close to the Sambian norms) continued up to the 5th century AD.

C. The most difficult problem is that of tracing the ethno-cultural sources for the tradition of placing horse remains in the graves of the inhabitants of the Amber Lands from late Roman times onwards. This tradition from phase C_2 became the main attribute of western Balt antiquities, firstly in eastern Mazurs, and, later, from phase D_3 in Sambia. Here the rites associated with horse burials returns as a result of the "Balt reconquista" of the Amber Coast, made, probably, by the population of eastern Mazurs and provoked by the outflow of the former Sambian population to the south for participation in the Hunnic wars. An attribute of this "reconquista" in the beginning of the 6th century AD is the distribution along the Amber Coast of horse burials placed to the west of the actual grave of the horseman and biconical vessels of subtypes 1.1-1.3 (Kulakov V.I., 1994, s. 42). Beforehand, this was only characteristic of the Suwalkian group of the western-Balt culture (Okulicz J., 1973, s. 415, 416, ryc. 200, e, h, j). This mass migration was promoted by the network of related ethno-cultural connections which developed at the end of the 4th/beginning of the 5th century AD between the people of the SNG and those of the Sudawian culture in the Prudz' phase of its existence. Already by then, at the beginning of the great migration, the Sudins started their movement westwards, occupying the Goldap micro-region (Bitner-Wróblewska A., 1998, s. 308, 309).

From the 1st century BC to the 2nd century AD, at settlements and sacrifice sites of north German tribes, there are special places for the concealment of horse carcasses and their parts; these places are fairly interpreted as sacrificial sites (Müller-Wille M., 1970/1971, S. 231-233). The special religious attitude of the Germans to horses in phase B_{1a} is well known (Tatsit, 1969, s. 357, 358). From the 2nd century AD there is a military tradition of horse sacrifices. Orosius in "Historiae adversum Paganus" described in detail these religious activities, during which the Germans (Teutons and Cimbri) "... threw into the river gold and silver, cut to pieces the military armour, broke horse harnesses..." (Pennik N., Dzhons P., 2000, s. 213). Identical results of such sacrifices have been noted in Vimose, Nydam, Kragehus, Ejsbøl, and Thorsbjerg. They are not only evidence for the reverence of gods, but also symbols of the authority of leaders-donors (Hedeager L., 1992, p. 285, 286). In these sacrificial spots, horse skeletons are not only accompanied by riding equipment, but also by weaponry and feasting vessels. By its composition, this set of goods (and horses) was obviously intended not so much for the gods, but rather for the warriors who had set out for the "Empire of those killed in action" - Valhalla. Probably, objective circumstances did not allow the surviving warriors to place these sets of objects into the graves of their brothers-in-arms who had fallen far from their ancestral native lands. Single military burials (but not those of chieftains) containing horse skeletons and weapons are known (in phases C_1-D_1) in areas of the Marcomanni (Preidel H., 1937, S. 590, Taf. 234,1; Müller-Wille M., 1970/1971, S. 227) and Quadi (Erdeli I., 1986, s. 290). As shown above, it was these peoples (more correctly together with them) that the tradition of Proto-Vimose-type headpieces penetrated into Sambia during phases B_1-B_2. Obviously, the idea of placing a horse skeleton or a horse skull in the grave of a soldier, as a demonstration of his occupation and social status, was of Old German origin and appeared in Europe prior to the beginning of wider contacts with the nomadic world. Subsequently the nomads only strengthened this autochthon tradition. In the burials of the Marcomanni, Quadi, and the "Veneti and Velti" of Sambia, the remains of a horse should be interpreted as an element of additional (necrological) structure connected with the action of sacrifice (Smirnov J.A., 1997, s. 76), and/or as the special nature of grave goods (above all in cases where headpieces were found without horse remains) of high-status character. Just on the Sambian borders, as well as at German sacrifice locations, arrangements of phase-B_1-B_2 horse burials (fig. 71) were connected with water and especially emphasized the sacral character of these assemblages. The sets of items from military sacrifices in eastern Denmark, without bodies of soldiers and indistinguishable from the grave goods of kenotaphos, correspond as a whole to the grave goods of the military burials from Sambia in phases B_2/C_1-C_1. However, the special conditions of the sacrifices, in view of all the above, allows us to interpret these actions not only (and perhaps not widely) as sacrificial ones, but as the creation of the kenotaphos phenomenon: the bodies of the dead are absent and a sufficiently complete set of objects is "sent" to the souls of the deceased heroes in Valhalla - not through an actual grave, but via the mysteries of a watery sacrifice. Thus the presence of horse skeletons in Sambian cemeteries should not be regarded as the means of transporting the deceased along a path to the next world, but as a symbol that the dead warrior is worthy to dwell in the transcendental halls of that other great god who was killed in action - Odin. In other words the horse was the subject of sacrifice according to Old German religious traditions.

In phase C_2 the rites of early German "followers" were adopted by the western Balts and activated in an initial stage of formation of Old Prussian culture. In phase D_3 the skeleton of a horse provides a link to the Old Prussian culture, placed in the lower level of a grave. A horse, properly bridled and even correctly posed (the back ready to accept the horseman, the head extended forward and upwards) would show its readiness to travel the long and transcendental journey; it is no longer a sacrifice for the lost soldier, but represents a chthonic image of the way the deceased's spirit must now travel. Recognized by the participants of the "Balt reconquista", the Old German ceremony is reinterpreted and becomes an integral feature of the Old Prussian spiritual culture. Among the eastern Slavs of the early Middle Ages, the barrow tradition, which was uncommon for them, began to be distributed in the same way, and was later to be replaced by the (totally alien for Slavic peoples) Christian rite of inhumation in a coffin orientated to the west. Nowadays these ceremonies are, by right, part of a set of attributes of historical and modern Slavic ethnography.

IX. The Amber Coast after the Marcomannic Wars

The modern level of our knowledge of the outlying north-eastern districts of Oikumene in the 1st-4th centuries AD allows us with a certain level of confidence to reconstruct the processes that took place on these boundaries, far from Rome, in Empire times. The travels of merchants and military campaigns expanded the limits of Oikumene, and at the beginning of the 1st century AD the Romans considered that its northern borders were not limited by the Alps, as it was before the epoch of Alexander the Great, but washed by the cold waves of the Sveb sea (the Baltic Sea) and the German Ocean (the North Sea). As has been already shown above, Roman knowledge of Barbaricum solum, and above all the Amber coast, developed considerably in the time of Nero. From this date the Great Amber Route (fig. 8) began to function fully, connecting as it did, via the Vistula and the Danube, the Aestian territory with the Roman Empire. It ran between Aquileia and Sambia and was served by people of the Wielbark culture – the Lubovidz phase (Kulakov V.I., 2001, S. 45). This is testified to by a significant amount of Wielbarsk jewellery taken from the burial sites of Western Sambia (phases B_1 and B_2), as well as by finds from the cemeteries at Dollkeim and Lauth (Kr. Samland - Bol'shoye Isakovo, Guryevsk District) of Kumpf-type vessels with a certain roughness to the lower body sections. Moreover, the absence in Amber Coast antiquities of barrows, and those jewellery derivatives that are characteristic of western Balt cultures from Roman times, indicate that "the western Balts from the 1st century AD did not comprise the ethnic majority of the Amber Coast" (Kulakov V.I., 2000, s. 380). As the evidence of the Roman authors summarized by Nowakowski (Nowakowski W., 1995, ryc. 2) shows, the antique world knew the inhabitants of the Amber Coast of the 1st century AD under the name of "Veneti". This name in the Julius-Claudius epoch might have been assigned to "provincial" dealers and handicraftsmen for whom there was no place within the limits of the power of Octavian Augustus (Shchukin M.B., 1998, s. 206). Searching the best share, they penetrated into a related area of the Luguii, and, by migrating northwards, reached Sambia – driven by their aspiration to take full control of the main Amber Route.

Intensive barter developed between the eastern provinces of the Empire and the lands of the Lower Vistula, and as a result objects of antique use began to appear frequently in the local environment. The Lugii of the Przeworsk area were intermediaries in the amber trade to the north of the Carpathians, as were the Marcomanni to the south of them (Hedeader 1983, p. 193, 197).

In the third quarter of the 2nd century AD, Svebia was to deal the next stroke to Rome: under Marcus Aurelius' government the Marcomannic Wars broke out; the "barbarians" tried to put under rigid control the southern extremity of the Great Amber Route - Aquileia (Budanova V.P., 1999, s. 43). Ending in 179 AD with the defeat of the "barbarians", these wars highlighted a new tendency in Barbaricum whereby the German tribes aspired not to attack the Empire but to settle in its provinces (Budanova V.P., 2000, s. 30, 31). However, during the final stages of the wars, the Roman troops, including the X Legion, occupied the basin of the River Morava, thus blocking a pass used by amber caravans along the Moravian route (Tejral J., 1997b, S. 130, 131). That is why, soon after the end of the Marcomannic Wars, the inhabitants of Jutland, the geomantic centre of Barbaricum (Jackson T.N., Podosinov A.V., 2000, s. 116), situated "to the north of some sea" and a legendary ancestral home of the Germans (Podosinov A.V., 1999, s. 344), tried to establish their own variant of the Amber Route (Wielowiejski J., 1980, s. 94), as shown in the left of figure 73, marked by a dotted line of triangles. Jutland, as well as Sambia, had amber deposits although considerably less than the lands of the Aestians. A significant part of the Roman "imports" that were received in exchange for amber, and seized during the military campaigns of phase B_2 (Ilkjær J., 1994, p. 133), were intended as sacrifices to Wodan and the goddess Nerthus, whose sanctuaries were located on the islands of Fine and Zeland and were honoured by all the German tribes in the early Iron Age, through to Roman times (Janssen H.-L., 1942, S. 205, 206). It is probable that trophies from the Marcomannic Wars, and the goods received in exchange for Jutland amber, were insufficient for these sacrifices, and the Jutlanders began in the 3rd century AD to attack the Rhineland provinces of the Empire: the weapons and equipment so obtained found their way into the sacrificial lake-sites (von Carnap-Bornheim C., 1997, S. 227). However, already by the final stage of the wars, new sources for the sacrificial material appeared on the islands of the Western Baltic. Shchukin has made great efforts to understand this problem. He has shown the establishment (in the 3rd century AD) of steady contacts between the Jutland population and the peoples of the northwest of the Black Sea area up as far as Dobrudja (Ščukin M.B., 2000, S. 352-354): he cites trading links as the reasons for these contacts. However the same author earlier determined a Pontic origin for the Typ Havor gold filigree neck-rings from the first half of the 1st century AD. These items were intended for sacrifices (Hagberg U.E., 1984, p. 77, 78), and especially emphasized the Scandinavian-Thracian connections which arose even during the time of the campaigns of the Cimbri and Teutons from 113-101 BC (Shchukin M.B., 1999, s. 171, 173). As shown, the linking of these connections to the time of Flavius can also be very persuasive.

It is known that in the Wielbarsk area of the Lower Vistula (phase B_2/C_1), the Lyubovidz stage of local antiquities was replaced by the Cecele phase, and the local inhabitants started their movement to the southeast (Godłowski K., 1994a, s. 166, Abb. 7). Among the taxons of this migration are wire rings with tied ends (taxon 1), the S-shaped connectors of the string-ends of necklaces (taxon 2), pole-axe-shaped (taxon 3) and bucket-shaped (taxon 4) pendants. It is true that taxon 2 and the prototype of taxon 4 (small-amphora-shaped pendant) are already encountered during the Lyubovidz stage of the Wielbark culture, in phases B_1 and B_2 (Wołągiewicz R., 1981a, tabl. XXIII, 7, 20, 21, 32, 39, 41). This phenomenon is, probably, a consequence of the above-mentioned Scandinavian-Thracian connections. With regard to the bucket-shaped pendants, the occurrence has already been ascertained of their prototypes in the cities of the northern Black Sea area in the 2nd-1st centuries BC, and their subsequent distribution in certain Sarmatian and Przeworsk areas at the beginning of the 1st century AD. It is thought that the tradition of the manufacture of bucket-shaped pendants passed from these areas to the Germans – the people of the Wielbark antiquities (Bazhan I.A., Kargapol'tsev C. Ju., 1989, s. 169). It is true that the wide circulation of late Hellenistic prototypes of similar small amphora-shaped pendants in the Roman provinces of Moesia Superior et Inferior and Thracia is not taken into account. In the 1st, to the beginning of the 2nd, century AD, these artefacts, together with wire bracelets with tied ends (variant Ruseva-Slokovska I_2) and pole-axe-shaped pendants, were specific attributes of the provincial-Roman culture of these provinces. The above-mentioned bracelets, characteristic above all of children's burials from the 2nd century AD, refer back to the imperial tradition of "dona militaris" (Ruseva-Slokovska L., 1991, p. 58, 64). In the Wielbark area they were transformed to rings with spiral knots (Wołągiewicz R., 1981, tabl. XXIV, 30). Pole-axe-shaped pendants from the 1st and 2nd centuries AD in Moesia and Thracia (for example hoards from the vicinities of Vidin, Northwest Bulgaria), and to a lesser degree in Dacia (type 1 - Kokowski A., 1998, s. 101), derive their form from equal-sided Thracian pole-axes of the early Iron Age. The spontaneous appearance of similar jewellery items among the antiquities of the Wielbarsk people testifies to the significant and simultaneous contact of the northern Germans with the culture of the Romanized Thracians, occurring not later than the 2nd century AD. This contact might take place only in the course of the Marcomannic Wars: their military theatre covered Illyria. In 170 AD the Sarmatians and "Germans" attacked the province of Pannonia Inferior, between Sirmium and Singidunum, and the Costoboci - Moesia Inferior (Gudea N., 1994, S. 373, Abb. 5). It is only possible to assume that one of the troops of "barbarians", after an unsuccessful campaign, might well not have retreated to the north, along the Morava, but southeast along the Danube, intending to escape from the limits of Empire to the Black Sea area. The opportunity for such a course of events is splendidly proven by the spontaneous occurrence of monetary hoards of coins of Antonine Pius and Marcus Aurelius within the limits of Moesia Superior and Inferior – the extent of the right bank of the Lower Danube (Gerov B., 1977, Karte I). The northerners plundered the coastal cities and settlements and then safely crossed the imperial border into Dobrudja. As the way to the north (along the Vistula) was closed to them by the outposts of the X Legion (see above), the "barbarians" most likely took advantage of the Dniester, Sereta and Vistula basins. Therefore, by then a *rocade* or transversal military communications route was probably open, far from the *limes* – along the northeast edge of Barbaricum – and connecting the north of Europe with the lower reaches of the Danube. Similar safe contact routes were well known to the Roman military engineers of the beginning of the 1st century AD; in later European military science, the military route was used during the Thirty Years' War by the Swedish king Gustav Adolf II.

Around 180 AD on the Amber Coast, the "barbarous" newcomers from the west imported the first indicators of the use of this military route. They brought with them the traditions of cremations, with and without urns, accompanied by Typ Dollkeim accessory vessels and weapons (as at Godłowski 6 - Godłowski K., 1992, s. 74), as well as the custom of constructing stone circles and "stelae" over the graves. Thus the obligatory presence at the burial sites of the newcomers of socketed iron axes (fig. 27), the tradition of which in Europe was retained up to the beginning of the 1st century AD in the Zarubinetsk area, hypothetically identifies these soldiers with the lands of the Upper Dnieper. As the occurrence of the newcomers coincided with the beginning of the south-easterly drift of the people of the Wielbark culture, it is possible to assume a certain connection between these two phenomena. Probably the new and bellicose inhabitants of the Amber Coast took control not only of the above local amber mines, but also of the above local population and their nearest neighbours. The outcome, which resulted in the creation of "the power of Germanaric" (Kulakov V.I., 2000, s. 378), is marked in the Wielbark area by the change from the Lyubovidz to the Cecele phase. The above-mentioned and simply manufactured spiral-ended rings, bucket-shaped pendants, and, to a lesser degree, the pole-axe-shaped pendants, are characteristic of this stage as well as of the Sambian antiquities of phases C_1-C_2/D_1. The subtle changes in the cultural and ethnic evidence (in phase B_2/C_1) of the western-Balt culture are to be seen in examples of buckles with extended rectangular frame and double tongue, which, between 150-220 AD, "... are frequently found together in closed associations with "Prussian series" brooches (Bazhan I.A., Gertseger D.S., 1993, s. 107). This statement by our St. Petersburg colleagues is rather open to argument as "Prussian series" brooches are characteristic of phase B_2. R. Madyda-Legutko attributes these buckles (fig. 22, Pr. Mus. (371) to type G36. As well as many taxons, the area found by the inhabitants of the southeast Baltic in east Europe is united by an impressive number of ancient German place names and hydronames, directly pointing to the ethnos dominating in these new lands.

The distribution of some categories of finds characteristic of the Chernyakhovsk antiquities indicates the steady connections that existed from the beginning of the 3rd, to the third quarter of the 4th, century AD, between the Dniester area and Jutland, or more exactly, with the islands to the west of the Baltic Sea area; connections that also exist with the Baltic coast (and Sambia) for phase C_2 (Ilkjær J., 1994, p. 132). This is confirmed by arrangements of finds of artefacts on both ends of the "barbarian" military route: brooches of Typ Monstruosa (fig. 74), belonging to the aristocracy; runic inscriptions (fig. 75) and iron combs connected with religious activities (Levada M., 2000, fig. 1). The amazing similarity of groups of finds of bronze coppers of Typ Eggers 44-49 and glass goblets Eggers 189-194 in the grave goods of the micro-regions of Gepedojos and Halibo/Hrains Halba (the south-western border area of Sambia), and on the lake sanctuaries of the island of Zeland, led J. Okulicz to conclude that there were constant contacts between the inhabitants of these parts of the Baltic coast at the turn of the 2nd-3rd centuries AD (Okulicz J., 1992, s. 145). A similar conclusion was made concerning the distribution (in phases C_1-C_3) of the mounts of drinking horns in Sambia and Jutland (Wielowiejski J., 2000, S. 147, Abb. 4). It is indicative that finds of cheap jewellery (taxons 1-4, characteristic of the Wielbark culture), together with the luxury items mentioned above, were not made in the western Baltic or along the south-eastern coast of the "Sueb Sea".

The presence of these contacts is quite natural; they were a component of the realization of one of many objectives of the "barbarian" route - they provided an uninterrupted flow of elite Roman "imports" intended for use in sacrifices at the heart of Barbaricum, sacred to the Germans, on Jutland, Fyn and Zeland. In Roman times, one of the major purposes of distant military campaigns was the extraction of trophies that were intended beforehand for sacrifices, confirming the high social status and legitimacy of the power structures of the German "followers" (Müller-Wille M., 1999, S. 63). It is clear, therefore, that the cheap, hand-made articles of the Gothic masters (taxons 1-4) were not wanted for sacrifices, and that high-quality Roman products were preferred. It is true that this fact was earlier incorrectly interpreted as an indicator of the Roman appearance of assemblages of weaponry belonging to early German "followers" (von Carnap-Bornheim C., 1991, S. 50). Lake sanctuaries of phase B_2/C_1 originated in the Wielbarsk area in the immediate proximity of the Amber Coast - at lake Wolka-See (Kr. Rastenburg - Raddatz C., 1992/1993, S. 127) and lake Alte-See (Ost-Pommern - Raddatz C., 1994, S. 231) – and marking the junction point of the communications route in the southeast Baltic. It is possible to assume that this area became a focal point for the settlement of the northerners, those veterans of the Marcomannic Wars, who passed from Dobrudja along the Dniester (or Dnieper) route in 170-180 AD. As well as in the later case of the micro-region of Halibo/Hrains-Halba (see below), Sambia - a convenient crossing of waterways and the centre of amber extraction - became a link between northern Barbaricum and eastern Europe.

The second objective of the great migration into eastern Europe, probably initiated by veterans of the Marcomannic Wars, was the prospect of taking possession of the rich, black arable land. Thirst for fertile land was one of the main stimuli of "barbarian" migrations in late Roman times. This was especially apparent in the campaigns of the Visigoths into Italy, and the Vandals into Africa (Skrzhinskaya E.Ch., 1999, s. 126); they were eager both for the immediate benefits of agriculture and future "annona" - land taxes.

Moving along the valleys of the Vistula, Pripyat and Dnieper, in part repeating the probable route of the veterans of the Marcomannic Wars (see above), immigrants began to appear on the fertile lands of the future Kiev area in the first half of the 3rd century AD. It was here, on the open spaces and black earth of the forest-steppe, that the new international community, known to archaeologists as the "Chernyakhov culture", began to develop. One of the chief motives of the northern Germans was realized: far from the troubled *limes*, on the distant boundaries of Oikumene, they created a reliable alternative to the established military line of communications (fig. 73), allowing them quickly and efficiently to unleash their military might from southern Scandinavia into the richest, and at that time still poorly protected, east Roman provinces, and to supply the north of Europe with foodstuffs and objects for sacrifice. In eastern Europe the military route was undoubtedly supervised by a union of various tribes known by late antique writers as "Goths". At the time of the formation of this ethnos the exchange operations over the wide spaces of the regions of the Dnieper and Oka rivers were adjusted by the population of the Kiev region and, later, by the peoples associated with the Moshchino antiquities. From the beginning of the 3rd century AD the craftspeople of these areas made clasp derivatives of type AIV, 92 with small amounts of enamelling. These artefacts are very well connected with the western-Balt area, where, at the turn of phases B_1/B_2, local imitations of provincial Roman clasps of type AIV, 92 were associated in the grave goods with articles from the Celtic-Roman enamel workshops of the Lower Rhein. It is possible to assume, with some caution, that the Balt (Veneti?) newcomers from the Amber Coast also tried to set up a trans-national route, and rather simple pieces of enamel jewellery were selected as a means of barter. Traces of a similar sort of trading operation are encountered earlier on the Amber Coast by finds of details of legionary equipment from the time of Nero in the graves of the Aestians (Veneti?). At a later date, the indicators of such trade are widely known, both in the Old World (brooches of Balt types made by German masters in the Mazurian Lakes area in the 6th century AD in exchange for amber), and in America (glass beads and bright fabrics brought to the natives by conquistadors in exchange for gold). Besides the active trading operations on the outlying eastern districts of Barbaricum after the termination of the Marcomannic Wars, there was intensive cultural exchange: for example the Heruli brought the written runic written language along the Great Amber Route, or along the Dnieper military route, from the area of the southern Germans on the north of the continent in the 3rd century AD (Makaev E.A., 2002, s. 43, 44, 52).

The ethnos of the Goths is connected with the Chernyakhov culture. Its origins were in Masovia and Podlyasie, and even towards the Middle Dnieper area, within the limits of phases C_{1b}-C_2, as is confirmed by the introduction of the main attribute of Gothic historical ethnography – female costumes featuring a pair of plate brooches (Tempelmann-Mączyńska M., 1989, s. 222-224). It is true that the assemblages of arms in eastern Europe, known from finds in the cemeteries of the Amber Coast from phases B_2/C_1 and C_1, are absent. Of course, this does not presuppose that Gothic society was especially peaceful. Begun by the Goths in 255 AD, the sea campaigns against the Roman cities of Pithyus, Phasis and Trebizond were, probably, carried out according to a concerted strategy (von Carnap-Bornheim C., 1997, S. 237), and were focused on the strengthening of the "barbarian" positions on the southern section of the Jutland-Dobrudja route. As a result of these actions, at the end of the phase C_2 the culture of Sintana de Mures (Bierbrauer V., 1994, S. 123) developed. It is evident that the intention of the "barbarians" to damage and plunder the trade routes leading to the imperial borders, on this occasion and also during the Marcomannic Wars, did not promote trade stability. The final disruption of the Jutland-Dobrudja route dates to 376 AD, the start of the period when the Huns began in earnest to force the Goths into the provinces of Moesia and Thracia. The breakdown of Gothic dominion in the territory between the Danube and Dnieper (the international "power of Germanaric") is witnessed by a series of finds of plate brooches and buckles in the assemblages of phases C_3 and D_1 (fig. 76), showing the Gothic migration at the end of the 4th century AD into the areas of the Middle and Upper Danube.

At the end of the antique era, the military route, the major unit of which was Sambia, exhausted its possibilities. In 369, as a result of the active military operations of the Emperor Valens, the Danube Goths were powerfully repulsed and lost any opportunity to trade with the Empire. They were also beginning to be pressed by the Huns, who ceased their organizational activities not only along the established military communications line but along the Great Amber Route as well. Admittedly, the beginnings of the Hunnic Wars, which in the 4th-5th centuries AD shook practically all the regions of the *limes*, provided the "barbarians"/Germans with an opportunity to take revenge on Rome. There was a fresh chance for the multi-tribal military troops who, undoubtedly, also included the inhabitants of the Amber Coast; they were able not only to attack the Roman provinces and resettle there, but also to capture and plunder them completely. Although some of these plundered goods were sacrificed in the sacred lakes of the north, many impressive trophies were brought back by the veterans of the Hunnic Wars to the southeast Baltic. This is illustrated, in particular, by the hoard (400-450 AD) at Młoteczno (former Hammersdorf, Kr. Heiligenbeil) in the micro-region of Hailibo/Hrains-Halba (Kulakov V.I., 1998б, p. 101), which also continued to be an important communication centre along the Jutland-Dobrudja line in Hunnic times, even though in the 5th century these contacts ran mainly along the Vistula and Tissa rivers. Sambian warriors, having long ago abandoned their former foot-battles, the consequence of which was usually a spear thrust, changed their tactics under the influence of their allied relations with the steppe-nomads/Huns. At the beginning of the 5th century AD, the Sembi (the name of the inhabitants of Sambia in the early Middle Ages) became lightly armed riders (actually the forerunners of dragoons). Their main weapon was a special dagger fastened by a belt behind the back that imitated the Roman sword belt of the balteus type (fig. 77, on the right). The single-edged blades, typologically looking back to the swords with ring-shaped tops of type Biborski IV, V, which appeared in Barbaricum at the time of the Marcomannic Wars (Biborski M., 1994, s. 89), had a long serrated edge that chain and plate armour could not withstand (Kulakov V.I., Skvortsov K.N., 2000, s. 47); the horsemen/ clibanarians, the Roman foederati of Hunnic times, were equipped with such blades (Taratorin V.V., 1999, s. 136).

The final breakdown of Hunnic power, after the encounters at Mauriak (451) and on the River Nedao (454/455), ended all hopes of the revival of the Jutland-Dobrudja line and the Great Amber Route. On the islands of the western Baltic, which were sacred for the Germans, the ethnic situation changed. In Hunnic times, the hypothetical Juts were probably replaced by newcomers from Skone - the Dans (Khlevov A.A., 2002, s.45). On the above-mentioned islands a story was told of the power of the legendary King Dan the Proud, who, apparently, took part in the Hunnic campaigns (Shuvalov P.V., 1999, s. 28). In the 5th-7th centuries, the sacred and political centre of the Dans moved from Zeland to Fyn (Gudme) (Thrane H., 1991, p. 261). From the end of the 5th century AD, the northern Germans aspired to recreate the former functions of the military route from Jutland to Dobrudja, within the limits of the Baltic waterways, establishing what would become the bases of the next main eastern route (the Avstrvegr of the Viking era), and, together with the Old Prussians, supervising the amber trade along the Sambia-Emajõgi line (Kulakov V., 2000, p. 285-289, fig. 1). That the religious aspect of these contacts was maintained is illustrated by various Balt-Scandinavian mythological analogies (Toporov V.N, 1970, s. 537).

X. Halibo, the cradle of Prussian culture

The ending of the late Roman era in the southeast Baltic was marked by serious cultural and ethnic changes that shook the foundations of local society. These changes were a consequence of the processes connected with the great migration of the peoples and the ethnic migrations touching the Amber Coast in particular were reflected in the local legends (Kulakov V.I., 1998b, s. 98).

In 1529, Simon Grunau wrote the "Prussian Chronicle", substantially founded on Prussian oral tradition. (The author said that the text was based on the records of Bishop Christian, who was held captive by the Prussians between 1231-1238. "In the book, which was written by Christian, the first bishop in Prussia, and called by him Liber filiorum Belial cum suis superstitionibur Bruticae factionis, he writes that he received the work (in anno Domini 110) in der Masau (from the Moscow area or Masovia) from one Jaroslav, a priest from Polotsk (in Byelorussia or Poland). The book was written in Russian, but with Greek letters, by a certain Dywoynis..." (Grunau S., 1876. S. 55). He is represented under the name of Dywoynis in the text of Grunau Dion Cassius from Vifinia (about 150 - after 229), and mentioned in "Gethica" of Jordanes, testifying to the acquaintance of Grunau (Christian?) with the work. The "Prussian Chronicle" is key to the understanding of the actual historical events of the time, comparing the oral tradition reflected in the book with the archaeological data (Šimėnas V., 1994. p. 56, 57; Kulakov V.I., 1997a, s. 152).

Some of the riddles of Grunau's text lie in certain river and place names of the Prussian lands. In particular, the chronicler informs that "Vistula... fliest in Crono, das gesaltzene mehr" ("The Vistula...flows into the Chrono, the salty sea"), "Brudeno und sein bruder Witowudo... qwomen durch Crono, das wassir Haillibo..." ("Bruten and his brother Videvut...came onto the Chrono, by the waters (sea, gulf?) of Haillibo...") "Vitowuto im baute ein schlos zwischen Crono und Hailibo... und ist auff der Neringhe..." ("Videvut constructed a castle between the Chrono and Hailibo...on Neringue (=spit)..." (Grunau S., 1876, S. 56, 61, 62)). It was considered from the context of the "Chronicle" that the names "Crono" and "Haillibo/Halibo", mentioned in connection with the description of the events of the great migration, correspond to modern geographical locations - "the River Pregolya" (most probably being the Vistula Lagoon or the Baltic sea) and the "Curonian, or Baltic Spit" (Grunau S., 1876, S. 56, Anm. 1; Ibid., S. 69, Anm. 5). In the modern literature, the opinion is that the name Κρονωσ/Chronos, according to Ptolemy's data, should correspond to the River Neman (Šimėnas V., 1994b. p. 29) or to the River Svezhaya (see above). However, in that case the Scandinavians Videvut and Bruten should have reached Ulmigeria (one of the names for the Amber Lands offered by Grunau) not from the west (as follows on from the text of the "Chronicle"), but from the north or from the northeast, making a very wide detour around the Baltic. In Grunau's text, the names of "Crono" and "Halibo" correlate (and are obviously connected) with the coast of the present Vistula (Kaliningrad) Lagoon (ehem. Frisches Haff). In particular the chronicler designates Natangia as follows: "...das lant zwischen Pergolla, Alle, Bassaro und Halibo den wassirn..." (Grunau S., 1876, S. 72). By this context, Halibo corresponds to the Vistula Lagoon or its coast. The archaism "Haillibo/Halibo" (hereafter Halibo) was maintained indirectly as the place name of Heiligenbeil (nowadays the town of Mamonovo, in the Bagrationovsk district of Kaliningrad). Here, according to written sources, Anselm the Varmian bishop in 1301 cut down the sacred oak that was the object of worship of the Old Prussians at the sanctuary of Swentomest (in Prussian "Sacred city") (Guttzeit E.J., 1966, S. 82). To the northeast of the town of Mamonovo, in the area of Ladushkin (ehem. Ludwigsort, Kr. Heiligenbeil), grows the oldest oak on the Amber Coast, its age is estimated at around one thousand years. The veneration of this oak at the beginning of the second millennium AD, and the religious significance of the largest and most defended Prussian site at Lipovka (4kms to the east of Mamonovo - ehem. Grünwalde, Kr. Heiligenbeil) are quite feasible. All this focuses special attention on the early medieval history of the stretch of the Vistula Lagoon coast, limited from the southwest by the River Ilfing (now the River Elblong) and from the northeast by the River Frisching (now the River Svezhaya) (fig. 78). For a fuller study of this micro-region, to which the name "Halibo" (and repeatedly mentioned by Simon Grunau) is connected, it is necessary to examine all the local archaeological sites from the least investigated period of Old Prussian history - the 5th century AD. (For the catalogue of archaeological sites of the Halibo micro-region see Appendix 3.)

The finds from the territory between the Ilfing and Frisching, dating from the time of the great migrations, are in some cases unique for the Baltic. Over recent decades they were treated completely differently depending on the scientific and political views of the experts. La Baume and Bolin considered that the abundance of finds of late Roman solidii in the territory between the Rivers Ilfing and Frisching is connected with the groups of the German population who deposited their treasures here in the 5th and (even) 6th centuries; in the latter case, as a result of the dangers posed by Slav movements along the left bank of the Vistula (Petersen E., 1936, S. 20, 21). Engel considered that, while the majority of the Gothic population moved in the 5th century, there was a Gothic aristocracy supervising the amber trade (see the example of "prince" burials at Warnikam and Hammersdorf. Engel C., 1942, S. 160-162) conducted by the early Prussian society along the

German/Balt border area. (The ancient German river name Ilfing was maintained by local inhabitants up to the Reformation.)

K. Godłowski has undertaken a full analysis of the antiquities of this micro-region. He has come to the conclusion that there were two waves of Old Prussians who populated the territories to the southwest of the River Paslyonka, the traditional western border of the Balt world. These waves are dated to the second half of the 5th century and the middle of the 6th century (Godłowski K., 1981b, s. 106, 110, 111, 118). The well-known Polish scholar considered that indicators of the first wave of migration were the cemeteries at Podguzhe and Paslyonka, founded by the Old Prussians in the middle of the 5th century, forcing out from these lands the last of the people of the Wielbark culture, called the "Vidivarii" by Jordanes. The very last of these, receding in 450-500 AD to the southwest, marked their way with hoards of solidii. Gąssowska, on the contrary, considers that these hoards contain solidii deposited by the inhabitants of "the northwest Slavic lands", acquired as a result of trade in Scandinavian furs to east-Gothic Pannonia in the times of Theodosius II and Justinian I. In the opinion of the Polish researcher, the re-immigration of the soldiers serving as mercenaries in the Byzantine army played a considerable role there (Gąssowska E., 1979, s. 166). The present author has already expressed his opinion that in the middle of the 5th century, the Aestians, passing the River Pasłęk, arrived at the eastern outlying district of "The Gepedojos Island" (the Elbląg Heights) into direct contact with the remnants of the German-speaking people of the Wielbark culture and generated with them (450-500 AD) a unit of the Vidivarii (Kulakov V.I., 1997b, s. 363).

Analysis of the distribution of the local archaeological sites (fig. 78) allows us to create a picture of the events that happened at the time of the great migration of the peoples - stages D_2-D_3 (horizon of Unterzibenbrunn-Sösdala) - to the east of the River Nogata. In the present work, J. Tejral's system is used. He revealed phases D_1 (360/370 - 400/130), D_2 (380/400 - 440/450), D_2/D_3 (430/440 - 470/480), D_3 (450 - 480/490) and D_3/E 9470-500/510) from the material of the Middle Danube area (Tejral J., 1997a, S. 351).

By the beginning of the 5th century AD, the last Wielbarsk cemetery - Ladekopp (Lubieszewo, woj. Gdańskie) – had already stopped functioning (La Baume W., 1924, S. 76) in the delta of the Vistula to which the micro-region under investigation adjoins from the east.

It is true that, in spite of the fact that the Wielbark population of "the Gepedojos Island" migrated at the end of the 3rd century to the south and southeast, according to the evidence from the Pruszcz Gdański and Wielbark cemeteries, the remnants of the Wielbark population supported the Vistula delta's function as "the fundamental 'window on the world' for the peoples of the Baltic" (Okulicz J., 1992, s. 145).

Single finds of late Roman and early Byzantine solidii were discovered in a concentration of Wielbark cemeteries of the 2nd-3rd centuries in the vicinity of the modern town of Elblong (Appendix 3: points 2-7, single hoard - point 3). The arrangement of these hoards, dating from the first half of the 5th century and the boundary of the 5th-6th century (phases D_2 and D_3/E), marks the 5th-century exit from the lower reaches of the Vistula to the calm extents of its wide delta (Okulicz J., 1992, ryc. 2). The location of the finds infers that these solidii were offerings of thanks from grateful travellers who had reached the threshold of home, and not, as Godłowski suggests, deposits buried by the last "Wielbarians" as they receded to the southwest under pressure from the Aestians. On the contrary, loaded with gold stolen from Rome, the barbarous "followers" made their way back to their northern European lands by way of routes then open along the Vistula (Baltic) Spit (Bertram H., 1924, S. 36). As finds from the River Paslyonka basin show, not all the participants in the Hunnic Wars passed via the coast of the Vistula Lagoon. There, on the former border between the Balt and German worlds (Gaerte W., 1932, Abb. 1), in the vicinity of the village of Młotezno, the richest of all Baltic hoards (A-C) were revealed. These unique finds, which in former times raised lively debates over their ethnic identities and dates, are nowadays forgotten (Nowakowski W., 1985, s. 99).

Hoard A (fig. 79), despite the early minting date of a Constans II solidus that was transformed into a medallion, was buried between 400 and 450 AD. Such a chronological interval of deposition is established by the specific features of a three-beamed brooch found in the hoard that appertained to stylistic group III of polychrome jewellery of Hunnic date and of Danube-Pontus origin (Ambroz A.K., 1992, p. 23, 93). According to the complicated detailing of the foot of this brooch, it was made at the late stage of stylistic group III (by A.K. Ambroz - in the first half of the 5th century).

Hoard B (fig. 80) represents two (or more) fragments of a large silver dish ("luterius", deposited together (and perhaps in) with a small silver dish (fig. 81). The large silver fragments probably represented a share of the military spoils received within the limits of the Empire and shared between the brothers-in-arms before they returned to their native lands. True, this assumption holds only if these fragments were not the result of the activities of the unknown discoverers of Hoard B. It is not obligatory that Hoard B was buried in the same period (corresponding to the date of manufacture of the objects found with it) as Hoard A. In the western Balt region of stage D, the long-term accumulation of fragments of silver objects for recycling into jewellery is well known (Kaczyński M., 1989, s. 194). A mark "×" on the inside of a small dish from Hoard B attracts attention. From Celtic times this mark was one of the important symbols of European paganism - the sign of the Sun. In the all-German runic inscriptions, the mark on the "small" dishes from Hoard B would correspond to the rune "gebu". The presence of this rune on the object from Hoard B may specify the sacral reason for its concealment as a sacrifice to the German deity.

Hoard C (fig. 82) - two gold rings - is a unique phenomenon for the Balt world. In Baltic hoards, gold, which was so popular among German tribes from late Roman times, is poorly represented. In Scandinavia (especially to the south), gold neck-rings of the same type as the finds from Hoard C - with overlapping bulged ends - are plentiful in hoards, and are dated in bulk by the 6th century. Admittedly, such Scandinavian neck-rings are decorated with the deep impressions of sickle-shaped stamps, which seem to be later derivatives of the ring ornamentation in Hoard C. The closest find by form to these rings occurs in the hoard from Piltene, Ventspils raj. (Latvia), conditionally dated to the 5th century (Urtāns V., 1977, p. 138); it is more accurate to attribute Hoard B to this time. The purpose of these finds is clear. The presence of pairs of rings, or neck-rings, in Celtic grave goods or hoards from the 4th-3rd centuries BC, and in particular in the German world of the 5th century, was connected with religious rites. Among the men of the German "followers" in the epoch of the great migration, the gold neck-ring was a symbol of a leader's prestige (Hauck K., 1954, S. 147-149), and so the attribution of Hoard C as a sacrificial offering, brought in the 5th century by one of the local leaders of German origin, does not cause doubt. It is necessary to note that gold chains with medallions/bracteates, and neck-rings were material indicators of social status for Germans of the 5th-6th centuries.

The presence of the plate brooches and neck-rings was obligatory for the "barbarian" aristocracy in various regions of east Barbaricum in late-, and post-Hunnic times (horizon Untersiebenbrunn). The above-mentioned artefacts correspond to Bierbrauer's category IA, distributed in the middle and second half of the 5th century AD in the Danube area and the northern Caucasus (Mastykova A.V., 2001, s. 60), and equating to the very highest stratum of the German elite (Bierbrauer V., 1987, S. 81, 82). This social stratum was the dominant one in the "barbarous kingdoms" that rose from the wreck of "Attila's power" on their outlying districts (Tejral J., 1997a, S. 349-352). If Hoards A and C are to be interpreted as genuinely intended horded objects, they should be regarded as "royal sacrifices", brought back from the German realm to preserve the good luck of their owners. In Viking times such actions were carried out by sacrificing a king to the gods (Steinland G., 1992, S. 747, 748). The radical influence of the prestigious Hammersdorf-type "prince" costume in those regions far from the Baltic Bosphorus and north Caucasus attracts attention (Kazansky M.M., Mastykova A.V., 1998, s. 112); it especially emphasizes an increased ability for contacts between members of the collective who deposited hoards at Młotezno.

The association of sites at Mlotezno from the time of the great migration, is supplemented by the cemetery of flat graves investigated by Ziemlińska-Odojowa. She has no doubt in its belonging to the Sambian-Natangian group of western-Balt culture. Horse burials, characteristic of the antiquities of early medieval Prussians, occur at Młotezno no earlier than the first decades of the 6th century.

Special interest is caused by a raised bank at Młotezno-Rogity (point 19) that limits the association of sites at Młotezno from the east. If we accept the option of a Prussian derivation for all of this association, or even for the cemetery where hoards A-C were found, the protection of this site from the main territory of the Prussians looks senseless. The possibility of a later dating for the Młotezno-Rogity bank (H. Crome suggests that the bank was part of the fortifications of medieval Braunsberg) is not credible. According to the context of a document of 4th September 1291, there were only reminiscences about this feature of fortification by the date of the early Order (Crome H., 1937, S. 82). Dating the bank to the time of the great migration appears more preferable, above all by the geomorphological associations with synchronous objects. It is necessary to note that the Germans, occupying the new lands in the 5th-7th centuries, protected themselves from the native population with banks. (The banks of Offa's Dyke, created by the Anglo-Saxons between Mercia and Wales in the 7th century, served a similar purpose.) In the same way the first settlers of the lands of the Upper Schmitze might have secured themselves against possible attack by the Aestians-Prussians. As well as a practical security measure, the feature served a religious function as a place for making sacrifices to the gods (cf. Hoard B at Młotezno). The newcomers' choice of this particular landscape was based only on geographical factors. From the palaeo-geographical data (Zhindarev L.A., Kulakov V.I., 1996, s. 62), the centre of the Vistula Lagoon coast was a peninsula limited by the marshy river mouths of the Passarge and Möcker (fig. 78) at the time of the great migration. A convenient bay (at the mouth of the River Schmitze) led to the high ground that nowadays distinguishes this landscape and is the location for the village of Młotezno. It is indicative that it is only possible to interpret Grunau's place names Chrono and Haillibo by recourse to the basics of the Gothic language. The words Hrains and Halba (Braune W., 1905, S. 152, 153) form a combination of words for a "distinct part/side/half". This corresponds well to the characteristics of the above-mentioned landscape in the middle of the 5th century at a time when the Wielbark population had left, but the Prussians were not then the occupiers. It is possible that the reconstructed place name was preserved by the Prussian oral tradition. In local legends this place name, in its variations Chrono and Haillibo, passed into the "Prussian Chronicle" of Grunau, who was a native of the Halibo region (point 3 - Gronowo Górne, woj. Elbląg/ehem. Grunau, Kr. Braunsberg/).

The evolution of the 5th-century sites around Młotezno helps us to restore the sequence of events that developed in the Halibo micro-region at the time of the great migration. The cemetery at Podgórze, 2.5km to the west of Młotezno on the left coast of the River Passarge, has revealed Prussian material dating not earlier than the first decades of the 6th century (Kulakov V.I., 1990b, s. 62). Before this period there are no features of western-Balt early medieval rites at this cemetery (horse burials, "temporary urns", etc.). From the second half of the 5th century, Grave 26 is a typical example. It contains within its grave goods a fragment of an animal-headed brooch (less possibly a buckle). This artefact appertains to clasps of group 1 (Kulakov V.I., 1990a, s. 211, 212). Using the Gotland analogies that appertain to stage VI.2 (Nerman B., 1935, Taf. 49, Nr 510), it is possible to date securely the find from Podgórze Grave 26 to the second half of the 5th century. As the antiquities of the southeast Baltic show, similar early forms of animal-headed brooches appear in the third quarter of the 5th century in the micro-region of Halibo. As accessories of female costume, the brooches of this type combine the attributes of the antiquities of the Frisians, Saxons, and the inhabitants of the middle Danube area. In the Baltic these brooches are a marker of the archaeological horizon of the Vidivarii (Kulakov V.I., 1990d, s. 212, 213). "Gathered from various tribes" (Jordan, 1997, s. 67), they brought to the Amber Coast not only features of their rites, but also some specific examples of the decorative arts of post-Hunnic times. These phenomena of material culture are most appreciable in the material from the cemetery at Warnikam (point 25). There, 16kms to the northeast of the cultic landscape feature (a sanctuary?) at Młotezno, are found the earliest mass indicators of the 5th-century penetration of foreign elements into the local environment. If the outline information on the Warnikam burial customs we have allows us to assume cautiously that there was a sharp increase in the attributes of these rites in late Hunnic times (Kulakov V.I., 1997a, s. 148), the question of the finds is clearer. Ten years ago, Bitner-Wróblewska engaged for the first time in a thorough analysis of the Warnikam antiquities. The Polish researcher created a helpful typology of star-shaped brooches (types I-VII[2]), based mainly on the Warnikam finds. However, the author's thesis about the genesis of this type of brooches gives rise to doubt. Type I is seen as the prototype (Bitner-Wróblewska A., 1986-1990, s. 52), where the crossbow-shaped brooches of variants Schulze IIx AF3b, Ix AF2a (Schulze M., 1977. Taf. 11, 12), found at the turn of the 4th century at various sites in Barbaricum, are united only by the arc-shaped top edge of trapezoid widening on a foot. In view of this approach to the material, a rather indistinct chronology has arisen for this star-shaped brooch. For example, type I is linked "throughout phase D" (according to Bitner-Wróblewska this is the boundary of the 4th-5th centuries, up to the beginning of the 6th century), type II equates to phase E, and type IV "up to the beginning of the early stage of the era of the great migration" (Bitner-Wróblewska A., 1986-1990, s. 61, 63). It is, perhaps, possible to solve the problem of the chronology of brooches (above all star-shaped ones), found in the 5th century in the grave goods of the population on the western border of the Balt world, not so much within the framework of formal typological research but by their associations with other artefacts.

Figure 83 shows the basic early types of star-shaped brooches and similar clasps of stages D_1 - E from the Baltic and northern Europe. Bitner-Wróblewska considers type I to be the earliest for star-shaped brooches (Bitner-Wróblewska A., 1986-1990, s. 56, tab. 2). Actually this clasp variant appertains to a typological line of development of crossbow-shaped brooches with a trapezoidal foot, and connects with the family of clasps from group AVI type 181. Classifying them in one typological line with star-shaped brooches within the limits of types I-VII, as suggested by Bitner-Wróblewska, looks rather strange methodologically, and only confuses the complex picture of the development of the material culture of the western Balts. Brooches with a trapezoidal foot, retained in Aestian antiquities to some late Wielbarsk examples of phase D_1 (fig. 83, 1), in the first half of the 5th century, are decorated on the foot with a compass ornament. This decorative element is known in the technological repertoire of late provincial-Roman masters. It is possible that it is a reproduction of the inlays of semiprecious stones and coloured glass characteristic of a similar form of Scandinavian brooch foot from phase D_1 (Jørgensen L., 1994, Abb. 121, 6). This idea was first put forward at the beginning of the 20th century by H. Kemke (Åberg N., 1919, S. 31). The final stage of the brooches, relatives of "type I", is designated by a find in Do-258 (fig. 83,4). In this assemblage (that also includes a pair of brooches) there is a late clasp-form with ring elements, characteristic of the Wielbarsk antiquities of 300-400 AD (Nowakowski W., 1996, S. 51, 52). Thus, according to the typological features and accompanying material, the brooch from Do-258 should be dated to phase D_2, and the next one (fig. 83.3) to the boundary of phases D_1 and D_2. In the middle of the 5th century, in some instances brooches with trapezoidal feet, becoming at this time one of the indicators of Balt antiquities, acquired elements of the form and decoration of star-shaped brooches (fig. 83,7): the leading attribute of this clasp type, the bent back, is retained. Similar exceptions are found within the limits of type I (Bitner-Wróblewska A., 1986-1990, s. 56, tab. 1, 1, 4, 8).

Besides these brooches, which became ethnographic indicators of the Aestians in the middle of the 5th century, an appreciable amount of the innovations occurring from the tribal areas adjoining the Aestians is found in the material of the Amber Coast from this time. Thus, for example, the real predecessors and prototypes of the star-shaped brooches are those of groups AVI type 179 (fig. 83,8), dating from phase D_1. The process of

[2] Actually they are subtypes, because "the star-shaped brooch" itself is one type within the range of "brooches".

development of these brooches in the north of Europe was quite long, resulting in the occurrence of local imitations of these artefacts in the first half of the 5th century (fig. 83,9). These brooches became a paradigm for the creation of the first copies of the star-shaped brooches of the late Aestians (fig. 83,11). It is important to note the rectangular outlines of the brooch's arch section, which dates the clasp from Do-258 to the same time (fig. 83,4). These clasps are attributed by Bitner-Wróblewska to type V and are dated to the boundary of phases D_2/D_3. The characteristic features of this type of brooch are the result of the symbiosis of the brooches' traditions of various cultures and ethnicities. The end of the foot, which evolved into the form of a circle with "beams", imitates the sheen of a stone or a glass inlay of brooches from the Hunnic era. At the top of the arch of the foot there is a rectangular platform consistent with an appropriate fixing point for a palm fastener to the cloak. In the post-Wielbark antiquities from Sambia (fig. 83,10), and in the Gotland material of stage Nerman VI.1, such platforms have an oval form. The bend and cross-ribbed decoration of the foot of a brooch from Gora Velikanov and of a similar find from Uzornoye (point 26, both brooches are iron ones), hark back to principles of decoration of stage-D_1 "three-beamed brooches" (Bügelfibeln Typ Leipferdingen - Voß H.-U., 1994, S. 505, Abb. 107, 1). Similar attributes are found in the later brooches of types Glaston (fig. 83,12) and Duratón (fig. 83,13). These brooches fix the influence of the people of the western-Balt culture to the antiquities of the western part of the European continent, distributed as a result of the participation of a separate force of soldiers from the Amber Coast in the events of the middle phase of the period of the great migration of the peoples. This is particularly well confirmed by the evolution processes of the type Duratón brooches. In Kazanski's opinion this type (as well as the clasp types Rouilleé and Albias) is an indicator of the fact that representatives of the non-German tribes participated in the development of Gallia by the Visigoths. Our French colleague numbers the western Balts among them (probably the Galindians; Kazanski M., 1994b, p. 168, 173). The toponymics of western Europe allow V.N. Toporov to make similar assumptions about the movements of the Galindians' "druzhinas" in the 5th-6th centuries from their Mazurian ancestral lands to the east, south, and southwest (Toporov V.N., 1983, s. 129-140). The Prussian variant of the type Duratón brooches gave rise to clasps of the related type Schönwarling, dating the hoard at Frombork (point 9) to about 450 AD, i.e. a terminus ante quem for the brooch from Do-163 (108). Together with it in the burial assemblage was a buckle decorated in the Sösdala style (fig. 83,14). This is the earliest appearance in Baltic monuments of the decorative style that appear, as finds in Roman *limes* show, at the end of the 4th century in the environment of the legionaries of Romanic and Gaul-Romanic origin. Further concrete evidence is supplied by material from the cemeteries in the northeast of modern France (Kazanski M., 1994c, Fig. 2, 1; 3, 4; 5, 7; p. 41). In the "barbarian" environment of the legionaries in phase D_1, the traditions of provincial-Roman art (i.e. types of artefacts) overlapped with the traditions of German art (use of the stamp). These tendencies were perceived by the poly-ethnic (both Romanic and "barbarous") population of the Danube area, which, in the first half of the 5th century, eventually produced the Sösdala style. It is true that belt sets with "star-shaped" ornaments, popular among the legionaries of Gallia, were unknown in Pannonia and Noricum. The composition of the late examples of the Sösdala style might have imitated the Hunnic polychromic style and brought to us a glimpse of the greatness of Hunnic power on the eve of its destruction.

Monuments of this decorative style are found in the antiquities of the Amber Lands from the middle of the 5th century in Sambia (for example assemblage Do-163, with a brooch of the Schönwarling type - a derivative of type Duratón), and in the northeast part of the Halibo micro-region (the cemetery at Warnikam). Prussian finds are the most recent monuments of style Sösdala. As a rule, assemblages contain details of belts accompanied by star-shaped brooches; these burial assemblages belong to noble warriors. The closest to the prototype (fig. 83,11) is a brooch from former Heinriettenfeld, Kr. Gerdauen, which is a casual find, as are, unfortunately, a significant number of the brooches from the first two thirds of the 5th century in Prussian territories. It is important to note the "fish-scale" ornament that covers the foot of this brooch (fig. 83,15). In all northern Europe there are no analogues to it, although this decoration is well known in post-Hunnic antiquities. Magnificent examples of this ornamentation are found on gold bracelets from Badokpuszta, made in the "Post-Nedao" era (phase D_3). Their tops represent figures of dragons with open jaws and closely-set, pointed ears (Bóna I., 1991, Abb. 112, S. 292). Genetically, these artefacts are connected with gold bracelets from the 1st century BC to the 1st century AD, found in graves 2 and 6 at the cemetery of Tillatepe (at Shibergan, Northern Afghanistan). These assemblages are linked to noble women from the "Great Juedzhi" tribe, adjoining the Hunnu tribe from the west. Bracelets from Tellatepe, the tops of which were decorated with images of antelopes and "lion-headed griffins", date back to Assyrian and Bactrian antiquities (Sarianidi V.I., 1989, s. 64, 128, 172). In Hunnic art these characters assumed the shape of mythical reptiles that were relatively popular in steppe antiquities of the 4th century. According to the presence on these disjointed bracelets of fastening screws, they were made by Byzantine masters and guided by steppe traditions. The analogues of these bracelets, which were found in the village of Sennaya on the Taman (Zasetskaja I.P., 1994, tab. 18, 1) and in the vicinity of Kiev, are dated to the second half of the 5th century (Kazanski M., 1992, p. 140, fig. 1, 9). True, these eastern European finds display simplified and degraded representations of dragon's scales (or, indeed, none at all), showing the secondary importance of these artefacts in comparison with the bracelets from Badokpuszta, and, accordingly, their later date.

The analogue of the brooch from Heinriettenfeld is known only from the cemetery at Grunajki, woj. Suwałki (Bitner-Wróblewska A., 1986-1990, tab. III, 9). According to the widening of the foot inscribed in a circle (the main attribute of the prototypes in fig. 83,8,9, 11), the two specified brooches are the earliest in the group of star-shaped clasps, and are closest to the Scandinavian prototypes (fig. 83,8,9). They are attributed to the beginning of the third quarter of the 5th century. Such a dating is confirmed by the presence of a brooch from Heinriettenfeld with cut "washers" on the ends of the cross bar. This attribute is characteristic of earlier brooches from a localized series with trapezoidal feet (fig. 83,4), and certain Bügelfibeln from Gotland of the second quarter of the 5th century.

Brooches from Heinriettenfeld and Grunajki determined the principles of decoration of the last star-shaped brooches. Their decoration is in the form a dragon-scale foot, made by the use of a tiny, sickle-shaped stamp (fig. 83,18). At the turn of phases D_2/D_3, these brooch "scales" were made by means of alternating groups of cross flutes (fig. 83,18). This is an abstraction from the ornamentation of brooches from the second quarter of the 5th century (fig. 83,11), and it is also echoed in the decor of brooch feet of the Schönwarling type. In the Sambian antiquities of phase D_3, the magnificent belt sets that were earlier decorated in the Sösdala style (fig. 83,19) begin to degrade. By phase D_3/E, assemblages featuring an already complete range of the attributes of the formed Prussian culture (Kulakov V.I., 1994b, s. 65), contain very simplified clasp derivatives (with oval frames and rings with stamp decoration), and star-shaped brooches (fig. 83,20,21). Types III, VI and VII, classified by Bitner-Wróblewska as star-shaped brooches, are found, in the 6th century, only in the areas of the northeast neighbours of the Old Prussians. Many other artefacts from the 5th-8th centuries were also subject to similar western-Balt cultural innovations (Kulakov V., Šimėnas V., 1992, S. 183).

The conclusions reached after consideration of the finds of star-shaped (and other) brooches of the 5th century from the Halibo micro-region and its vicinity are:

1. One of the components of the range of attributes of late-Aestian finds is the number of variants of brooches with a trapezoidal foot. Adopted, at the turn of the 4th century, by those western Balts from the late Wielbarsk population of "the Gepedojos Island", these brooches subsequently became an ethnographic attribute of the early medieval Sembs, and then of some other Balt tribes. For a long time, up to the 9th century, the derivatives of these clasps were in use especially by the inhabitants of the territory of modern Lithuania (Tautavičius A., 1996, p. 198-201). Using a chronology based on the data of brooches with a trapezoid foot, it is possible, in particular, to date graves 1 and 20 from the cemetery at Młotezno to phase D_2/D_3 (ок. 440-470). This dating can be established according to the simplified form of brooches found in the above-mentioned assemblages and in view of the presence of the rudimentary decoration, based on the star-shaped brooches of type I (fig. 83,11) from an artefact from grave 1.

2. In the first half of the 5th century, there are traces in the archaeological material (for example the brooches) of the influence of Scandinavian antiquities on the western outlying districts of the Balt world. In particular, as a result of the north German impulse, there are direct prototypes of star-shaped brooches (fig. 83,11). Finds of such clasps show that, in the second quarter of the 5th century, these Aestian sites were especially important for foreign soldiers, in that they marked the location of possible incursion points for Balga Island and the Amber Lands of Sambia. In turn, some features of late-Aestian brooches are identifiable across a wide spectrum of west-European material from the middle/third quarter of the 5th century up to the middle of 6th century, showing the participation of the western Balts in the military campaigns of the post-Hunnic era in the western outlying districts of the continent.

3. From the middle of the 5th century, and simultaneously on the coast of the Vistula Lagoon and in Sambia, there is a marked intensification in the antiquities of the post-Hunnic era. It is important to note the exclusiveness of local phenomena, such as specific forms of star-shaped brooches (the development phase, type II) that had existed for a long time in the western part of the former tribal area of the Aestians.

4. A little later, from phase D_3 in the same territory, animal-headed brooches begin to be distributed (fig. 84, 85). They originated under the direct influence of the Jutland antiquities from the second half of the 5th century (Bakka E., 1958, fig. 28-30). True, according to finds of the same brooches in Moravia (Sokolnitse), categorized by J. Tejral to the middle of 5th century (Tejral J., 1997a, S. 350, Abb. 28, 14), certain tribes from the Danube area might also have participated in the creation of animal-head clasps. Coinciding within the limits of one Old Prussian cemetery (never in one single assemblage) with finds of star-shaped brooches - late type-II variants (for example, at "Gora Velikanov") - and in contrast to them, the animal-head brooch was an accessory of female costume. This phenomenon serves as an example of the various stages in the adoption of these items, which were in themselves indicators of the decline of Roman greatness in the environment of the local western Balts (Kulakov V.I., 1990c, s. 210-213).

5. The most important conclusion from the above is the possibility to put a date to the first Sösdala-style monuments in the Baltic (and the last in Europe) at around 450 AD. These artefacts occur in association with star-shaped brooches and magnificent belt sets. In other categories of objects (excluding the blades of knives/ daggers) in the area comprising Prussian culture, and among the other Balts, there are no such features of the

Sösdala style. They disappear from the repertoire of local goldsmiths at the beginning of stage E. Thus, N. Åberg's thesis about the dating of the Stempelornamentik in the southeast Baltic (Åberg N., 1919, S. 32, 50) is fair. It is true, as analysis of some types of brooches carried out above shows, that the Sösdala style appeared in the Amber Lands not from southern Scandinavia, as Åberg suggseted, but from the lands adjoining the Roman *limes*. It is important to note that an especially significant part of the provincial-Roman "imports" (not so much the products of trade, but the results of robbery) in the 3rd-5th centuries was accumulated by the bellicose inhabitants of the island of Seland (Lund Hansen U., 1994, S. 205), possibly for sacrifices on the island sanctuaries. This situation repeats a similar course of events around the Baltic just after the end of the Marcomannic Wars. It was thanks to the islanders from the west of Baltic that buckles with silver overlay arrived in Halibo and Sambia. These buckles were decorated in the Sösdala style and imitated the buckles of the late-Roman legionaries: these finds are unknown in the Danube area. However, from this region as far as the Aestian area, the decoration on the knives/daggers correlate with the decorations found on Sösdala-style objects. The knife/dagger evolved during the Hunnic wars and was absorbed into the German, Balt, Steppe, and late Roman traditions (Kulakov V.I., Skvortsov K.N., 2000, s. 40-44). It is indicative that, falling out of use among the Prussians at the turn of the 5th century, knives/daggers remained popular for some decades to come among the populations on the right bank of the Neman.

6. All the above leads to the conclusion that, in the middle of the 5th century, groups of an alien ethnic population arrived on the coast of the Vistula Lagoon, and, practically simultaneously, in Sambia. These groups sprang from the former imperial provinces of Pannonia and Noricum - the territory that witnessed the fiercest fighting in the "Post-Nedao" era. At the same time, it is possible to mention the connection of this group of soldiers with the island of Zeland, located on the outlying western district of the Baltic. These newcomers are quite comparable to the Vidivarii of Jordanes. They brought with them a variety of jewellery and weapons of poly-ethnic origin, and obviously possessed a high level of social and military organization. To a considerable measure they promoted the formation of Prussian culture, which began to date, according to the bulk of the attributes of the material culture and burial rites, from phase D_3 (Kulakov V.I., 1994, s. 65). The most representative antiquities connected to the genesis of this culture are available within the limits of the investigated micro-region (the cemetery at Warnikam, adjoining the synchronous defended site of Pillgarten). From the 3rd to the beginning of the 5th centuries, the burials of this cemetery represented a typical picture of the sites of western-Balt culture in late-Roman times. As the finds made by Klebs show, the members of the tribal community of Warnikam, alongside the other Aestians, participated to some extent in the amber trade. A significant presence of alien ethnic elements in the environment of the western Balts of the Vistula Lagoon coast in this epoch is not marked. The middle of the 5th century became a turning point in the history of the inhabitants of Warnikam. It was one of the first stopping points in the territory of the western Balts for the poly-ethnic "followers". Unfortunately, our western-European colleagues have overlooked this obvious historical fact that is based on direct evidence from the archaeological material of Halibo. Thus, for example, Bitner-Wróblewska, with some persistence, in her newest book discusses (based on Scandinavian and Prussian material from the 5th century AD) the similarity of the development of the appropriate societies. This shows, in the 5th century AD, the clear development of the "costume elements of the middle class (sic!)..." and "the strong competition between separate clans,...the distant connections between the elites of Scandinavia, the Baltic lands and Western Europe" (Bitner-Wróblewska A., 2001, p. 122, 123).

The occurrence of a system of fortifications of German origin adjoining the cemetery of the defended site at Pillgarten should be attributed to the middle of the 5th century. It is located 5kms to the northeast of the Balga peninsula. In the 1st millennium it was an island, and this plot of land was, most likely, an initial landing point for newcomers into the territory of the Balts. The first castle of the Order, founded in 1239, was located on the Balga, and this became a base for the eventual capture of the centre of the Old Prussian lands by crusaders. This castle was built over the Old-Prussian defensive site at Honeda (Batūra R., 1985, p. 365), which, according to local tradition, was considered to be the spot where Videvut's warriors first landed in the territory of the Old Prussians. It is possible that among the first Prussian "followers" (men from Warnikam) there was a force of Herulii who moved to the north of the Danube after 455 AD. Lightly armed warriors, hardened by their actions under the banners of Attila, became a serious threat to the surrounding Aestians. The abundance and luxury of the grave goods of the newcomers, and the obvious poverty of the synchronous Aestian graves (above all at "Gora Velikanov"), allow us to assume the possible dependence of the Aestians on the Vidivarii, at least in the first years of their contact in the middle of the 5th century AD. Modern interpretation draws a similar conclusion based on the well-known message from Jordanes about the Aestians: "pacatum hominum genus omnio" - not only "quite peaceful..." but also "...a subjugated people" (Bajer H.V., 2001, s. 368).

From this time there is a sense of the need to distinguish conditionally members of the local clan collectives from the warriors, who were opposite them socially, and the distinction "the Aestians - the Prussians" becomes apparent. For the century following the battle at Nedao the newcomers, gradually assimilating into the local environment, created a sufficiently stable social system with the institution of the leadership authority of the "followers" on the coast of the Vistula Lagoon and in Sambia. This was reflected in the formation of the early medieval antiquities of the Prussians, represented at the Warnikam cemetery by cremations without urns and with biconical accessory vessels, and the arrangement of horse burials below the graves, and by collective military graves. The Prussian-culture population, having a strongly pronounced "followers" character, as the material found at the Warnikam cemetery shows, had already adopted by the time of Theodoric those stable attributes that existed on the Amber Coast at the beginning of the 12th century (Kulakov V.I., 1997a, s. 149, 153, 154).

As discussed above, the earliest burials at the Młotezno cemetery date to the beginning of phase D_2/D_3 (about 440-460 AD) and had no features of Prussian culture. If the brooches from the oldest assemblages of this cemetery have a local shape, the wheel-made pottery is a continuation of the tradition of Wielbark accessory vessels from the Cecele phase (Kokowski A., 1988, s. 165, ryc. 5, g). Similar finds of the middle of the 5th century were discovered at the cemeteries of "Gora Velikanov" (H-19k) and Pasłęk, and emphasize not only the participation of a foreign military elite in the formation of the Prussian culture, but also that of the local community (Kulakov V.I., 1997b, s. 361, 363).

The country around Młotezno might well have seen the first settlements of the newcomers into Hrains Halba (in the Prussian tradition Chrono and Haillibo). The Podguzhe cemetery to the west of Młotezno shows the development by the newcomers of their surroundings. However the geomorphological conditions of the district did not lend themselves to long-term residence because of the great difficulty of establishing defended sites. The bank to the east of Młotezno afforded some protection against sudden attack only and would not have withstood a long siege. The only site available to the Prussians for erecting fortifications was at the mouth of the river Passarge, and these were not constructed until the 6th century. The development of this river basin, marked with sacrifices as they went (separate solidi and hoards at points 11-14, fig. 70), convinced the newcomers of the futility of creating there a network of long-term sites. Only to the northeast of Hrains Halba - on Balga Island and near by - was there the possibility of doing so. True, because of the events of 1239-1240 (the building and siege of Balga castle), and of April 1945 (the last bridge-head of the Wehrmacht in East Prussia outside Königsberg), the archaeological researches on the former island of Balga are extremely complicated. However, there is no doubt that the military experience of the 5th-century settlers suggested the possibility of using the island of Balga as a springboard. Its neighbourhood, according to the find at Uzornoye of an archetype of a star-shaped brooch, was surveyed by the foreigners shortly before the middle of the 5th century on their way to war in Pannonia. The use by the newcomers of Balga Island as a springboard is indirectly marked by the existence of a cemetery at Keimkallen, near a ferry from the island to the continent. Numerous military burials of the middle of the 5th century, with knives/daggers - an attribute of the antiquities of the Vidivarii (point 30)- were found at this cemetery; the defended site at Warnikam may definitely be considered as their stronghold. The principles of fortification at this defended site have analogies in the German antiquities from the time of the great migration. Among them there is an inverted and extended external bank, as well as at Młotezno-Rogity to the east. A cemetery (point 31) adjoins the external side of the bank, giving mass evidence of the presence there of a non-Balt population in 450-500 AD. For example, buckles from Warnikam belt sets, decorated in the Sösdala style, originate from the Hunnic buckles of groups III, department I (a subgroup 3, type A), categorized by Zasetskaya as an "homogeneous group of objects of western origin distributed in the first half of the 5th century." (Zasetskaya I.P., 1994, s. 90, 19в). This "homogeneous group" of artefacts represents a stage in the development of Haillot-type buckles, and dates back to the production of the provincial-Roman workshops known from the *limes* of northeast France (Böhme H.W., 1974, S. 302, 303). In the southeast Baltic "tongue-shaped" belt-ends that decorated belts with buckles derived from the Haillot type, were fashioned in the late-Wielbarsk environment around the extent of the Vistula basin (Bitner-Wróblewska A., 1989, mapa 3, s. 168) - that is along the route of the military "followers" from Scandinavia to the battlefields of the Hunnic Wars and back again. In addition, the route is dotted with hoards of the same type as Hoard A from Młotezno (Petersen E., 1936, S. 51). The above-mentioned characteristic may also be afforded to many other artefacts from the antiquities of Halibo from the second half of the 5th century. In particular, gold rings from Hoard B at Młotezno reflect the same decorative model as post-Hunnic bracelets from Badokpuszta of the middle of the 5th century (see above). And so the bulges at the ends of these bracelets (but already without open dragon jaws) are maintained on the neck-rings. The triangles at the ends of a small ring, complete with the prints of various stamps, represent rudimentary dragon ears (fig. 82,1). The dragon scales on this ring are imitated by cross ribbing, while on the large ring they are recreated with a sickle-shaped stamp. In the antiquities of the Untersiebenbrunn horizon (Kazanski M., 1991, p. 80, 81), this particular stamp was not employed, but it is encountered in the last Sösdala-style phase in western-Balt material (Åberg N., 1919, Abb. 44, including on the blades of knives/daggers), in the antiquities of the early Merovingi era in Scandinavia and on the continent. This allows us to assume, cautiously, that the tradition of manufacturing gold neck-rings (rings) with bulging ends occurred in the Halibo micro-region in period D_3. This thesis is confirmed by the full (in comparison with other artefacts of this type) set of stamps found on the rings from Hoard C, as well as by the wide circulation of the derivatives of this type of decoration in the adjoining

Baltic zone as far as modern Estonia (Lincke B., 1938, Karte 2-4). Traces of jewellery manufacture (including gold items) can be seen in a hoard from Frombork (point 9), and the tendency in the western-Balt area of manufacturing jewellery for the German 'market' is evidenced by the find of a mould from the southern border of the Balt world (Pupki, woj. Mazursko-Warmińskie, ehem. Klein-Puppen, Kr. Ortelsburg).

A principal function of the activity of the poly-ethnic newcomers along the Amber Coast over the period of "the great migration" (Bitner-Wróblewska A., 1992, p. 261, 262) was the maintenance of the essential needs of the various military groups who were returning to Scandinavia at the end of the Hunnic wars. The veterans/"barbarians", weighed down with Roman gold, were little interested in this specie, willingly brought by them as sacrifices to the gods (points 1 - 2, 11, 14). Collective sacrifices, emphasizing the social status of the "followers" and the power of their leaders (Geisslinger H., 1970, S. 208, 212), were brought as massive gold neck-rings (Hagberg U.E., 1984, S. 77, 78), according to the German tradition. The ritual character of this type of jewellery is stressed by finds in Sorte Muld (Bornholm) of plates of gold foil, representing a deity, with a short beard and moustache, sporting a neck-ring, goblet, and staff (Freyr?) (Hauck K., 1994, S. 438). Interestingly, later carved stone figures representing the legendary Prussian leaders Bruden and Videvut are shown with just the same attributes. Returning to the gold neck-rings of the 5th century, it is necessary to note that similar decorations, according to the Roman military customs that the "barbarians" also observed, marked service in battle. This aspect of neck-rings, and also their use as means of payment ("ancient gold of ring-treasures") is reflected in the ancient German epos (Beowulf, 1975, s. 86, 94). As well as neck-rings, the earliest forms of decorations included gold medallions (with portraits of the emperors), chains, and even the gilded bridle bits of battle horses (Jordan, 1997, s. 246). The distribution of gilded and bronze bits in late and post-Hunnic Barbaricum is obviously connected with the last fact. All this points to the desire of those soldiers, deprived of decorations on the battlefield (especially after the defeats at Mauriac and Nedao), to receive their equivalents (or, rather, their "barbarous" imitations) at the first safe point on their way home. They tried, by acquiring these external attributes of recognition of their military valour, to reproduce the elements of the value system of the great Hunnic era. Along the entire length of waterway from the Danube to the Vistula the most convenient point for this purpose was Halibo, reconnoitred by the inhabitants of the islands of the western Baltic on their way to Pannonia. The symbiosis of elements of the applied arts of the provincial Romans, Germans, and Balts was one of the specific features of the material culture of the population of this historical landscape (a landscape unique for the Baltic in the 5th century), occupied, according to Jordanes, by the Vidivarii. The distribution, about 450 AD (most likely right after the arrival of the soldiers defeated at Nedao and forced to leave the inhospitable Danube area), of star-shaped brooches and belt sets in the Sösdala style, became one of such phenomena. The present author has already noted the possibility of a reflection of this unique phenomenon in northern Barbaricum in the legend of "Brisings' belt" (brisinga girdr) (Kulakov V.I., 1997c, s. 59). Taking this into account, it is hypothetically possible to connect the activity of the legendary goldsmith Veland with Halibo ("The Song of Veland" is brought to mind by memories of the real events connected with the battle at Nedao - Wamers E., 1987, S. 85). According to the Gothic language, the word "weiha" ("priest" - Braune W., 1905, S. 166) in combination with the word "land" is close to the sense of "the country of priests/wise men" (a variant "veh" /"sanctuary"/ - "the land of a sanctuary"). The western part of the Prussian area is known by the same derivative from the 9th to the 14th centuries (Witland/Weydelant - "the country of wise men"). The micro-region, with a sacred site near the modern village of Młotezno, was one of the first points in the southeast Baltic where, in the middle of the 5th century, the Old Prussian culture began to develop. Later, the area of Prussian-culture influence received the name of "Vitland". In addition, there is the following reference in the German epos: "...Veland alone, sitting in Ulvdalir, ...rings as snakes, skilfully twined..." (Beovulf, 1975, s. 243). A fascinating analogy to these lines is to be found in the snake-shaped neck-rings from Hoard C in Młotezno. Other references in the epos to the "Valley of the Elves" (Ulfdalir) and the "Lake of the Elves" (Ulfc'ar), in all northern Europe, have their parallels only in Halibo: Halibo is limited from the southwest by the river Ylfing (the "River of the Elves"?), flowing out from Lake Druzhno (in the 5th century a section of the extensive delta of the Vistula/Nogata). The name of the Vistula is comparable to the ancient Icelandic "kvísl" ("mouth" - Dzhakson T.N., 1993, s. 60). Taking into account that antique authors mention (from the 1st century BC) the name of the greatest river in the southeast Baltic in the form of the ancient German hydro-name Vistula-kvísl (first mentioned by M. Vipsanius Agrippa - Podosinov A.V., 2002, s. 46), its occurrence in such a form should be connected with the activity of the Goths. Appearing there at the end of the 1st century BC/beginning of the 1st century AD, the newcomers from Scandinavia paid special attention to the broadest delta of the Vistula - a huge funnel to be negotiated by the northern warriors on their way to finding Roman treasure. It is probable that the Vidivarii gave the name Ylfing to the eastern part of the Vistula delta-kvísl in the 5th century.

It is necessary to note that within the limits of Halibo there is the largest congestion of hydro-names of ancient German origin in the Baltic. They are concentrated in the north-eastern part of this micro-region and are represented by such names as the streams Banava and Stradik (Pėteraitis V., 1992, p. 311). However, the names of the adjoining rivers - Frisching, Jarft (from the ancient

Icelandic "jorđ" - "land") and the island of Balga (from the Gothic "balgs" - "passage"), also assume the appropriate origin. Prussian hydro-names of obviously late origin cover only the large rivers of Halibo (Passarge, Baude, Wogenapp), evenly distributed throughout Halibo (fig. 78). The first, traditionally considered to be a border between the Balts and the Germans during the time of Roman influence, may actually have a Gothic root (in Gothic, "saúrga" - "alarm" (Braune W., 1905, S. 160)), and thus completely referring to its function as a boundary. The hoards at Młotezno were intended to assuage this "alarm" (i.e. the mutual fear of the representatives of various tribes), which was reflected in the hydro-name Passarge. Such a custom is characteristic of the intertribal zones of the Baltic in the early Middle Ages (Šimėnas V., 1995, S. 153). A similar sense is linked to the hydro-name, Sorge, which is located to the southwest of the southern extremity of Lake Druzhno, on the later border between the possessions of the Order and the territory of the Pomorze Slavs. Here, finally, one of the numerous names from the Prussian lands, Ulmerigia (known by 15th-century sources - Grunau S., 1876, S. 58, Anm. 1), has a Gothic origin, deriving from "holmr" - "island" (Jordan, 1997, s. 187) and "reiki" - "kingdom" (Braune W., 1905. S. 159). If the meaning of Ulmerigia is not a direct borrowing from Jordanes' "Gethica" ("...soon they (the Goths) moved...to the areas of the Ulmerugii that were then found on the ocean's coast..." - Jordan, 1997, s. 65), it might have derived from the environment of the Vidivarii, designating their springboard to Aestian lands - Halibo and Balga Island.

The archaeological finds (above all the stylistics of those objects of decorative art), and hydronimic data, confirm the presence of some groups of non-Balt populations in the centre of the Halibo micro-region, (mainly) in its northeast, in the 5th century. This presence appeared in the earlier unpopulated territory about 450 AD and was designated Vidivarii by Jordanes. Their traces (particularly the early forms of animal-headed brooches) are found in the antiquities of the first decades of the 6th century. Being involved in the sphere of activity of the Vidivarii, the western Balts arrived in the course of the "Balt reconquista" from Upper Pregolya in the second half of the 5th century, and begin to master first the left-bank of the Frishing, and then all the territory of Halibo. This is testified by the hydronimic data and distribution of Prussian cemeteries and defended sites[1] (Kulakov V.I., 1994, fig. 2, 43). At these burial sites the following attributes of Prussian culture, new to the Baltic, are characteristic. At the upper level of the grave there is a lens with the remains of a funeral pyre among which was a vessel of biconical form, serving not as an urn, but for carrying bones from the pyre to the grave, where it was located nearby. In the lower level of the graves there are no skeletons, but only the horse skull, oriented to the west and often with the remains of the headpiece. The basic features of later Prussian cremations from the 8th-11th century are as follows. The deceased were burnt on a pyre outside the cemetery; the burnt bones were gathered in a small vessel (a "temporary urn") and carried to the grave, where, earlier, the skull or, less often, the legs of the horse had been buried (but not the whole horse as practised before). Taking into account the abundant organic material found in the lower levels of the burials, it is possible to assert that the skin, head, and some body parts were also included.

Thus, in the 5th/first half of the 6th century AD, in their search of raw materials for trade (above all furs), groups of early Prussians (together with separate bands of Vidivarii?), on their migration to the east and northeast, penetrated "...as far as the lands of the local Finnish Balt/Slav population forest zone, and in particular to the long-barrow cultures – the Tushemlya, late Djakovo, Moshchino, and Ryazan-Oka area cemeteries" (Kazanski M.M., 1999, s. 414).

It is remarkable that in the vicinity of Młotezno in the 6th century, there is only one Prussian defended site (Podguzhe) to protect the western Balts' oldest and most revered Halibo complex. At the beginning of the 6th century, one of the latest indicators of the presence of Vidivarii in this territory is the "second wave" of solidii (for stage E - no further hoards, but only separate coin finds). This is indirectly confirmed by the discovery at the defended site of Sonnenburg bei Braunsberg (Ehrlich B., 1923. S. 200, 201) of a fingered brooch of Regio-Emilia style (500-550 - Kulakov V.I., 1990b, s. 180). It is probable that the defended site of Grünwalde was established (point 21) during the hostile activity of the Vidivarii and Prussians, and that it adopted then the sacral functions of the main Halibo centre (points 15-19), becoming the new religious centre of the Prussians and known by Peter fon Dusburg and Simon Grunau by the name Romove. Correspondingly, the idea of Halibo itself was broadened. Grunau was already using the name Halibo for a sacral territory extending to other areas of Balga Island.

[1] Prussian settlements from the 6th-13th cent. in Halibo: 1 - Janovo; 2 - Myslętin; 3 - Weklize; 4 - Weklize-2; 5 - Weklize-3; 6 - Janów Pomorski; 7 - Beliany Welke; 8 - Łęcze; 9 - Tolkmizko; 10 - Bogdany; 11 - Grunenberg; 12 - Mamonovo; 13 - Vesioloe; 14 - Moskovskoye; 15 - Timiryazevo; 16 - Lipovka; 17 - Pervomajskoye; 18 - Beregovoye; 19 - Ushakovo.

XI. Conclusions

The major conclusion to be drawn from this analysis of the antiquities of the Amber Lands of the first centuries AD, is the revealing of three chronological horizons for the era of Roman influence in the southeast Baltic. These horizons correspond to the great historical events occurring on the outlying western districts of the Balt world in antique times. These events may be plotted against the chronological framework of Aestian antiquities in the following way:

The Venetian horizon (B_1-B_2) is characterized by the dominance of cremations - in male sets, with a single brooch, knife, belt-buckle, and, less often, a shield-boss. Material evidence is found for the first contacts with Romans (Dollkeim, Ilischken). For Sambian female graves there are numerous brooches of groups AIII and AIV, details of Noricum-Pannonian belt sets, and bracelets of the Sambian Kameńczyk-type variant are characteristic. As a whole, there is a distinct impression of the "female nature of grave goods". In the area of the SNG, by the beginning of the 1st century AD, there are no traces of the anticipated quantities of Aestian/western-Balt antiquities, which, apparently, should have developed the tradition of the barrow culture of the western Balts in the Amber Lands. This leads Nowakowski to replace the "groups" comprising the western-Balt culture with various independent "cultures". It would seem that our Polish colleague's innovation here is not without merit. Besides the individual features of ceremonialism and several types of jewellery, the Sambian archaeological material presents the amazed spectator with a completely new ethno-cultural formation (named for ease the Sambian/Natangian group of western-Balt culture), which should be connected with those newcomers from the south, who brought with them their post-Celtic and provincial-Roman traditions. Thus, Tacitus' informants on the Amber coast encountered not "autochthons", 'but allochthons' - "the wandering Veneti" and Velti, whose language "was close to British". The earliest horizon for antiquities of the Amber Lands in Roman times is named after the first of the mentioned tribes, those neighbouring with the Germans of Aenigia (see below). This horizon is marked by the stabilization of the Great Amber Route and the appearance of its military escorts - the first professional military formation in the Baltic, and conditionally called the "Sambian ala". This irregular formation afforded protection for the amber caravans in the Vistula basin, considered part of the border between Germaniae and Skythiae as early as the 1st century AD (Podosinov A.V., 2002, s. 263, 264).

The Cimbrian horizon of phase B_2/C_1 corresponds to a transitional stage of the Dollkeim development. Inhumations are replaced by urn cremations, with finds of between 150-180 socketed axes, details of Vimose headpieces and iron "stiloses". Versions of above-grave stone constructions are also found. All these innovations testify to the appearance in former Aestian territories of groups of warlike Germans connected both with the south (stone rings as "symbols of unity, harmony and completeness" - Toporova T.V., 1994, s. 25), and western Scandinavia (some types of grave goods). These German tribes, connected by origin to the north Jutland peninsula, are conditionally designated by the ethnic name of "Cimbri". There are no direct indications that every Cimbrian tribe left their native land, setting out in 113 BC on the long and fatal campaign. On the contrary, many antique and even early medieval authors continue to call the inhabitants of the lowlands of northern Jutland by the collective name "Cimbri" (Budanova V.P., 2000, s. 251, 252). In the present research, the connection of this ethnic name with the German newcomers along the Amber Coast in phase B_2/C_1 is conditional. If a certain (and most likely straight) connection of these "Cimbri" with the Goths of Jordanes is obvious enough, their correlation with other German tribes in the mouth of the Vistula (the Skiri, Sarmatians, Veneti and Hirri, and, according to Pliny the Elder, inhabiting the country of Aenigia - Podosinov A.V., 2002, s. 104) is not clear. It is probable that the former name of Witland is also included within the designation of this "country" (see Chapter X). In phases C_1-C_2, the "Cimbrian" antiquities in Sambia assume a fully developed appearance; and the only form of their burial rites is the typical urn cremation. Such standardization is a consequence of the religious character of the German tribal associations (Budanova V.P., 1999, s. 17), in which non-Germans also participated (Budanova V.P., 1999, s. 139). The presence in military graves of Roman bronze coins, which were symbols of the soldiers' leading role in the control of the amber trade, is an innovation in burial assemblages of phases C_1-C_2. Provincial-Roman brooches and their derivatives replace brooches with tied feet, including those with ring elements. Details of female costume included a pair of brooches and an insignificant number of glass, copper alloy, and amber beads. The bead necklace was supplemented by iron bell-pendants, which, in phase C_2, were replaced by bucket-shaped pendants. One of the few indicators of the presence in Sambia of western-Balt material of phases C_1-C_2, are single brooch-derivatives of clasps of types AIV, 72 and AV, 98, related to "Mazurian brooches". The close of the Cimbrian horizon fell in phases C_2/D_1 - D_2. The distribution of urns with horizontally applied decoration, known previously in the island antiquities of the western Baltic, is an innovation. From phase C_2/D_1, the main attribute of female Aestian assemblages becomes the large "crossbow-shaped" brooch. At the late stage of phase-C_2/D_1 - D_2 burials, brooches with star-shaped feet are found. There are also assemblages containing a plentiful number of military grave goods belonging to optimates.

The horizon of the Vidivarii (phases D_2-D_3) in 400-450 AD is characterized by changes in the burial rites: the urn cremation is replaced by groups of cremated bones (variant 1.2). The main type of clasp at this time is the Duratón type. The historical phenomenon of this horizon consists in the creation (on the former border of the Germans and Balts, in the Halibo micro-region and, probably, western Sambia) of a "barbarous kingdom" under the leadership of Vidja-Videvut, glorified in the legends and mythologized according to the traditions of the ancient German onomastics (Toporova T.V., 1996, s. 132). The genesis of this community is similar to that of the Germans at the outset of the Cimbrian period. They, as well as the "people of Vidja" (designated by Simon Grunau as "natives of Cimbria"), came to Vitland, escaping from the military defeats of the Danube area. The remnants of the "barbarous" troops, arriving in the Amber Lands from the south, were prepared to subject themselves to the military authority of Vidja. These remnants were the ethnic components that gave rise to the community of the Vidivarii in Halibo in phase D_3. In effect, Prussian culture developed in the middle of the 5th century AD as a result of their search for new trade routes, the sources of raw materials/goods, and tributaries. Its second component was the Balt "reconquista" (p. 106), due to which, at the end of antiquity, the descendants of the autochthons of the Amber Lands returned from Sudawiania to the native country of their ancestors; they were subsequently called "Prussians". This "reconquista" was probably provoked by the above-mentioned activity of the Vidivarii, and might well have been responsible for carrying the original "virus" of migration to Witland (Budanova V.P., 2000, s. 108).

Therefore, the era of Roman influence in the Amber Lands witnessed the incessant movements of various (essentially German) tribes. The instability of the cultural and ethnic situation in the regions discussed in this present work resulted in the original "scrappiness" and diversity of the local material culture – features that relate the southeast Baltic to other areas of Barbaricum. On the right bank of the Neman there was another, and much more consolidated, material culture in the years of the Roman Empire: one that directly indicates the mono-ethnic population of the area.

Appendix 1

Catalogue of Sambian "barrows" from the first centuries AD.

45. DUBROVKA, Zelenogradsk District, Kaliningrad Region, Russia (former Regehnen, Kr. Fischhausen). In 1887, Georg Bujack excavated 22 stone coverings (overlapping urn cremations) from the cemetery. These coverings, stones arranged in concentric circles, were no more than 0.33m high. Horse remains (more exactly horse skins) were found in the lower parts of the grave pits. By elements of the grave goods, this cemetery is dated to the beginning of the 3rd century AD. To the north of this cemetery there are barrows dating from the early Iron Age (Bujack G., 1887-1888c, S. 122, 123).

49. KLEVERNOYE, Zelenogradsk District (former Drusker Forst, Kr. Fischhausen). In 1889, G. Bujack investigated here 21 stone coverings with concentric structures, being fully analogous to the coverings in Dubrovka (see # 45) and categorized as 1st century AD. Within the limits of this cemetery there are barrows (NrNr IV, V, VI, VIII, XII, 1.5m high). In one of these barrows there was a stone axe of the late Bronze Age (Bujack G., 1889, Taf. V). Sections of these barrows (probably dating from the early phase of western-Balt barrow culture) were damaged during construction of the 2nd/3rd century AD burials (Bujack G., 1889, S. 113-116).

60. GEROJSKOYE, Zelenogradsk District (former Goithenen, Kr. Fischhausen). In 1906, A. Bezzenberger and E. Hollak excavated here a flat grave cemetery of the 4th/5th century AD, the finds of which have not been fully published (fig. 86). According to local residents, the arable land near the village of Goithenen was "... earlier covered by woods and low stone barrows (of small size), grouped together and in which there were urns lying between flat stones" (Bezzenberger A., 1914a, S. 132). However, this early Iron Age barrow cemetery was destroyed by ploughing at the beginning of the 20th century, and was 30m from a site investigated by A. Bezzenberger and E. Hollak. It bore no relation to Aestian antiquities from the time of Roman influence.

95. YAROSLAVSKOYE, Zelenogradsk District (former Schlakalken, Kr. Fischhausen). In 1893, Karl Krechman, the manager of the East Prussian provincial museum, excavated here 38 burials (barrow-shaped, covered earth eminences, consisting of 2 or 3 layers of stones, overlapping the cremations (urn and without-urn) as well as the inhumations. The diameters of the coverings reached 5m, with a height of no more than 0.40m. Of special interest is Grave 1, where the stone covering overlapped the circle to a diameter about 5m. In the northern part of this circle an inhumation with a northerly orientation is revealed, and the walls of the grave were arranged with stones. Among the grave goods of this military burial there was a copper-alloy brooch of type AV, 125, dating Grave 1 as stage B_2/C_1 or 150-200 AD. To the west of this inhumation were found the teeth of a horse and two harness buckles (Jankuhn H., 1939, S. 246, 247). This assemblage (fig. 87) has no analogies in terms of its burial rites to the local material of the previous era; it is considered as one of the earliest horse-burial assemblages in the southeast Baltic (Nowakowski W., 1996, S. 63).

256. RADUZHNOYE, Ozersk District (former Rominten, Kr. Darkehmen). In 1895, A. Bezzenberger excavated here a flat grave cemetery dating to the 1st-3rd centuries AD, and containing stone coverings of various forms: "...the site of the cemetery was a large forest and the constructions stood out as groups of stones with longitudinal apertures, (appearing) as round; the area was overgrown, grassy, and low rising" (Bezzenberger A., 1896, S. 36). The upper parts of the stone coverings were probably destroyed when the wood was cleared at the end of the 19th century; these features gave rise to the "low rising" nature of the site.

320. The former WARGENAU-KUNTERSTRAUCH site, actually - WOSEGAU, Kr. Fischhausen (and now the village of Klintsovka, Zelenogradsk District). In 1899, G. Heydeck investigated here 18 flat graves with log-inhumations ("sarcophagi") and urn cremations. The barrows of "late pagan times" (10th-11th centuries) are located within the modern boundary of Dubki, 0.7kms to the southwest of this cemetery. The grave goods (fig. 88) date the burials that were opened in the natural boundary of Kuntershtrauch to phases B_2-C_1/C_2 (Heydeck J., 1909, S. 207-216).

325. The former WENGERIN, Kr. Insterburg (Chernyakhovsk District). The cemetery was investigated by Frolih, C. Engel (1931), Walter Grunert and Walter Gronau (1936). They opened 60 burials, mainly urn and without-urn cremations of the 2nd-3rd centuries AD. Grave 35, before excavation, was an eminence 5m in diameter and 0.40m high. After removing the grass covering, there was a further layer of stones with a ring of large boulders around the perimeter. Under the stone covering there was a stone stela. The cremations (without urns) were found below, in 2 pits 0.7 and 0.8m deep. Directly from the northwest, the urn cremation, containing grave goods of the 3rd-4th centuries AD, adjoins the covering of this assemblage. Only Assemblage 60, located at the east section of the cemetery, represented a real barrow, the mound of which included three layers of stones. Under them were found two stone boxes (cists), one with small walls 0.30m high. Unfortunately, the investigation of this Bronze Age assemblage was not completed (Grunert W., 1935, S. 50, 51, 55).

Thus, as the catalogue data shows, the "barrows" marked by J. Jaskanis in the Aestian regions as being of early Roman date are actually either mounds from the second half of the 1st millennium BC, or flat graves above which stone coverings overlapped. Theoretically these may be faint rudiments of a barrow tradition but obviously do not represent barrows. It is necessary to examine further the published assemblages:

26. CHRUSTAL'NOYE, Zelenogradsk District (former Wiekau, Kr. Fischhausen). In August-September 1884, J. Heydeck, with the assistance of the graphic artist Ekkart, dug 58 burials here. In the first publication of this excavation there was no reference to barrows; Heydeck concentrated on the description of the inhumations in the wooden "sarcophaguses" found during the excavations. True, the unburned bone remains, as well as those from the sites of the southeast Baltic from the 1st millennium AD as a whole, were not retained because of the features of ground (Heydeck J., 1900-1904, S. 217). In addition, 48 cremations were found when excavating in Wiekau the same season. The burnt bones were mainly from urns found under round stone coverings and accompanied by accessory vessels, copper-alloy brooches, bracelets with overlapping ends, and round-framed buckles of the Dollkeim-Rominten type, all dating to the 1st-3rd centuries AD (Bujack G., 1887-1888b, S. 276-283). One area of the cemetery contained a group of 10 inhumations in grave pits; layers of brown colouring from wooden coffins were traced (Bujack G., 1887-1888a, S. 273). The most noteworthy burial on the site was Burial XIIII (according to J. Heydeck's numbering). In this pit (about 0.5m below the level of the ploughed soil) Heydeck and Ekkart found only "white traces of bones, together with the brown traces left by the (wood) bark of the sarcophagi". To the north of the edge of Grave 4, accessory vessels were placed in a cist, one of which contained the remains of the funeral pyre (fig. 89,3). At the east edge of the grave the remains of a shield and two broken spearheads (fig. 90, h, q, i, l) were found. Other grave goods - iron buckle, brooch, axe and battle knife, copper-alloy brooch (fig. 90, p, n, o, k) - were not described in detail in the publication (Heydeck J., 1900-1904, S. 218). The designation of the accessory vessels from this assemblage ("mug of Wiekau type") and the dating of the assemblage to stage B_2/C_1 = 150-200 AD are worthy of attention (Nowakowski W., 1996, S. 29, 30). For a long time this burial, because of Heydeck's reconstruction (fig. 89,6), was considered as an under-barrow. Actually, according to Ekkart's field sketch copied in 1931 by Heinrich Sommer and kept in the Samland-Museum (Pinneberg), the stones protecting the grave from above hardly broke the surface (fig. 89,7). This burial assemblage may be reconstructed as follows:

1) In the mound of the early Iron Age barrow (the nearest analogue to which is near Burial XIIII) a trench was dug through on a north/south (fig. 89,1,2) axis before the "sarcophagus" was buried. As a result, the stone covering of the barrow (from north to south) was disturbed.

2) The urns were withdrawn from the cist in the barrow and replaced at the east edge of the new grave by accessory vessels ("Wiekau-type mugs") with the remains of the ritual food and pyre. On the base of one of these "mugs" (fig. 89,4,5), and obviously from early Roman times, an ancient potter has imitated a late La Tène sacral image (Okulicz J., 1973, ryc. 155, a), as though echoing the grave goods from the end of the 1st millennium BC. The style of this ornament was probably discovered by the Aestians on the urns taken from the barrow cists.

3) The skeleton in the wooden "sarcophagus" in the trench was then oriented towards the north. The deceased had two brooches of different sizes in the "sarcophagus", indicating the clothes of a soldier - cloak and jacket (von Richthofen J., 1994, Abb. 12). To the east of the "sarcophagus" a pair of spears, covered by a shield, was placed at the bottom of the newly dug trench.

4) Then the trench (containing the "sarcophagus" with the body of the deceased) was unsystematically filled with the earlier extracted stones of the barrow's covering. Subsequently, after the "sarcophagus" top had rotted, these stones fell below the level of the bedrock within the limits of the grave pit (fig. 89,7).

5) The last stage of the burial process was to top up the surface level of the stone covering with earth – a completely uncharacteristic action for true antiquities of the western-Balt barrow culture.

Burial XIIII is dated by a pair of brooches (late variants of clasps of type AV, 131) to stage B_2/C_1, according to the chronological system of Ulla Lund Hansen (Lund Hansen U., 1987, Abb. 13, 20). Shield-bosses with a long central axis, and accessory vessels with S-shaped sections, find good parallels in the Przeworsk antiquities of the second half of the 2nd century AD (Godłowski K., 1981a, s. 115), illustrating the frequent contacts between the inhabitants of the Amber Coast and Masovsha (Dąbrowska T., 1988, s. 84). Ceramic forms of Przeworsk origin are found in the 2nd century AD on sites in Sambia and the Lower Nogata (Gaerte W., 1929, Abb. 125, 126). For the Aestians' southern neighbours such pottery is considered an indication of contacts between the "Przeworskians" and the population of the Jutland peninsula (Dąbrowska T., 1988, s. 196).

92. IZOBIL'NOYE, Polessk District (former Klein Fließ, Kr. Labiau). At this point (at the end of the 19th century) in the history of the ploughed plot, fragments of Megar bowls (inscribed "CINNAMI") were found (Brinkmann A., 1896-1900, S. 73). There was a stone covering that was destroyed by forester Bommel's activities in the area. In this place in 1897, J. Heydeck opened, according to the field sketch (fig. 91,3), a split-level burial [I]. An urn with three accessory vessels was found (one of the vessels - fig. 91,2) under the remains of the stone covering (a small area of rising ground no more than

0.40m high - fig. 91,3). Below, at a depth of about 2.5m from the modern ground surface, a horse skeleton (south (?) orientation of the skull) was revealed. It was accompanied by a headpiece with copper- alloy details and by a bit with copper-alloy rings. The stone covering (no more than 0.5m high) was located 20m from the burial described above. When excavating, Heydeck found Burial 3 under the stone covering (consisting of 2 or 3 layers of boulders). This contained a group of burnt bones with fragments of a vessel, as well as a spearhead, iron celt, shield-boss, copper-alloy brooch of group AV, 192-203 (Bezzenberger A., 1904, S. 85), copper-alloy details of a belt set (fig. 92) and copper-alloy slag (Heydeck J., 1896-1900, S. 57-59). According to the brooches and accessory vessels, both investigated by Heydeck, the cemetery at Izobil'noye is dated as stage B_2/C_1 = 150-200 AD.

Appendix 2

Catalogue of horse headpiece details, 1st-4th centuries AD, characteristic of Sambia.

Phases B_1-B_2.

4 (hereinafter NrNr - according to the S. Wilbers-Rost's catalogue - Wilbers-Rost S., 1994, S. 164-213) - **Glane-Visbeck**, Ldkr. Osnabrück: a chain link similar to type (Form) Z1; a chance find on a road.

13 - **Saalburg**, Kr. Homburg: a chain link similar to type type (Form) Z1, provenance unknown.

22 - **Gudbjerg**, Gudme sn.:links of type Z1 and straight snaffle of type Ks3, found in composition of headpiece in cremation together with two copper-alloy brooches, fragment of blade of one-edged sword or knife, vessel, fragment of shears, tongue of copper-alloy buckle, tooth of a wild-boar and other finds (phase B2).

41 - **Klein-Wangen**, Kr. Nebra: links of type Z1 and straight snaffle of type Ks1, found (in composition of headpiece) under a stone during ploughing.

45.1,2,3 - **Pantlitz**, Kr. Ribnitz-Damgarten: links of types Z1 and Z3 in composition of headpiece, as well as strap holders of type Rh1, found in the River Reknitz.

52 - **Mödring**, Bez. Horn, Niederösterreich: strap holders of type Rh1 and links of type Z4c,f, straight snaffle of type Ks4a in composition of headpiece with plates in style opus interrasile, found on Mt. Wechselberg together with silvered mounts.

61 - **Cekanów**, woj. Kaliszskie: a casual find of links of type Z1, strap holders of type Rh1 and straight snaffle of type Ks3 in composition of headpiece.

64 - **Jeźów**, woj. Jeleniogórskie: in the grave goods of two (?) burials - links of type Z3, strap holders of type Rh1 and straight snaffle of type Ks1 in composition of headpiece accompanied by: spearhead with gold mounts, 4 spearheads, 2 bosses of type J7b, 4 knives (or details of shears), whetstone, axe with lug.

71 - **Ostrówek**, woj. Bydgoskie: a find in an unnumbered burial of links of type Z3, straight snaffle of type Ks1b in composition of headpiece fragment accompanied by pieces of a knife and iron buckle.

110 - **Lugovskoye**, Zelenogradsk District (ehem. Lobitten, Kr. Königsberg) - Grave 21 (point 2): links of type Z1, strap holders of type Rh1 and straight snaffle of type Ks2b in composition of headpiece in association with brooches of types AII, 42 and AIII, 60/61 (phase B_{1a}-?).

112 - **Mel'nikovo**, Zelenogradsk District (ehem. Rudau, Kr. Samland): links of type Z1 and straight, covered by copper-alloy iron snaffle in composition of headpiece chain.

115 - **Nikėlai**, Šilutė raj., Lietuva - links of type Z1 and strap holders of type Rh1 in composition of a headpiece fragment, a chance find in the cemetery.

120.2 - **Polesye**, Guryevsky District (ehem. Gr. Ottenhagen, Kr. Königsberg) - Grave 23 (1931): links of type Z2 and strap holders of type Rh1 in composition of headpiece in association with brooches of types AIV, 72 and AV, 120 (phase B_2).

121.2 - **Povarovka** (fig. 56), Zelenogradsk District (ehem. Kirpehnen, Kr. Samland): Grave III, in the urn there are two headpieces, in the composition of one of them there were links of type Z1, strap holders of type Rh1 and straight snaffle of type Ks2b accompanied by brooch of type AIII, 61 and objects for arms (phase B_{1a}-?).

125 - **Ramenskoye**, Zelenogradsk District (ehem. Norgau, Kr. Samland): links of type Z1, strap holders of type Rh1 and straight snaffle of type Ks3 in composition of headpiece, found in inhumation (according to H. Jankuhn).

127 - **Russkoye**, Zelenogradsk District (ehem. Germau, Kr. Samland): links of type Z1 and strap holders of type Rh1 in composition of a headpiece fragment with head mount in style opus interrasile, a chance find in the cemetery (?).

130 - **Soldatovo**, Gvardeysk District (ehem. Friedrichstal, Kr. Wehlau): links of type Z3 and the strap holders close to type Rh1 in composition of a fragment of headpiece (of belt?), a chance find in the cemetery.

135 - **Pervomajskoye**, Bagrationovsk District (ehem. Warnikam, Kr. Heiligenbeil) - Grave 29: links of type Z1 in composition of headpiece with iron details.

Phases C_1-C_2

01 - (in S. Wilbers-Rost's catalogue is absent) - **Vizsoly**, kep. Borsod-Abaúj-Zempelén (Végh K.K., 1969, p. 76, 77, 79, XII. Tab., 1-3): links of type Z4 in composition of fragmented headpiece, provenance unknown.

14.1-3 - **Thorsberger Moor**, Kr. Flensburg: links of type Z4 in composition of three headpieces, found as sacrifices (hereinafter signifying marsh/lake sacrifice sites).

19 - **Bissinge**, Stege sn.: links of type Z4 in composition of a fragment of headpiece, found during ploughing (according to H. Jankuhn).

21.1-3 - **Ejsbøl Mose**, Haderslev Amt: strap holders of type Rh4 in composition of three headpieces, found as sacrifices.

23.3.4 - **Illerup Ådal**, Skanderborg sn.: links of type Z4 and strap holders of type Rh2 in composition of two headpieces, found as sacrifices.

25.1 - **Maltbæk Mose**, Malt sn.: links of a type Z4 and strap holders of type Rh2 in composition of headpiece, found as sacrifice.

30 - **Sanderumgård**, Davinde sn., Odense Amt.: link of a chain of type Z4 in composition of grave goods of female inhumation. Composition of grave goods: bucket of type E58, gold brooch of type AVI, 184, two silver brooches of type AVII, 206, silver washer-shaped brooch of type A, 235, a silver "Baltic" pin of type Beckmann 124, an S-shaped silver clasp of a necklace, pendants (?), glass and amber beads, bone and iron combs, shears and other finds (phase B_2/C_1).

31 - **Sørup Mose**, Østofte sn.: links of type Z4 in composition of copper-alloy chain found as sacrifice.

32 - **Tranebær Mose**, Skeerup sn.: links of type Z4 and Rh3 in composition of copper-alloy chain found as sacrifice.

33 - **Trinnemose**, Torslev sn.: links of type Z4 and strap holders of type Rh3 in composition of one (?) headpiece, found as sacrifice.

35.1 - **Vimose**, Allesø sn., Odense Amt: links of type Z4 and strap holder of type Rh3 in composition of headpiece, found as sacrifice.

39 – **Dresden**: links of type Z4 and strap holders of type Rh3 in composition of a fragment of headpiece, provenance unknown.

40 - **Ketzin**, Kr. Nauen: strap holders close to type Rh2 in composition of a fragment of headpiece, found in the River Havel.

42 - **Langsdorf**, Kr. Ribnitz-Damgarten: links of type Z4 and strap holders of type Rh3 in composition of headpiece, found in the River Trebel.

49 - **Schagen** ("Lagedjik"): link close to type Z4, found at a settlement (phase B_2/boundary of phases C_1-C_2).

52 - **Mödring** (see the first section of the catalogue).

65 - **Kamieńczyk**, woj. Ostrołęka: links of type Z4, strap holders of type Rh2 and straight snaffle of type Ks4a in composition of headpiece, found in the River Bug.

73 - **Sikucin**, woj. Sieradzeckie: strap holder close to type Rh2, provenance unknown.

80 - **Żabin**, woj. Koszalińskie: strap holders of type Rh4 in composition of headpiece, provenance unknown.

81 - **Zlaków** Koscielny, woj. Skierniewickie: links of type Z4 and straight snaffle of type Ks4a in composition of headpiece - from burial (according to Wilbers-Rost).

90.12-15 - **Skedemose**, Öland: links of type Z4 in composition of headpieces, found out sacrifices.

91 - **Skee**, Bohuslän: links of type Z4 in composition of a fragment of copper-alloy chain, a single find (?).

92 – **Småland**: links of type Z4 and strap holders of type Rh2b in composition of headpiece, found as sacrifice (?).

94 - **Stora Katthagen**, Västergötland: fragment of strap holder of type Rh4, found during ploughing.

98.1 - **Chrustal'noye** (fig. 67, 68), Zelenogradsk District (ehem. Wiekau, Kr, Samland): links of type Z4 and strap holders of type Rh2a in composition of headpiece, found in wooden sarcophagus (INH) Nr. 35; the grave goods consisted of a sword, spearheads, shield-boss, socketed axe, silvered brooches of types AV, 29 and AV, 130 and other goods (phase B_2/C_1).

99 - **Galich-Lovachka**, Lvov Region: links of type Z4a in composition of copper-alloy chain, found when excavating the settlement (the end of the 1st mil. BC/1st-2nd cent. AD).

109 - **Logvino**, Zelenogradsk District (ehem. Gr. Medenau, Kr. Samland): links of type Z4 and strap holders of type Rh4e in composition of a fragment of headpiece, found in burial (?) with pair of spurs of group Godłowski II.

121 - **Povarovka**, Zelenogradsk District (ehem. Kirpehnen, Kr. Samland) (see the first section of the catalogue).

131 - **Kulikovo** (fig. 61), Zelenogradsk District (ehem. Sorthenen, Kr. Samland): links of type Z4, strap holders of type Rh2a and straight snaffle of type Ks1a in composition of headpiece, found in numberless grave together with brooch of type AV, 145, bronze coin of Lucilla, and other goods (phase B_2/C_1).

138 - **Hungary**, "comitat Turoc": links of type Z4a and strap holders of type Rh2a together with a straight snaffle of type Ks4a in composition of headpiece, provenance unknown.

139 – **Hungary**: links of type Z4h1 and Rh2a in composition of headpiece, provenance unknown.

140 – **Hungary**: links of type Z4h1 and Rh2a in composition of headpiece, provenance unknown.

Appendix 3

Catalogue of archaeological sites from the Halibo micro-region (5th cent. AD).

1 - **Komorowo**, woj. Elbląg: solidus of Theodosius I (361-395) (Godłowski K., 1981b, s. 120).

2 - **Dębina**, woj. Elbląg (ehem. Klein Eichen, Kr. Elbing): solidus of Honorius (395-423) (Haftka M., 1977, s. 265).

3 - **Gronowo Górne**, woj. Elbląg (ehem. Grunau, Kr. Braunsberg): 2 (?) hoards of solidi, the latest coins - Livius Severus (461-465) (Haftka M., 1977, s. 265).

4 - **Elbląg** (ehem. Elbing): 11 separate solidi (odd hoard?), the latest coins - Leon I (457-471) and Anastasius (491-518) (Haftka M., 1977, s. 261).

5 - **Czechowo**, woj. Elbląg (ehem. Böhmischgut, Kr. Elbing): 2 solidi, including - Theodosius II (408-450) (Haftka M., 1977, s. 265).

6 - **Dębice**, woj. Elbląg (ehem. Dambitzen, Kr. Elbing): solidus of Leon I (457-471) (Haftka M., 1977, s. 265).

7 - **Przezmark**, woj. Elbląg (Preußisch Mark, Kr. Elbing): solidus of Valentinian III (424-425) (Haftka M., 1977, s. 274).

8 - **Lęcze**, woj. Elbląg (ehem. Lenzen, Kr. Elbing): solidus of Anastasius I (491-518) (Haftka M., 1977, s. 270).

9 - **Frombork**, woj. Elbląg (ehem. Frauenburg, Kr. Braunsberg): hoard consisting of whole copper-alloy jewellery and their fragments (raw material for bronze foundry), including solidus of Theodosius II (408-450) (Godłowski K., 1981b, s. 120). From the material of the cemetery at Kobbea (Bornholm Island) the hoard is dated to 450-470 (Jørgensen L., 1989, p. 181).

10 - **Podgórze**, woj. Elbląg (ehem. Huntenberg, Kr. Braunsberg): cemetery of about 450 AD (Peiser F.E., 1914, S. 335-356; Åberg N., 1919, S. 94). Opened by F.E. Peiser, 56 flat graves in some cases were overlapped by stone coverings. Some of the bones were found mixed with the remains of their funeral pyres. Several graves adjoined horse burials. Only in Grave 23 was a horse (according to the Prussian burial rites of the end of the 5th cent.) placed in the lower level (Peiser F.E., 1914, S. 342). In Grave 26, under the stone covering together with burnt bones and charcoal, were found two accessory vessels, fragments of iron objects, copper-alloy small buckle with a frame of oval form, fragments of rings, part of twisted neck-ring, belt mount, fragment of the foot of a brooch (or buckle?) with an animal head and amber washer-shaped bead (Peiser F.E., 1914, S. 343). As with most of the burials opened at the cemetery, Grave 26 is dated to about 450-500 AD. It is remarkable that the "animal-headed" brooch found is connected by its stylistics to the antiquities of the Vidivarii, as mentioned by Jordanes, and finds strong analogies in the antiquities of the island of Gotland from the second half of the 5th cent. (Nerman B., 1935, Taf. 49).

11 - **Garbina**, woj. Elbląg: solidus of Theodosius II (408-450) (Godlowski K., 1981b, s. 122).

12 - **Trąbki**, woj. Elbląg (ehem. Klein Tromp, Kr. Braunsberg): hoard of solidi, 44 coins of which were struck under Theodosius II (408-450) (Godłowski K., 1981b, s. 121).

13 - **Trąbki**, woj. Elbląg: hoard of solidi, 11 coins of which were struck under Theodosius II (408-450) (Godłowski K., 1981b, s. 121).

14 - **Nałaby**, woj. Elbląg (ehem. Nallaben, Kr. Braunsberg): solidus of Theodosius II (408-450) (Godłowski K., 1981b, s. 121).

15 - **Młotezno**, woj. Elbląg (ehem. Hammersdorf, Kr. Heiligenbeil): hoard (fig. 79), consisting of a gold brooch (close to group AVI type 184) with traces of repair, decorated by granulation and garnets, three (?) bucket-shaped pendants, a part of neck chain and medallion of Constantius II, struck in the interval between September 335 and May 22, 337 (Ebert M., 1923, S. 161-163). This assemblage conditionally called "Hoard A" was found in stages between 1913 and November 1917. When F. Ben, between November 22-24 (1917), drew up the report on this hoard there was an obvious problem when plotting its component finds. According to his report the medallion was found 1.5km to the east of the village, on its western outlying district (Ebert M., 1923, Abb. 13). One of bucket-shaped pendants was melted soon after detection by a jeweller from Braunsberg. Later, at an exhibition in the Prussia-Museum, three pendants were nevertheless exhibited as parts of the hoard (Engel C., um 1934, S. 4). It is important to note that the brooch from this assemblage was found on a slope of the hill situated in the western suburb of the village. On the top of this hill in 1915 the remains of a funeral pyre and the destroyed urn were found (Ebert M., 1923, S. 155, 156). There is an opinion, therefore, that this assemblage of goods belonged to the burial from the beginning (Åberg N., 1919, S. 67), or the first half, of the 5th cent. AD (Kulakov V.I., 1990b, s. 204). It is probable that the pair of late-antique disc brooches with portrait images, found in the vicinity of present-day Młotezno at the end of the 19th cent., also belongs to this cemetery (Hollack E., 1908a, S. 56).

16 - **Młotezno**, woj. Elbląg: hoard (conditionally "Hoard B"), consisting of an axe cut into two pieces, a silver dish with gilt and partly nielloed (Niello method) pictures of hunting scenes, and six fragments of a second silver dish of smaller size with a central medallion featuring an acanthus rosette (Hirschfeld G., 1886. S. 77-82. Taf. VI, VII, VIII). On the inside of this vessel there is a scored cut mark in the shape of a "slanting cross". These finds were made during ploughing (in the autumn of 1873) to the east of the village. One year ago a metal object was also found there. From the way the finder rolled the object into a ball (Hirschfeld G., 1886, S. 77), it is possible to assume that it was part of an item made of thin silver, or gold, foil. Details of both the Młotezno dish fragments find analogies in material from the northern Black Sea area, dated to the 4th/beginning of the 5th cent. (Bank A.V., 1977, s. 40). Vessels from northern Europe, with outlines close to the Młotezno "Rosette" dish, appertain to type Eggers 115 and are dated to stage B_1 (Eggers H., 1951, S. 170).

17 - **Młotezno**, woj. Elbląg: a hoard with two gold rings simultaneously dug up by a plough (Ebert M., 1923, S. 155-160, Taf. 3, 4). Dimensions of the rings: the smaller one - 13 cm (weight 549,4 gram), the larger one - 15 cm (weight 991,5 gram). The term "neck-ring" used for these artefacts (Petersen E., 1939, S. 37, 38) is rather conditional because of the effects of actual and repeated use. Nils Åberg dated these rings of "von germanischen Typus" to the beginning of the 5th century (Åberg N., 1919, S. 67). Max Ebert considered that these rings, in view of the traces of attrition on the lower parts, were "worn for a long time", and this became for the researcher an additional argument in the definition of the date of this hoard (conditionally "Hoard C") as not less than 150 years younger than other local hoards with brooches and fragments of vessels (i.e. 550 AD, and not later than 400 AD) (Ebert M., 1923, S. 171).

In recent decades, finds from Młotezno have not attracted much attention from experts. Regarding their dating, it is only necessary to mention that Hans Bott considered all the Młotezno finds as parts of one hoard, buried by someone in the 6th cent. as a result of the perceived danger of Slavic penetration into the Paslenka basin. Bott attributed the "neck-rings" (in his opinion "Ringgeld") as fragments of "bowls" and bucket-shaped pendants from the 3rd cent. AD, and the chain and brooch, made, in the opinion of the researcher, for the Goths in the workshops of Pontus about 335 AD (Bott H., 1976/1977, S. 142, 145, 152).

18 - **Młotezno**, woj. Elbląg: a cemetery where, in 1974 and 1975, W. Ziemlińska-Odojowa dug 94 flat graves that she referred to as Prussian culture. The earliest assemblages at the cemetery to be dated to the advanced stage of the era of the great migration (about 450 AD) are Graves 1 and 20, containing the remains of adult men (Ziemlińska-Odojowa W., 1991, s. 113). A particular attribute that could link these assemblages to Prussian antiquities is the presence of urn cremations. However, the form of the urn from Grave 20 is far from those urns typical of type Grebieten. The typical bent copper-alloy nails and the ring found in the grave assume the presence of a casket, and this would be characteristic of German tradition (Ziemlińska-Odojowa W., 1991, ryc. 4).

19 - **Młotezno-Rogity**, woj. Elbląg (ehem. Hammersdorf-Regitten, Kr. Heiligenbeil): a bank, which, from the medieval written sources, limited the east and southeast vicinities of the village Młotezno (Crome H., 1937, S. 82).

20 - **Zelenodol'skoye**, Bagrationovsk District (ehem. Preußisch Bahnau/Carben, Kr. Heiligenbeil): a cemetery from the 4th/beginning of the 5th cent. (Bezzenberger B., 1914b, S. 138, 139). The excavator dated part of the finds to stage C, although in 1724 urns were found in the area of the cemetery territory that might be attributed to Prussian antiquities of the 5th cent. (Bezzenberger A., 1900a, S. 337, 338).

21 - **Lipovka**, Bagrationovsk District (ehem. Grünwalde, Kr. Heiligenbeil): the defended site of Jungfernberg. It is considered as a religious centre because of its unique size (for the Baltic) and the absence on the platform of any habitation material (Kulakov V.I., 1988, s. 129, 130).

22 - **Vesioloye**, Bagrationovsk District (ehem. Balga, Kr. Heiligenbeil): a defended site known in the Prussian oral tradition as "Honeda" (Kulakov V.I., 1990b, s. 49). This was the name Simon Grunau gave to the first "castle" constructed by the legendary princes/brothers, and which subsequently became the centre of the administrative authority of the Prussian lands (Grunau S., 1876, S. 61).

23 - **Vesioloye**, Bagrationovsk District (ehem. Scheckenberg, Kr. Heiligenbeil): a cemetery of the 4th/beginning of the 5th cent. (Reusch Chr., 1725, S. 551, 552). In 1705, on the defended site of Shnekkenberg, an urn cremation containing a turned amber bead was revealed by chance. The urn, from a representation of Christoff Reusch, has four handles, and finds a full analogy in the material of the Wielbark cemetery ehem. Pettelkau (Kr. Braunsberg) from phase C_1 (Bezzenberger A., 1909, Abb. 113, 122). True, the amber bead found in the urn at this defended site belongs, according to Reusch's representation (Reusch Chr., 1725, Abb. 4), to the Pauckenperle-type, dated in the Aestian area to 400-500 AD (Kulakov V.I., 1994, fig. 28, a position 77). (It is obvious that there were a flat grave cemetery in existence before the hill fortifications were constructed). Thus, it is possible to assume that the burial opened in 1705 on the Balga peninsula appertains to one of the latest Wielbark assemblages on the Amber Coast (from the beginning of the 5th cent.).

24 - **Pervomajskoye**, Bagrationovsk District (ehem. Warnikam, Kr. Heiligenbeil): the defended site at Pillgarten, built, from the evidence of the adjoining

cemetery, not later than 450 AD (Kulakov V.I., 1997a, s. 143).

25 - **Pervomajskoye**, Bagrationovsk District: the cemetery at Kapnizerberg. There were Aestian burials here from the beginning of the 4th cent. About 450 AD, the assemblages containing "druzhina" antiquities appeared at this cemetery, finding analogies both in the late Hunnic Danube area, and on the island of Bornholm (Kulakov V.I., 1997, s. 144, 148, 149). Considerable interest is raised by the find (Wa-65) of a fragment of gold folia showing reliefs of "Erots struggling with wild boars and panthers" and a "group of hunters among trees" (Kulakov V.I. 1997, s. 147). This find, dated to the beginning of the 6th cent., has much in common with finds from Hoard B at Młotezno.

26 - **Uzornoye**, Bagrationovsk District (ehem. Jacknitz/Rosen, Kr. Heiligenbeil): a cemetery of stages B and D. The first phase of the cemetery is reflected in Aestian urn cremations complete with Roman coins of the 2nd cent. AD. The second phase of the cemetery (the second quarter of the 5th cent.) includes Grave IIb which contains part of urn, an iron buckle, a knife, and a brooch with star-shaped foot, categorized by Bitner-Wróblewska as type V (Bitner-Wróblewska A., 1986-1990, tabl. IV, 18). Burial assemblage B at the same site has 13(!) brooches ("bent foot" type), an iron bucket-shaped pendant, burnt glass beads, a bit, and several horse teeth (Bezzenberger A., 1900b, S. 126-128).

27 - **Beregovoye**, Bagrationovsk District (ehem. Patersort/Domniksruh, Kr. Heiligen-beil): an Aestian cemetery of the 4th-5th cent. (Jaskanis J., 1977, s. 309, 310).

28 - former **Tengen**, Kr. Heiligenbeil: an Aestian cemetery of stage D (Klebs R., 1877, S. 51-62).

29 - **Berezovka**, Bagrationovsk District (ehem. Groß Sausgarten, Kr. Heiligenbeil): a hoard containing a silver neck-ring (16.4cm - 16.6cm diameter). The bulging ends of the neck-ring are decorated with stamps close in style to the small ring from Hoard C at Młotezno (Lincke B., 1938, S. 80, 81).

30 - **Krasnodonskoye**, Bagrationovsk District (ehem. Keimkallen, Kr. Heiligenbeil): a cemetery that revealed (at the beginning of the 1930s) about 100 urn cremations of Grebieten type (4th-5th cent. AD), in some cases accompanied by short swords (knives/daggers?). Several horse burials were also found there (Engel C., 1933b, S. 259, 260).

Bibliography

Åberg N., 1919. Ostpreussen in der Völkerwanderungszeit, Uppsala-Leipzig.

Adler W., 1993. Studien zur germanischen Bewaffnung. Waffenmitgabe und Kampweise im Niederelbegebiet und im übrigen Freien Germanien um Christi Geburt, Bonn.

Almgren O., Nerman B., 1923. Die ältere Eisenzeit Gotlands, H. 2, Stockholm.

Andrzejowski J., 1992. The Roman Period brooches made of decoratively coiled wire // Barbaricum'92, t. 2, Warszawa.

Andrzejowski J., 1994. Powiązania kultur przeworskiej i wielbarskiej w świetle znalezisk bransolet // Kultura przeworska, t. I.

Astrauskas A., Gleizniené G., šiménas V., 1999. Perpetés baltū kraptuose I tūkstantmečio viduryje // Archaeologia Lituana, t. I, Vilnius.

Bakka E. 1958. On the beginning of Salin's Style I in England // Universitetet i Bergen. Årbok 1958, Historisk-antikvarisk rekke Nr. 3.

Baranowski T., 1973. Rząd koński z wodzami łańcuchowymi na terenie Europy środkowej w okresie wpływów rzymskich // Archaeologia Polski, t. XVIII, z. 2.

Batūra R.,1985. Paaiškinimai // Petras Dusburgietis. Prūsijos žemės kronika. Vilnius.

Beran J., 1997. Carl Engel 1895-1947 // Alteuropäische Forschungen, Bd. I, Weissbach.

Berendt G., 1874. Zwei Gräberfelder in Natangen und zwar Tengen bei Brandenburg und Rosenau bei Königsberg // SPÖG, Bd. XIV.

Bertram H. 1924. Die physikalische Geschichte des Weichsel-Nogat-Delta // Das Weichsel-Nogat-Delta. Danzig.

Bezzenberger A., 1896. Das Gräberfeld bei Rominten // Prussia, Bd. 20.

Bezzenberger A. 1900a. Das Gräberfeld bei Pr. Bahnau und Carben // Prussia. 21. H.

Bezzenberger A. 1900b. Gräberfeld bei Jäcknitz // Prussia, Bd.XXI.

Bezzenberger A., 1904. Analysen vorgeschichtlicher Bronzen Ostpreussens, Königsberg.

Bezzenberger A. 1909. Gräberfeld bei Pettelkau, Kr. Braunsberg // Prussia, Bd. 22.

Bezzenberger A., 1914a. Fundberichte (VIII. Gräberfeld bei Goithenen) // Prussia, Bd. 23, I. Teil.

Bezzenberger A. 1914b. Zerstörtes Gräberfeld bei Carben, Kr. Heiligenbeil // Prussia, Bd. XXIII, H. 1.

Biborski M., 1994. Typologie und Chronologie der Ringknaufschwerter // Markomannenkriege. Ursache und Wirkungen, Brno.

Bierbrauer V., 1987. Ostgermanische Oberschichtgräber der römischen Kaiserzeit und des frühen Mittelalters // Peregrinatio Gothica, t. VIII, Łódź.

Bierbrauer V., 1994. Archäologie und Geschichte der Goten vom 1.-7. Jahrhundert. Versuch einer Bilanz // Frühmittelalterliche Studien, 28. Bd., S. 51-171.

Bitner-Wróblewska A. 1986-1990. Zapinki z gwiazdistą i lopatkowatą nózką z poludniowo-wschodnich wybrzezu Bałtyku // Wiadomosúi Archeologiczne. T. LI, z. 1.

Bitner-Wróblewska A., 1989. Elementy bałtyjskie w kulturze wielbarskiej // Kultura wielbarska w młodszym okresie rzymskim. Lublin.

Bitner-Wróblewska A., 1991-1992. Z badań nad ozdobami emaliowanymi w kulturze wielbarskiej na marginesie kolekcji starożytnosci Paula Schachta z Malborka // Wiadomoúci Archeologiczne, t. LII, z. 2.

Bitner-Wróblewska A., 1992. The Southeastern Baltic Zone and Skandinavia in the Early Migration Period // Barbaricum'92, t. 2, Warszawa.

Bitner-Wróblewska A., 1998. Suwalszczyzna w okresie wędrówek ludów // Ceramika zachodniobałtyjska od wczesnej epoki żelaza do początku ery nowoýytnej, Białystok.

Bitner-Wróblewska A., 2001. From Samland to Rogaland. East-West connections in the Baltic basin during the Early Migration Period, Warszawa.

Bliujienė A., 2001. Lithuanian amber artifacts in the middle of the first millenium and their provenance within the limits of Eastern Baltic region // Acta Academiae artium Vilnensis, vol. 22.

Böhme H.W., 1974. Germanische Grabfunde des 4. bis 5. Jahrhunderts zwischen unterer Elbe und Loire. Textteil, München.

Bóna I., 1991. Das Hunnenreich, Stuttgart.

Bohnsack D., 1940. Ostgermanische Gräber mit Steinpfeilern und Steinreifen in Ostdeutschland // Gothiskandza, H. 20.

Bolin St., 1926. Die Funde römischer und byzantinischer Münzen in Ostpreußen // Prussia, H. 26.

Bott H. 1976/1977. Zu Datierung der Funde aus Hammersdorf (Ostpreußen) // Jahrbuch der Römisch-Germanischen Zentralmuseums. Festschrift Hundt. T. 3. Text. 23.-24. Jg.

Braune W. 1905. Sammlung kurzer Grammatik germanischer Dialekte. Bd. I. Gotische Grammatik, Halle.

Brinkmann A., 1896-1900. Funde von Terra Sigillata in Ostpreussen // Prussia, Bd. 21.

Bujack G., 1887-1888a. Accessionen des Prussia-Museums (Baumsärge zu Wiekau, Kr. Fischhausen) // Prussia, Bd. 14.

Bujack G., 1887-1888b. Accessionen des Prussia-Museums (Brandgräber mit Achangefeßen aus Wikau, Kr. Fischhausen) // Prussia, Bd. 14.

Bujack G., 1887-1888c. Das Gräberfeld zu Regehnen, Kr. Fischhausen (Palve des Besitzers Lestin) // Prussia, Bd. 14.

Bujack G., 1889. Das Gräberfeld in der Drusker Forst, Schutzbezirk Espenhain, Jagen Nr 209 // Prussia, Bd. 14.

von Carnap-Bornheim C., 1991. Die germanische Gefolgschaft. Zur Interpretation der Mooropfer der jüngeren römischen Kaiserzeit in Südskandinavien - Ein archäologischer Diskussionsbeitrag // Peregrinatio Gothici, Bd. III, Fredrikctad.

von Carnap-Bornheim C., 1997. Zur Bedeutung der militärischen Seefart bei den Barbaren im 3. Jahrhundert n. Chr. - Skandinavien, die Nordsee und das Schwarze Meer // Military Aspects of Scandinavian Society in a European Perspective, AD 1-1300, Copenhagen, p. 226-237.

Connoly P., 1981. Greece and Rome at War, London.

Crome H. 1937. Längwälle in Ostpreußen // Mannus, Jg. 29, H. 1-2.

Dąbrowska T., 1988. Wczesne fazy kultury przeworskiej. Chronologia-zasiąg-powiązania, Warszawa.

Deschler-Erb E., 1991. Römische Militaria der 1. Jahrhunderts aus Kaiser-Augst. Zur Frage des frühen Kastels // Deschler-Erb E., Markus P., Deschler-Erb S. Das Frühkaiserzeitliche Militärlager in der Kaiseraugster Unterstadt. Augst.

Ebert M. 1923. Neuwerbungen des Prussia-Museums // Prussia, 24. H.

Ehrlich B., 1939. Das preußische Gräberfeld von Hochschnakeinen, Kreis Pr. Eylau // Prussia, Bd. 33. 1.-2. H.

Eggers H.J. 1951. Der römische Import im Freien Germanien. Bd. 1. Text. Hamburg.

Ellegård A., 1987. Who were the Eruli? // Scandia, Bd. 53:1.

Engel C., 1931. Das Gräberfeld bei Sanditten // Prussia, Bd. 29.

Engel C., 1933a. Das Samland als altbaltisches Kulturzentrum und seine vorchristlichen Beziehungen zu den Nachbargebieten // Altpreussische Beiträge, Festschrift, Königsberg.

Engel C. 1933b. Altprussisches Gräberfeld im Kreise Heiligenbeil // Unsere Heimat. Jg. XV, Nr 22.

Engel C. um 1934. Silber- und Goldfunde aus ostpreußischer Vorzeit. Königsberg.

Engel C., 1942. Die Ostgermanische Stämme in Ostdeuschland, die gotische Ostseeherrschaft und das Gotenreich in Osteuropa // Deutsche Ostforschung. Ergebnisse und Aufgaben seit dem ersten Weltkrieg. Bd. 1, Leipzig.

Engel C., 1962. Typen ostpreußischer Hügelgräber, Neumünster.

Engel C., La Baume W., 1937. Kulturen und Völker der Frühzeit in Preussenlande, Königsberg.

Engström J., 1992. Skandinaviskt krigsväsen under mellersta järnaldern // Medellande 52 Armémuseum, Årsbok 1992, Stockholm.

Ethelberg P., 1990. Hjemsted 2 - tre gravpladser fra 3. og 4. årh.e.Kr. // Skrifter fra Museumsr¿det for Sønderjyllands amt, 3, Haderslev.

Filip J., 1951. Pradzieje Czechosłowacji, Poznań.

Gaerte W., 1928. Skelettgräber zwischen Weichsel und Memel aus der römischen Kaiserzeit // Festgabe für den 70jährigen Gustav Kossina von Freunden und Schülern. Mannus VI. Ergänzungsband, Leipzig.

Gaerte W., 1929. Urgeschichte Ostpreußens, Königsberg.

Gaerte W. 1931a. Ostpreussische "Schatzgräber" des 15. und 16. Jahrhunderts // Prussia, Bd. 29.

Gaerte W., 1931b. Wittwenverbrennung im vorordenszeitlichen Ostpreußen // Prussia, Bd. 29.

Gaerte W., 1932. Die Ostgrenze der gotischen Weichselmündung-Kultur in den Römischen Kaiserzeit // Mannus, Bd. 24.

Gaerte W., 1939. Das Hackendreieck // Zeitschrift der Altertumsgesellschaft Insterburg. H. 22, Insterburg.

Ganzelewski M., Slotta R., 1996. Bernstein - Tränen der Götter, Bochum.

Gàssowska E. 1979. Bizancjum a ziemie pólnocno-zachodnio-słowiańskie we wczesnym úredniowieczu. Wrocław-Warszawa-Kraków-Gdańsk.

Gebühr M., 1996. Fürstengräber. Römische Kaiserzeit // Reallexikon der Germanischen Altertumskunde, Bd. 10, Lief. 172, Berlin-New York.

Geisslinger H. 1970. Soziale Schichtungen in den Opferorts der Völkerwanderungszeit // Vorgeschichtliche Heiligtümer und Opferplatze in Mittel- und Nord-Europa, Göttingen.

Gerov B., 1977. Die Einfälle der Nordvölker im Lichte der Münzfunde // Aufstieg und Niederung der römischen Welt. Geschichte und Kultur Roms im Spiegel der neueren Forschung. T. II. Prinzipat, Sechster Band, Berlin-New York.

Gimbutas M., 1963. The Balts, London.

Gimbutas M., 1983. Die Balten, München-Berlin.

Godłowski K., 1981a. Kultura przeworska // Praehistoria ziem polskich, Wrocław-Warszawa-Kraków-Gdańsk, t. V.

Godłowski K. 1981b. Okres wędrówek ludów na Pomorzu // Pomorania Antiqua, T. X.

Godłowski K., 1986. Gegenseitige Beziehungen zwischen der Wielbark- und Przeworsk-Kultur, Veranderunngen ihrer Verbreitung und das Problem der Gotenwanderung // Peregrinatio Gothica. Polonia 84/85. Vol. VII, Łódź.

Godłowski K., 1992. Zmiany w uzbrojeniu ludnoúci kultury przeworskiej w okresie wpływów rzymskich // Arma et ollae, Łódź.

Godłowski K., 1994a. Die Synchronisierung der Chronologie der germanischen Funstoffes zur Zeit der Markomannenkriege // Markomannenkriege. Ursachen und Wirkungen, Brno.

Godłowski K., 1994b. Vorrömische Eisenzeit und Römische Kaiserzeit im östlichewn Mitteleuropa und in Osteuropa // Fibel und Fibeltracht. Reallexikon der Germanischen Altertumskunde, Bd. 8, Lieferung 5/6, Berlin-New York.

Grabarczyk T., 1983. Metalowe rzemioslo artystyczne na Pomorzu w okresie rzymskim, Wrocław-Warszawa-Kraków-Gdańsk- Ùódý.

Grunau S,. 1876. Preussische Chronik. Bd. I, Leipzig.

Grunert W., 1935. Nadrauer Grabungen // Zeitschrift der Altertumsgesellschaft Insterburg, H. 20, Insterburg.

Gudea N., 1994. Dacia Porolissensis und die Markomannenkriege // Markomannenkriege. Ursachen und Wirkungen, Brno.

Günther C., Voss A., 1880. Photographisches Album der Prähistorischen und Anthropologischen Ausstellung zu Berlin, Berlin, Bd. I.

Guttzeit E.J. 1966. Heiligenbeil // Handbuch der historischen Stätten. Ost- und Westpreussen. Stuttgart.

Hagberg U.E., 1984. Opferorte der Kaiser- und Völkerwanderungszeit in Schweden // Frühmittelalterliche Studien. 18. Bd., Berlin-New York.

Haftka M., 1977. Elbląg we wszesznym średniowieczciu i problem lokalizacji Truso // Pomorania Antiqua, t. VI.

Hauck K. 1993. Die bremische Überlieferung zur Götter-Freiheit Altuppsals und die bornholmischen Goldfolien aus Sorte Muld // Frühmittelalterliche Studien. 27. Bd., Berlin-New York.

Hedeager L., 1983. A Quantitative Analysis of Roman Imports in Europe North of the Limes (0-4000 A.D.), and the Question of Roman-Germanic Exchange // New Directions in Scandinavian Archaeology. Studies in Scandinavian Prehistory and Early History, vol. I, 191-216.

Heydeck J., 1888. Der südliche Teil des Gräberfeldes von Grebieten, Kreis Fischhausen // Prussia, Bd. 13.

Heydeck J., 1900-1904. Gräberfeld bei Wiekau, im Volk Kucksberg genannt // Prussia, Bd. 21.

Heydeck J., 1909. Fundberichte (Gräberfeld Kunterstrauch) // Prussia, Bd. 22.

Hirschfeld G. 1886. Bruchstücke von zwei silbernen Geräten // Prussia, Jg. 41.

Hoffmann J., 1991. Otto Tischler. W śtulecie smerti // Borussia, z. 1, Olsztyn.

Hoffmann J., 1992. Adalbert Bezzenberger - archeolog, językoznawza, historik i etnograf // Borussia, z. 1, Olsztyn.

Hollack E., 1908a. Die Grabformen ostpreussischer Gräberfelder // Zeitschrift für Ethnologie, Bd. 40, H. 2.

Hollack E., 1908b. Erläuterungen zur vorgeschichtlichen Übersichtskarte von Ostpreussen, Glogau-Berlin.

Ilkjær J., 1994. Skandinavien zur Zeit der Markomannenkriege // Markomannenkriege. Ursache und Wirkungen, Brno.

Jankuhn H., 1933a. Zur Besiedlung des Samlandes in der älteren römischen Kaiserzeit // Prussia, Bd. 30, H. I, 1933.

Jankuhn H., 1933b. Die Bevölkerung Ostpreußens in beiden ersten Jahrhunderten unserer Zeitrechnung // Ostdeutsche Monatshefte, Bd. 13, H. 11.

Jankuhn H., 1933c. Gürtelgarnituren der älteren römischen Kaiserzeit im Samland // Prussia, Bd. 30.

Jankuhn H., 1939. Ein Gräberfeld der ersten Jahrhunderte u. Zr. aus Schlakalken, Kr. Fischhausen // Prussia, Bd. 32, H. II.

Jankuhn H., 1950. Zur räumlichen Gliederung der älteren Kaiserzeit in Ostpreußen // Archaeologia Geographica, Jahrg. 1, H. 4.

Jankuhn H., 1968. Spuren von Anthropophagie in der Capitulatio de partibus Saxoniae? // Nachrichten der Akademie der Wissenschaften in Göttingen I. Philologisch-historische Klasse, Nr. 3.

Janssen H.-L., 1942. Die Toten in Brauchtum und Glauben der germanishen Vorzeit // Mitteilungen der Anthropologischen Gesellschaft in Wien, LXXII. Bd., I.-III. H.

Jaskanis J., 1968. Pochówki z końmi na cmentarzyskach Protojaćwieskich z okresu rzymskiego i wędrówek ludów // Rocznik Białystocki, t. 8.

Jaskanis J., 1974. Obrządek pogrzebowy zachodnich bałtów u schyłku starożytności (I - V w.n.e.), Wrocław-Warszawa-Kraków-Gdańsk.

Jaskanis J., 1977. Cmentarzyska kultury zachodniobałtyjskiej z okresu rzymskiego. Materiały do badań nad obrządkiem pogrzebowym // Materiały starożytne i wszesnośredniowieczne, t. IV, Warszawa.

Jentzsch A., 1896. Bericht über die Verwaltung des Ostpreußischen Provinzialmuseum in den Jahren 1893-1895 // SPÖG, Bd. XXXVII.

Jørgensen L. 1989. En kronologi for yngre romersk og aeldre germansk jernalder på Bornholm // Simlegård-Trelleborg, Kobenhavn.

Jıırgensen L. 1994. Jüngere römische Kaiserzet // Reallexikon der Germanischen Altertumskunde. Bd. 8, Lief. 5/6, Fibel und Fibeltracht, Berlin-New York.

Kaczyński M., 1976. Problem zróznicowania wewnætrznego "kultury sudowskiej" w późnym podokresie wpływów rzymskich i okresie wędrówek ludów // Zeszyty naukowe Uniwersitetu Jagellonskiego, Nr CCCCXXII, Prace archeologiczne, z. 22.

Kaczyński M., 1987. Die Balten, die nördlichen Nachbarn der Slawen // Die Balten, die nördlichen Nachbarn der Slawen, Freiburg-Braunschweig.

Kaczyński M. 1989. Z problematyki kontaktów zachodnich bałtów z prowincjami Imperium Romanum // Rocznik Białostocki, T. XVI.

Kazanski M., 1991. Les Goths (Ier - VIIe apres J.-C.), Paris.

Kazanski M. 1992. L'influence Dnubienne dans l steppe Ponyiue pendant l seconde motie du Ve siecle: la rolle des Angiskires // Medieval Europe. Death and Burial. Volume 4. York.

Kazanski M., 1994. Les éperons, les umbo, les manipules de boucliers et les haches de l'épogue romaine tardive dans la region pontigue: origine et diffusion // Beiträge zu römischer und barbarischer Bewaffnung in den ersten vier nachchristlichen Jahrhunderten. Lublin/Marburg.

Kazanski M., 1994b. À propos de quelques types de fibules germaniques de l'époque de Grandes Migrations trouvées en Gaule au Sud de la Loire // Antiquités nationales, T. 26, 1994.

Kazanski M., 1994c. L'équipement et le materiel militaires au Bas-Empire sen Gaule du Nord et de l'Est // Revue du Nord-Archeologie, T. LXXVII, N° 313.

Kazanski M., 1999. Les barbares en Gaule du Sud-Ouest durant la première moitie du Ve siècle // L'Ossident romain et l'Europe centrale au debut de l'époque des Grandes Migrations. Spisy Arch. Ústavu AV ÈR, t. 13, Brno.

Kemke H., 1901. Fundverzeichnis zu Tafel 7 - 15 der 1. (ostpreussichen) Section des Photographischen Albums der Berliner Anthropologischen Ausstellung im Jahre 1880 // SPÖG, Jg. XLII.

Kilian L., 1980. Zur Herkunft und Sprache der Prussen, Bonn.

Klebs R. 1877. Bericht über die neuen Ausgrabungen in Tengen bei Brandenburg (Natangen) ausgeführt im Sommer 1875 // SPÖG, Bd. XVII.

Klindt-Jensen O., 1957. Bornholm i folkevandringstiden og forudsætningerne i tidlig jernalder, København.

Kokowski A. 1988. Ze studiów nad ceramiką wykonaną na kole w kulturze wielbarskiej // Kultura wielbarska w młodszym okresie rzymskim. Lublin.

Kokowski A., 1995. Die Goten in den archäologischen Kulturen der älteren römischen Kaiserzeit // Schätze der Goten, Stuttgart.

Kokowski A., 1998. Metalowe wisiorki w ksztalcie topora na terenie Barbaricum na pólnoc i pólnocny wschód od limesu rzymskiego w okresie rzymskim i we wschesnym okresie wędrówek ludów // 20 lat archeologii w Maslomęczu, Lublin.

Kolendo Z., 1981. Wyprawa po bursztyn bałtycki za Nerona // Pomorania Antiqua. T. X.

Kolník T., 1980. Römerzeitliche Gräberfelder in der Slowakei, Bratislava, Teil I.

Koulakov V., 2000. Les contacts entre la cote sud de la Baltique et Rome a l'epoque de Neron // Les sites archéologiques en Crimeé et au Caucase durant l'Antiquité tardive et la haut Moyen-Age. Colloquia Pontica, vol. 5, Leiden-Boston-Köln.

Kowalski J., 2000. Chronologia grupy elbląskiej i olsztyńskiej kręgu zachodniobałtyjskiego (V-VII w.). Zarys problematyki // Barbaricum, t. 6, Warszawa.

Kulakov V.I., 1998. Die Kulturen und Stämme auf westlicher Grenze der Welt der Balten // Kraje stowiańskie w wiekach średnich. Profanum i sacrum. Poznań.

Kulakov V., 2000a. Dollkeim 1992: Neue Ausgrabungen des spätkaiserzeitlichen Gräberfeldes im Samland (Russland) // Archäologisches Korrespondenzblatt, Bd. 30, H. 4, Mainz.

Kulakov V., 2000b. La terre prusse entre Scandinavie et Orient // Les centres proto-urbains russes entre Scandinavie, Byzance et Orient, Réalités Byzantines, t. 7, Paris.

Kulakov V.I., 2001. Jütland-Dobrudscha: die Rochade der "Barbaren" vom 2. bis 4. Jh. Chr. // Archaeologia Bulgarica, Jg. V, H. 2, Sofia.

Kulakov V.I., Šimėnas V., 1992. Latgalische Schmucktypen in des Sammlung des Instituts für Ur-und Frühgeschichte der Universität Wien // Archaeologia Austriaca. Bd. 76.

Kulakov V.I., Valuev A.A., 1996. Veluva (Alt-Wehlau), ein heidnischer Friedhof im christlichen Nadrauen // Eurasia Antiqua, Bd. 2, Berlin.

La Baume W. 1924. Die Besiedlung des Weichsel-Nogat-Delta in vor- und frühgeschichtlicher Zeit // Das Weichsel-Nogat-Delta. Danzig.

La Baume W., 1940a. Ost- und Westpreussen in germanischen Zeit, Königsberg.

Laur W., 1954. Aisten/Esten, eine germanische Volksbezeichnung im baltischen Raum // Zeitschrift für Ostforschung. 3. Jahrg.

Lewada M., 2000. Metal combs of the second quarter of the first millenium AD in Eastern Europe // Die spätrömische Kaiserzeit und die frühe Völkerwanderungszeit in Mittel- und Osteuropa, Łódź.

Lincke B. 1938. Eine baltische Halsringform der Völkerwanderungszeit // Prähistorische Zeitschrift, Bd. 28-29.

Lund Hansen U., 1987. Römischer Import im Norden, København.

Mączyńska M., 1998. Ein spätkaiserzeitliches "Totenhaus" aus Ostpolen und seine Beziehungen zu Skandinavien // Studien zur Archäologie des Ostseeraumes von der Eisenzeit zum Mittelalter. Festschrift für Michael Müller-Wille, Neumünster.

Madyda-Legutko R., 1986. Die Gürtelschnallen der Römischen Kaiserzeit und der frühen Völkerwanderungszeit im mitteleuropäischen Barbaricum, Oxford.

Mannhardt W., 1936. Letto-preussische Göttenlehre, Riga.

Martin M., 1995. Tradition und Wandel der Fibelgeschmückten frühmittelalterlichen Frauenkleidung // Jahrbuch des Römisch-Germanischen Zentralmuseums Mainz, 38. Jg. 1991.

Maxfield V.A., 1981. The military decorations of the Roman Army. London.

Mažiulis V, 1981. Apie senovės vakarų baltų bei ju santyklus su slavais, ilirais ir germanais // Iš lietuviu etnogenezės, Vilnius.

Mesterhazy K., 1989. Ethnische- und Handelsbeziehungen zwischen der Weichselmündung und der ungarischen Tiefebene in der römischen Kaiserzeit // Peregrinatio Gothica. Archaeologia Baltica, vol. VIII, Šódź.

Mihelbertas M., 2000. Zu den Handels- und Kulturbeziehungen der Bewohner Litauens mit Skandinavien in der römischen Kaiserzeit // Archaeologia Baltica, Bd. 4, Vilnius.

Mortensen H., Mortensen G., 1937. Die Besiedlung des nordlöstlichen Ostpreußens bis zum Beginn des 17. Jahrhunderts, T. I // Deutschland und Osten, Bd. 7, Leipzig.

Müller-Wille M., 1970/1971. Pferdegrab und Pferdeopfer im frühen Mittelalter // Berichten van de Rijksdienst voor het Oudheidkundig Bodemonderzoeg, Amersforst, vol. 20-21.

Müller-Wille M., 1999. Opferkulte der Germanen und Slawen, Stuttgart.

Nerman B., 1934, Zur Kenntniss der ältesten Eisenzeit in Schweden // Acta Archaeologica, Vol. V, Stockholm.

Nerman B., 1935. Der Völkerwanderungszeit Gotlands. Stockholm.

Niewęgùowski A., 1981. Obrządek pogrzebowy ludności kultury przeworskiej na przełomie er (II wiek p.n.e. - II wiek n.e.). Wrocław-Warszawa-Kraków-Gdańsk-Łódź.

Nowakowski W., 1985. Rzymskie importy przemysùowe na terytorium zachodniobałtyjskogo krągu kulturowego // Archeologia, t. XXXIV.

Nowakowski W., 1986-1990. Osiedla kultury bogaczewskiej - próba podsumowania stanu badań // Wiadomości Archeologiczne, t. LI, z. 1.

Nowakowski W., 1989a. Studia nad ceramiką zachodniobaùtyjską z okresu wędrówek ludów. Problem tzw. pucharów na pustych nożkach // Barbaricum'1989, Warszawa.

Nowakowski W., 1989b. Kultura wielbarska a zachodniobałtyjski kręg kulturowy // Kultura wielbarska w młodszym okresie rzymskim, Lublin.

Nowakowski W., 1990. Ludy na północno-wschodnich skrajach Barbaricum. "Germania" Tacyta w świetle analizy źródel archeologicznych // Meander, Bd. 2, H. 3.

Nowakowski W., 1994a. Tintinnabula auf den Ostseeinseln. Die römischen Bronzeglocken auf den Inseln Gotland und Bornholm // Fornvännen, vol. 89.

Nowakowski W., 1994b. Krieger ohne Schwerter - Die Bewaffnung der Aestii in der Römischen Kaiserzeit // Beiträge zu römischer und barbarischer Bewaffnung in den ersten vier nachchristlichen Jahrhunderten. Lublin/Marburg.

Nowakowski W., 1995a. Σουδινοι - lud między Bałtykem a Morzem Czarnym // Nunc de Svebis dicendvm est, Warszawa.

Nowakowski W., 1995b. Od Galindai do Galinditae. Z badań nad pradziejami bałtyjskiego ludu z Pojezieria Mazurskiego // Barbaricum, t. 4, Warszawa.

Nowakowski W., 1996. Das Samland in der Römischen Kaiserzeit und seine Verbindungen mit den Römischen Reich und der barbarischen Welt. Veröffentlichungen des Vorgeschichtlichen Seminars Marburg. Sonderband 10, Marburg-Warszawa.

Nowakowski W., 1999. "Książeca" zapinka ze zniszczonego cmentarzyska kultury przeworskiej w Karwowie, pod Sochaczewem // COMHLAN, Warszawa.

Odoj R., 1970. Dzieje Prusów do czasów krzyýackich // Komunikaty mazursko-warmińskie, Nr 1 (107).

Okulicz J., 1973. Pradzieje ziem pruskich od póznego paleolitu do VII w. n. e., Wrocław-Warszawa-Kraków-Gdańsk.

Okulicz J., 1976. Powiązania pobrzeża wschodniego Bałtyku i centrum sambijskiego z południem w podokresie wczesnorzymskim // Prace archeologiczne, Nr CCCCXXII, Z. 22.

Okulicz J. 1992. Centrum kulturowy z pierwszych wieków naszej ery u ujścia Wisły // Barbaricum'92. T. 2, Warszawa.

Okulicz J., 2000. Polska północna w młodszym okresie przedrzymskim i w okresie wpływów rzymskich // Archeologia i prahistoria polska w ostatnim pólwieczu, Poznań.

Okulicz Ł., 1970. Kultura kurhanów zachodniobałtyjskich we wchesnej epoce żelaza, Wrocław -Warszawa-Kraków.

Okulicz Ł., 1976. Osadnictwo strefy wschodniobałtyckiej w I tysiącleciu przed naszą erą, Wrocław-Warszawa-Kraków-Gdańsk.

Orosius, 1861 // Scriptores rerum prussicarum, Leipzig, Bd. I.

Oxenstierna E., 1945. Die Urheimat der Goten, Leipzig.

Peiser F.E. 1914. Gräberfeld bei Huntenberg, Kr. Braunsberg // Prussia, H. 22, T. 1.

Peiser F.E., 1919. Fundberichte (V. Zerstörtes Hügelgrab mit Nachbestattung im Skardelies Wald) // Prussia, Bd. 23, II. Teil.

Pėteraitis V., 1992. Mažoji Lietuva ir Tvanksta prabaltų, pralietuvių ir lietuvininkų laikas, Vilnius.

Petersen E. 1936. Fragen der germanischen Besiedlung im Raume zwischen Oder und Weichsel in der Völkerwanderungszeit // Mannus, 28. Jg., H. 1.

Petersen E. 1939. Der ostelbische Raum als germanisches Kraftfeld im Lichte der Bodenfunde des 6.-8. Jahrhunderts. Leipzig.

Preidel H., 1937. Die Markomannen und Quaden // Vorgeschichte der deutschen Stämme. Germanische Tat und Kultur auf deutschem Boden, Bd. 2, Leipzig-Berlin.

Raddatz C., 1957. Der Thorsberger Moorfund. Gürtelteile und Körperschmuck, Neumünster.

Raddatz K., 1976. Bewaffnung der Germanen in der jungere römische Kaiserzeit // Reallexikon der Germanischen Altertumskunde, 2. Auflage, 2. Bd., Berlin-New York.

Raddatz K., 1992/93. Der Wolka-See, ein Opferplatz der Römischen Kaiserzeit in Ostpreußen // Offa, Bd. 49/50.

Raddatz K., 1994. Der "Alte-See" bei Butzke, ein Opferplatz urgeschichtlicher Zeit im östlichen Pommern // Jahrbuch des Römisch-Germanischen Zentralmuseums, Bd. 41.

Reusch Chr. Fr., 1725. Nachricht von Preußischen Grab-Hügel und Aschen-Töpfen // Erleutertes Preßuen, Bd. III.

von Richtchofen J., 1994. Gebrauchspuren an Silber- und Bronzefibel der ältesten römischen Kaiserzeit Norddeutschlands // Offa, Bd. 51.

Ruseva-Slokovska L., 1991. Roman jewellery. A collection of the National Archaeological Museum - Sofia, London.

Schultze E., 1992. Zu den Grab- und Bestattungssitten in Mitteleuropa während der ersten Jahrhunderte n. Chr. // Praehistorische Zeitschrift, 67. Bd.

Schuster J., 1995. Ein Werkstattkreis der Fibelherstellung in der Altmark? // Ethnographisch-Archäologische Zeitschrift, 36. Jg., H. 2.

Sedovs V., 1992. Balti senatnē. Rīga.

Steinland G., 1992. Die mythologische Grundlage für die nordische Königsideologie // Germanische Religionsgeschichte. Quellen und Quellenprobleme, Berlin - New York.

Steponaitis V., 2000. Bronze et early Iron Age. 18th c. B.C. up to the time of Christ // Prehistoric Lithuania. Archaeology exposition guide, Vilnius.

Steuer H., 1998a. Fürstengräber // Reallexikon der Germanischen Altertumskunde, Bd. 10, Berlin - New York.

Steuer H., 1998b. Gefolgschaft // Reallexikon der Germanischen Altertumskunde, Bd. 10, Berlin - New York.

Ščukin M.B., 2000. Three ways of the contacts between the Baltic and the Black Sea littorals in the Roman period // Die spätrömische Kaiserzeit und die frühe Völkerwanderungszeit in Mittel- und Osteuropa, Łódź.

Šimėnas V. 1994b. Legenda apie Videvuti ir Bruteni (legendos šaltinių klausimų) // Prūsijos kultūra. Vilnius.

Šimėnas V., 1995. Die Wildnis und baltische Stämmegrenzen // Archaeologia Baltica, Vilnius.

Šturms Ed., 1954. Die sowjetrussische archäologische Forschungen in Ostpreußen // Zeitschrift für Ostforschung, Jg. 3, H. 1.

Tautavičius A., 1987. Baltų gentys I-IV amžiuje. Archeologijos duomenys // Lietuvių etnogenezė, Vilnius.

Tautavičius A., 1996. Vidurinis geležies amžius Lietuvoje (V-IX a.), Vilnius.

Tejral J., 1997a. Neue Aspekte der frühvölkerwanderlichen Chronologie im Mitteldonauraum // Neue Beiträge zur Erforschung der Spätantike im mittleren Donauraum. Brno.

Tejral J., 1997b. The Amber route and the Roman military campaigns north of the Middle Danube area during the first two centuries A.D. // Peregrinatio gothica. Supplementum ad Acta Musei Moraviae. Scientia sociales, t. LXXXII.

Tempelmann M., 1985. Die Perlen der römischen Kaiserzeit und der frühen Phase der Völkerwanderungszeit in mitteleuropäischen Barbaricum // Römisch-Germanishe Forschungen, Bd. 43, Mainz am Rhein.

Tempelmann-Mączyńska M., 1989. Das "reduzierte" Trachtmodell der gotischen Frauen und seine Ursprünge // Peregrinatio Gothici. Archaeologia Baltica. Polonia'86, vol. VIII, Ùódý.

Thomsen P.O., 1993. Lundeborg - an Early Port of Trade in South-East Funen // The Archaeology of Gudme and Lundeborg, København.

Tischler O., 1879. Ostpreussische Gräberfelder, Teil III, Königsberg.

Tischler O., 1880a. Königsberg. Provinzial-Museum der Physikalisch-Ökonomischen Gesellschaft // Katalog der Ausstellung prähistorischer und anthropologischer Funde Deutschlands, Berlin.

Tischler O., 1886. Ostpreussische Grabhügel. Teil I // SPÖG, Bd. XXVII.

Tischler O., 1889. Das Gräberfeld bei Oberhof, Kr. Memel // SPÖG, Bd. XXIX.

Tischler O., Kemke H., 1902. Ostpreussische Altertümer aus der Zeit der grossen Gräberfelder nach Christi Geburt, Königsberg.

Thrane H., 1991. Om Gudmes funktion. En tolkningsskitse // Samfundsorganisation og Regional Variation. Norden i romersk jernalder og folkervandringstid. Jisk Arkæologisk Selskab Skrifter XXVII, Aarhus.

Trotzig G., 1991. Craftsmanship and function. A study of metal vessels found in Viking Age tombs on the island of Gotland, Sweden, Stockholm.

Urbanavičius V., Urbanavičienė S., 1988. Archeologiniai tyrimai // Obelių kapinynas. Lietuvos Archeologija, t. 6, Vilnius.

Urbańczyk P., 1998. The Goths in Poland - where did they come from and when did they leave? // European Jornal of Archaeology, vol. 1, Nr 3.

Urtāns V. 1977. Senākie depozīti Latvijū, Rīga.

Vaitkunskienė L. 1981. Sidabras senovės Letuvoje, Vilnius.

Vaitkunskienė L., 1995. Pagribio kapinynas // Lietuvos archeologija, t. 13, Vilnius.

Vandermeersch B., 1970. Une sépulture moustérienne avec offrandes découverte dans grotte de Qafzeh // Correspondences Rewiew Academie Sciences Paris, t. 270.

Vedel E., 1886. Bornholms oldtidsminder og oldsager. Christiania-Leipzig-Kjøbenhavn-Stockholm-London.

Végh K.K., 1969. Kelta leletek a Miskolci museumban // A Herman Ottó Muzeum Évkönyve, t. VIII.

Voß H.-U. 1994. Frühe Völkerwanderungszeit. Fibelformen // Reallexikon der Germanischen Altertumskunde. Bd. 8, Lief. 5/6, Fibel und Fibeltracht, Berlin-New York.

Walenta K., 1980-1981. Obrządek pogrzebowy w okresie póznolatenskim i rzymskim // Archaeologia Baltica. Vol. V, Łódź.

Wamers E., 1987. Die Völkerwanderungszeit im Spiegel der germanischen Heldensagen // Germanen, Hunnen und Awaren. Schätze der Völkerwanderungszeit. Nürnberg.

Waurick G., 1994. Zur Rüstung von frühkaiserzeitlichen Hilfstruppen und Verbündeten der Römer // Beiträge zu römischer und barbarischer Bewaffnung in den ersten vier nachchristlichen Jahrhunderten. Lublin/Marburg.

Werner J., 1970. Opus interrasile an römischem Pferdegeschirr des I. Jahrhunderts // Werner J. Spätes Keltentum zwischen Rom und Germanen. München.

Wielowiejski J., 1980. Główny szlak bursztyniowy w czasach cesarstwa rzymskiego, Wrocław-Warszawa-Kraków-Gdańsk.

Wielowiejski J., 2000. Die nördliche Strecke der Bernsteinstrasse zur römischen Kaiserzeit // Die spätrömische Kaiserzeit und die frühe Völkerwanderungs-zeit in Mittel- und Osteuropa, Łódź.

Wilbers-Rost S., 1994. Pferdegeschirr der römischen Kaiserzeit in der Germanis Libera, Oldenburg.

Wołągiewicz R., 1981a. Kultura wielbarska // Prahistoria ziem polskich, t. V, Wrocław-Warszawa-Kraków-Gdańsk.

Wołągiewicz R., 1981b. Kultura wielbarska – problemy interpretacji etnicznej // Problemy kultury wielbarskiej, Słupsk.

Wołągiewicz R., 1986. Die Goten im Bereich der Wielbark-Kultur // Peregrinatio Gothica. Polonia 84/85. Vol. VII, Ùódý.

Ziemlińska-Odojowa W. 1991. Wstępne wyniki badań na wielokulturowym stanowisku w Młotecznie (gmina Braniewo) // Archeologia Bałtyjska, Olsztyn.

Žulkus V., 1995, Vakarų baltai gotų-gepidų migracioje (I-IV a.) // Lietuvininkų kraštas, Kaunas.

Ambroz A.K. - Амброз А.К., 1966. Фибулы юга европейской части СССР. Свод археологических источников, вып. Д1-30, М.

Ambroz A.K. - Амброз А.К. 1992. Боспор. Хронология раннесредневековых древностей // Боспорский сборник. Вып. 1, М.

Anfertjew A.N. - Анфертьев А.Н., 1991.Иордан // Свод древнейших письменных известий о славянах, т. I, М.

Aun M. - Аун М., 1985. Взаимоотношения балтских и южноэстонских племён во второй половине I тысячелетия н.э. // Проблемы этногенеза и этнической истриии балтов, Вильнюс.

Bažan I.A., Gerzeger D.S. - Бажан И.А., Герцегер Д.С., 1993. Бронзовая пряжка римского времени из Ольвии // Петербургский археологический вестник, № 4.

Bažan I.A., Kargapol'zew S.O. - Бажан И.А., Каргапольцев С.О. 1989. В-образные рифлёные пряжки как хронологический индикатор синхронизации // КСИА, вып. 198.

Bajer Ch.-F. - Байер Х-Ф., 2001. История крымских готов как интерпретация Сказания Матфея о городе Феодоро, Екатеринбург.

Bank A.W. - Банк А.В. 1977. Искусство Византии в собраниях СССР. Т. 1. М.

Beowulf, Starshaja Edda - Беовульф. Старшая Эдда. Песнь о Нибелунгах, 1975, М.

Budanowa W.P. - Буданова В.П., 1999. Германцы в эпоху Великого переселения народов // Великое переселение народов. Этнополитические и социальные моменты, М.

Budanowa W.P. - Буданова В.П., 2000. Варварский мир эпохи Великого переселения народов, М.

Vitiaz' S.P., Kulakov V.I., Medwedev A.M. - Витязь С.П., Кулаков В.И., Медведев А.М., 2000. Эталонные археологические памятники и проблемы хронологии европейских древностей (I тысячелетие н.э.) // Вестник Фонда фундаментальных исследований, № 4, Минск.

Gamkrelidze T.W., Iwanov Wc.Ws. - Гамкрелидзе Т.В., Иванов Вяч. Вс., 1984. Индоевропейский язык и индоевропейцы, т. II, Тбилиси.

Gibbon E. - Гиббон Э., 1998. История упадка и разрушения Римской Империи, т. III, СПб.

Gurewith F.D. - Гуревич Ф.Д., 1960. Из истории юго-восточной Прибалтики в I тысячелетии н.э. // Материалы и исследования по археологии СССР, вып. 76, М.-Л.

Djakson T.N. - Джаксон Т.Н., 1993. Исландские королевские саги о Восточной Европе (с древнейших времён до 1000 г.), М.

Djakson T.N., Podosinov A.V. - Джаксон Т.Н., Подосинов А.В., 2000. Норвегия глазами древних скандинавов: к вопросу о специфике древнескандинавской ориентации по странам света // Другие средние века, М.-СПб.

Dini P.U. - Дини П.У., 2002. Балтийские языки, М.

Driachlov W.N. - Дряхлов В.Н., 1988. Янтарный путь // Вопросы истории, № 11.

Gindarev L.A., Kulakov V.I. - Жиндарев Л.А., Кулаков В.И. 1996. Уровенный режим Балтийского моря в голоцене // Известия РАН. Серия географическая, № 5.

Zasezkaja I.P. - Засецкая И.П., 1994. Культура кочевников южнорусских степей в гуннскую эпоху (конец IV - V вв.), СПб.

Jordan - Иордан, 1997. О происхождении и деяниях гетов, СПб.

Kazanski M. - Казанский М.М., 1999.О балтах в лесной зоне России в эпоху великого переселения народов //Археологические вести, № 6, СПб.

Kazanski M., Mastykova A.W. - Казанский М.М., Мастыкова А.В., 1998. Германские элементы в культуре населения Северного Кавказа в эпоху Великого переселения народов // Историко-археологический альманах, вып. 4, Армавир-Москва.

Kolobov A.W. - Колобов А.В., 1999. Римские легионы вне полей сражений (Эпоха ранней Империи), Пермь.

Kolosowskaja J.K. - Колосовская Ю.К., 1973. Паннония в I-III веках, М.

Kolosowskaja J.K. - Колосовская Ю.К., 2000. Рим и мир племён на Дунае I-IV вв. н.э., М.

Kulakov V.I. - Кулаков В.И. 1988. Дружинные элементы в могильниках пруссов VII - начала XI в. // Археологические памятники европейской части РСФСР, М.

Kulakov V.I. - Кулаков В.И., 1990а, Древности пруссов VI-XIII вв. САИ, вып. Г 1-9, М.

Kulakov V.I. - Кулаков В.И., 1990б. Могильники западной части Мазурского Поозерья конца V - начала VIII вв. /по материалам раскопок 1878-1938 гг./ // Barbaricum-1989, Warszawa.

Kulakov V.I. - Кулаков В.И., 1990в. "Звериноголовые" фибулы балтов (V-VII вв.) // СА, № 2.

Kulakov V.I. - Кулаков В.И., 1990г. Поселенческая ситуация в Самбии II-XIII вв. // Проблемы изучения древних поселений в археологии (социологический аспект), М.

Kulakov V.I. - Кулаков В.И., 1994. Пруссы (V - XIII вв.), М.

Kulakov V.I. - Кулаков В.И., 1994в. «Гора Великанов». Раскопки 1992 г. // Barbaricum, T. 3, Warszawa.

Kulakov V.I. - Кулаков В.И. 1997а. Варникам. Древности прусских вождей // Гістарычна-археалагічны зборнік. № 12. Мінск.

Kulakov V.I. - Кулаков В.И. 1997б. Эстии и видиварии // Балто-славянские исследования 1988-1996. М.

Kulakov V.I. - Кулаков В.И., 1997в. Brisingamen. Стилистика «Восточной страны» // Квадрат, № 2, Калининград.

Kulakov V.I. - Кулаков В.И. 1997г. Истоки культуры пруссов // Vakarų baltai: etnogenezė ir etninė istorjia, Vilnius.

Kulakov V.I. - Кулаков В.И., 1997д. Прусский "обол Харона" // Пятая Всероссийская нумизматическая конференция, М.

Kulakov V.I. - Кулаков В.И., 1997е. Янтарная лихорадка в эпоху Юлиев-Клавдиев // Памятники старины. Концепции, открытия, версии. СПб-Псков, т. I.

Kulakov V.I. - Кулаков В.И., 1998а. Зональный орнамент на сосудах малых форм из могильников Янтарного берега // 20 lat archeologii w Maslomęczu, t. II, Lublin.

Kulakov V.I. - Кулаков В.И., 1998б. Holibo. Междуречье Ильфинг и Фришинг в 5 в. // Гістарычна-археалагічны зборнік, № 13. Мінск.

Kulakov V.I. - Кулаков В.И., 1998в. Майцский регистр фибул из музеев Кёнигсберга // Гістарычна-археалагічны зборнік. № 13. Мінск.

Kulakov V.I. - Кулаков В.И., 2000. Интерпретация "курганных" погребений эстиев I-III вв. н.э. // Балто-славянские исследования 1998-1999. XIV, М.

Kulakov V.I. - Кулаков В.И., 2001а. Козельский микрорегион в среднем железном веке//Вопросы археологии, истории, культуры и природы Верхнего Поочья, вып. IX, часть I, Калуга.

Kulakov V.I. - Кулаков В.И., 2001б. Археология Калининградской области // Археологическая карта России. Очерки археологии регионов, кн. 1, М.

Kulakov V.I. - Кулаков В.И., 2002а. От Восточной Пруссии к Калининградской области. Исторический путеводитель, Калининград.

Kulakov V.I. - Кулаков В.И., 2002б. Руноподобные знаки и руны старшего футарка в древностях юго-восточной Балтии // Civitas et castrum ad mare Balticum. Baltijas arheoloīijas un vēstures problēmas dzelzs laikmetā un viduslaikos, Riga.

Kulakov V.I., Valujev A.A. - Кулаков В.И., Валуев А.А., 1999. Тевтонский крест и бог Перкуно // Наука в России, № 6.

Kulakov V.I., Skwortzov K.N. - Кулаков В.И., Скворцов К.Н., 2000. Клинки из Кляйнхайде // Гістарычна-археалагічны зборнік. № 15. Мінск.

Kulakov V.I., Puzakova G.S., Tepliakov G.N. - Кулаков В.И., Пузакова Г.С., Тепляков Г.Н., 2001. Остров Розиттен: история заселения, Калининград-Кёнигсберг.

Kushner P.I. - Кушнер П.И., 1951. Этническое прошлое Юго-Восточной Прибалтики // Этнические территории и этнические границы. Труды Института этнографии, Новая серия, т. XV.

Kucharenko J.W. - Кухаренко Ю.В., 1980. Могильник Брест-Тришин, М.

Makajev E.A. - Макаев Э.А., 2002. Язык древнейших рунических надписей. Лингвистический и историко-филологический анализ, М.

Mastykova A.W. - Мастыкова А.В., 2001. Социальная иерархия женских могил северокавказского некрополя Дюрсо V-VI вв. (по материалам костюма) // Историко-дрхеологический сборник, вып. 7, Армавир-Москва.

Meletinskij E.M. - Мелетинский Е.М., 1980. Имир // Мифы народов мира, т. 1, М.

Monthyńska M. - Мончыньська М., 1999. О так называемых "сарматских" фибулах в Средней и Восточной Европе // Сто лет черняховской культуре, Киев.

Moshkov N.N. - Мошков Н.Н., 2002. Янтарь в медицине и косметологии, Калининград.

Nowakowskij W. - Новаковский В., 1997, Земли над нижним Неманом и Самбийский полуостров в Римскую эпоху // Vakarų baltai: etnogenezė ir etninė istorjia, Vilnius.

Ozere I.A. - Озере И.А., 1986. Миниатюрные глиняные сосуды в куршских погребениях V-IX вв. // Latvijas PSR Zinatnū Akademijas vēstis, Nr 1.

Pashuto W.T. - Пашуто В.Т., 1959. Образование Литовского государства. М.

Pennik N., Dzons P. - Пенник Н., Джонс П., 2000. История языческой Европы, СПб.

Plinij Starshij - Плиний Старший. Естественная история // Древние германцы. М., 1937.

Podosinov A.W. - Подосинов А.В., 1999. Ex oriente lux. Ориентация по странам света в архаических культурах Евразии, М.

Podosinov A.W. - Подосинов А.В., 2002. Восточная Европа в римской картографической традиции. Тексты. Перевод. Комментарий, М.

Pronin G.N. - Пронин Г.Н., 1989, Два типа погребений в земле ятвягов // Vakarų baltų archeologija ir istorija. Klaipėda.

Rethi Waftrudnira - Речи Вафтруднира // Беовульф. Старшая Эдда. Песнь о Нибелунгах, М., 1975.

Rostowtzev M.I. - Ростовцев М.И., 2000. Общество и хозяйство в Римской Империи, т. I, СПб.

Sarianidi W.I. - Сарианиди В.И., 1989. Храм и некрополь Теллятепе. М.

Skrzynskaja E.Th. - Скржинская Е.Ч., 1999. Примечания // Олимпиодор, История, СПб.

Smirnov J.A. - Смирнов Ю.А., 1997. Лабиринт. Морфология преднамеренного погребения, М.

Taratorin W.W. - Тараторин В.В., 1999. Конница на войне. История кавалерии с древнейших времён до наполеоновских войн, Минск.

Tautawithus A.Z. - Таутавичюс А.З., 1980. Балтские племена на территории Литвы в I тысячелетии н.э. // Из древнейшей истории балтских народов по данным археологии и антропологии, Рига.

Tazit - Тацит Корнелий. О происхождении германцев и местоположении Германии // Тацит Корнелий, Сочинения в двух томах, т. 1, Л.

Tokarev S.A. - Токарев С.А., 1985. Погребальные обычаи, их смысл и происхождение // Природа, № 9.

Toporov W.N. - Топоров В.Н., 1970. К балто-скандинавским мифологически связям // Donum Balticum, Stockholm.

Toporov W.N. - Топоров В.Н. 1983. Галинды в Западной Европе // Балто-славянские исследования 1983, М.

Toporov W.N. - Топоров В.Н., 1984. Oium Иордана (Gethica, 27-28) и гото-славянские связи в северо-западном Причерноморье // Этногенез народов Балкан и Северного Причерноморья, М.

Toporov W.N. - Топоров В.Н., 2000. К реконструкции балто-славянского мифологического образа Земли-Матери *Zemia & *Mātē (*Mati) // Балто-славянские исследования 1998-1999, М.

Toporova T.W. - Топорова Т.В., 1994. Семантическая структура древнегерманской модели мира, М.

Toporova T.W. - Топорова Т.В., 1996. Культура в зеркале языка: древнегерманские двучленные имена собственные, М.

Torsson Ed. - Торссон Эд., 2002. Руническое учение. Введение в эзотерическую рунологию, М.

Fiz J. - Фиц Й., 1986. Римская Паннония // Археология Венгрии. Конец II тысячелетия до н.э. - I тысячелетие н.э., М.

Chlewov A.A. - Хлевов А.А., 2002. Предвестники викингов. Северная Европа в I-VIII веках, СПб.

Shuwalov P.W. - Шувалов П.В., 1999. Позднеантичный мир в первой четверти V в. н.э. и сочинение Олимпиодора // Олимпиодор, История, СПб, с. 10-41.

Ščiukin M.B. - Щукин М.Б., 1994. На рубеже эр. Опыт историко-археологической реконструкции политических событий III в. до н.э. - I в. н.э. в Восточной и Центральной Европе, СПб.

Ščiukin M.B. - Щукин М.Б., 1998. Янтарный путь и венеды // Проблемы археологии, вып. 4, СПб.

Ščiukin M.B. - Щукин М.Б., 1999. К проблеме происхождения североевропейского филигранного стиля I - II вв. н.э. // Скандинавские чтения 1998 года, СПб.

Erdely I. - Эрдели И., 1986. Даки, сарматы, германцы, гунны в I - VI вв. // Археология Венгрии. Конец II тысячелетия до н.э. - I тысячелетие н.э., М.

Archivals

Archiv IA RAN, Kulakov V.I. - Архив ИА РАН, Кулаков В.И., Отчёт о работе Балтийской экспедиции ИА РАН в 1992 г., № 17428.

Heym W., 1938. Das Gräberfeld Zohpen. Diss., Biblioteka Muzeum Archeologicznego w Poznaniu, Inv. Nr 9177.

Abbreviations

AS GE - Arkheologicheskyi sbornik Gosudarstvennogo Ermitazha (Archaeologic collection of the State Hermitage)

Å.31 - a find is published in: Åberg N., 1919, Abb. 31.

Gbf. V, 16 - a find is published in: Tischler O., 1881, Taf. V, 16.

IA RAS - Russian Academy of Sciences Archaeological Institute

KSIIMK - Kratkiye soobshcheniya Instituta istorii material'noi kul'tiry (Short Reports of the Institute of a history of material culture)

KSIA - Kratkiye soobshcheniya Instituta arkheologii (Short Reports of the Archaeological Institute)

L. - Leningrad

M. - Moscow

MWiM - Muzeum Warmii i Mazur (Olsztyn)

MZM/A - Muzeum Zamkowe w Malborku, dzial archeologiczny.

PM - Prussia-Museum

Prussia - Sitzungsberichte der Altetumsgesellschaft PRUSSIA, Königsberg.

RA - Rossiyskaya arkheologiya (Russian archeology)

Ra.5 - the find is published in: Raddatz K. 1992-1993, Abb. 11,5.

SAI - Svod arkheologicheskikh istichnikov SSSR (Code of archaeologic sources of the USSR)

SPb. - Saint Petersburg

SPÖG - Schriften der Physikalisch-Ökonomischen Gesellschaft zu Königsberg

I, 10 - a find is published in: Tischler O., Kemke H., 1902, Taf. I, 10.

Alphabetic indexations of cemeteries:

A - Althof-Insterburg (the western surburb of Chernyakhovsk-Sity); A-Bod - Nowa Bozwinka; E - Eisliethen/Geroyskoye-Roshchino; Detlevsruh - Detlevsruh, Kr. Friedland (destroyed, Pravdinsk District?); Do - Dollkeim, Kr. Samland (Kovrovo-1, Zelenogradsk District); Eis - Eisselbitten, Kr. Samland (Sirenevo, Zelenogradsk District); Fü - Fürstenwalde, Kr. Königsberg (Poddubnoye, Guryevsk District); G - Grebieten, Kr. Samland (destroyed, Zelenogradsk District); Gr. Ott - Gross Ottenhagen/Beriozowka; Grei - Greibau, Kr. Samland (Lyublino-Krasnolesye, Zelenogradsk District); Gru - Grunaiki, woj. Białystok; Grei - Greibau; Hen - former Henriettenfeld; Ho - Hochschnakeinen, Kr. Pr. Eylau (nowadays - Bagrationovsk District); H - Hünenberg bei Rantau-Neu Kuhren, Kr. Samland (natural boundary "Gora Velikanov", Pionerskyi); Ka - Kampischkehmen/ Sinyavino; Ki - Kirpehnen/Povarovka; Kly - Klycken; Kr - Kruglanki; Ku - Kunterstrauch/natural boundary Dubki; Mo - Moritten/Sibirskoye; Na - Nautzau/Kovrovo-3; Ost - Ostróda; Ostpr - Ost-Preussen; Po - Pollwitten/Rovnoye; Pr. M - Prussia-Museum; Palwe bei Rantau - flat natural boundary to the east from Pionerskyi (Zelenogradsk District); Re - Regehnen, Kr. Samland (Dubrovka, Zelenogradsk District); Samland - Sambia; Sc - Schlakalken, Kr. Samland (Yaroslavskoye, Zelenogradsk District); Se - Seerappen/Lyublino; St. L - Sct. Lorenz/Sal'skoye; Str - Stręgiel; Tra - Trausitten/Guryevsk; Trö - Trömpau/Lazovskoye; Tro - Tropitten/Kumachevo; Wa - Warnikam/Pervomayskoye; War - former Warengen; Wi - Wiekau/Hrustal'noye; Wo - Wogau/Lermontovo; Wack - Wackern, Kr. Pr. Eylau (nowadays - Yelanovka, Bagrationovsk District); Zo - Zohpen/Suvorovo (Gvardeysk District).

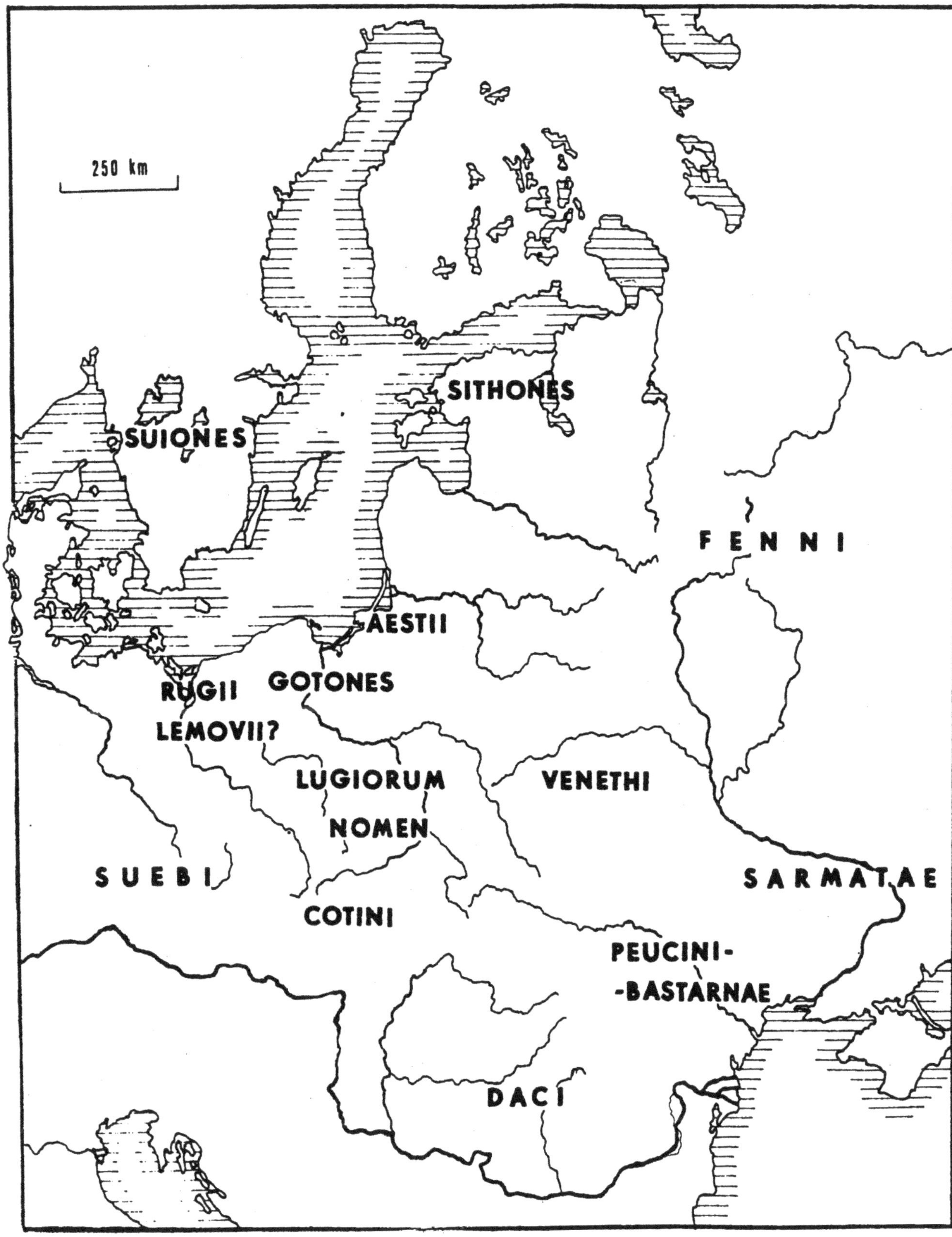

Fig. 1. The tribes of "Svebia" (after Tacitus).

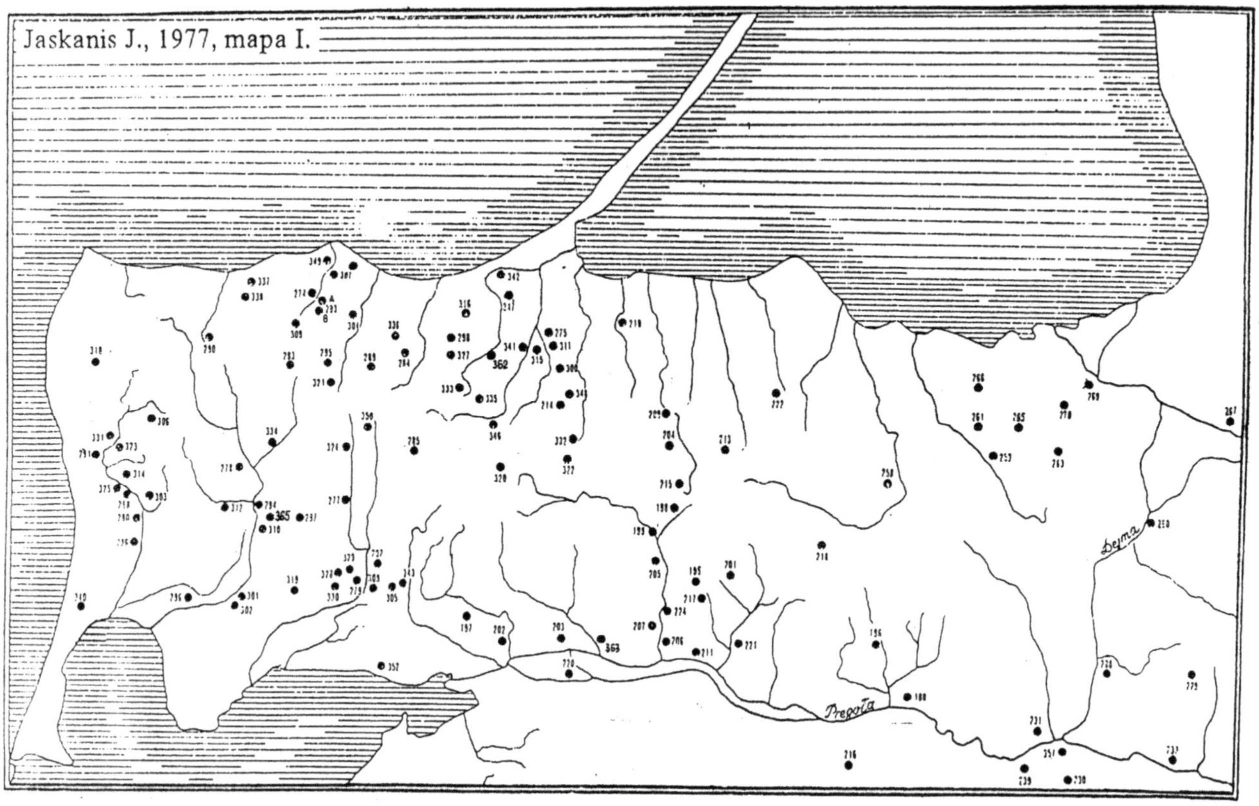

Fig. 2. Flat-grave Sambian cemeteries from Roman times.

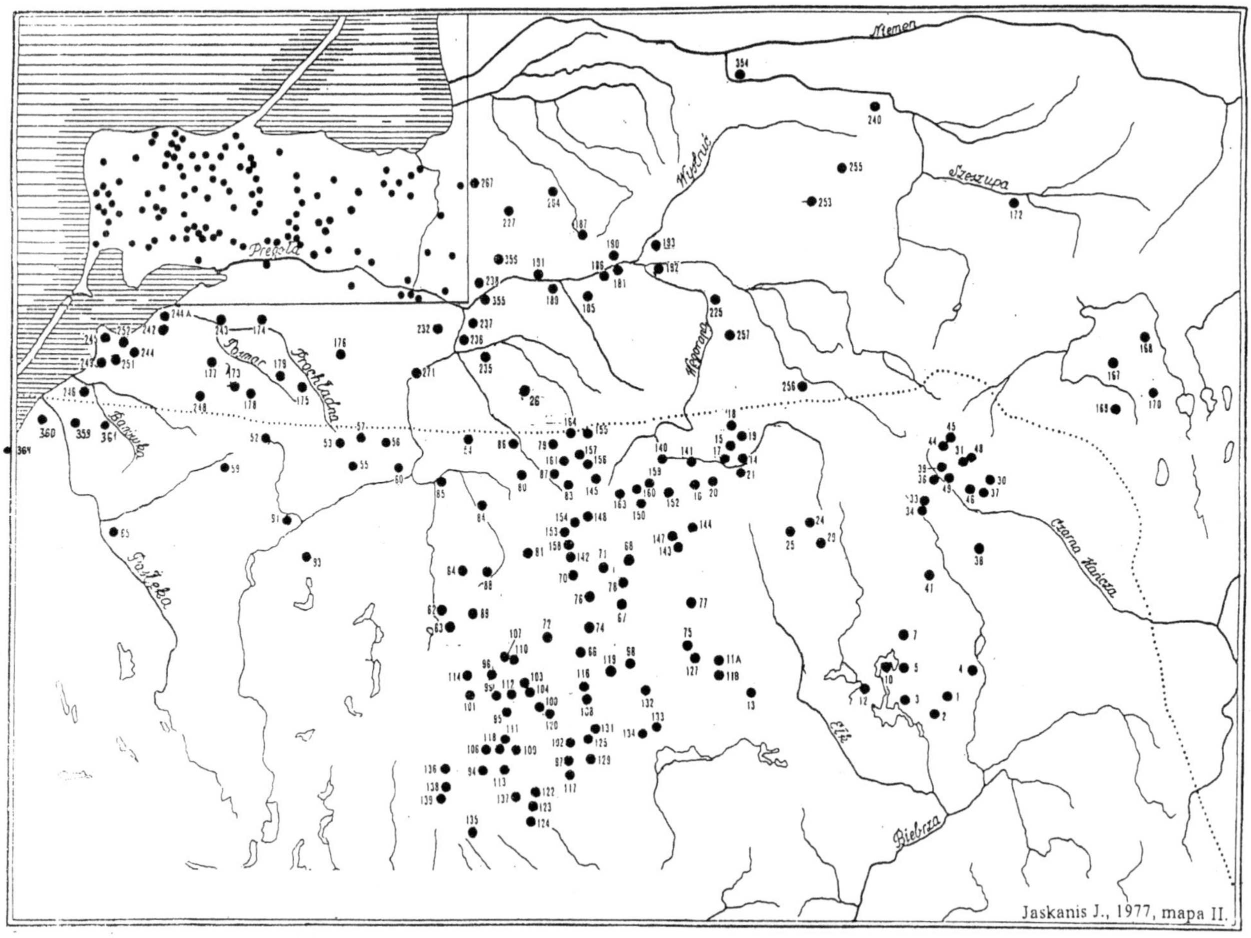

Fig. 3. Flat-grave cemeteries of the southeast Baltic region from Roman times.

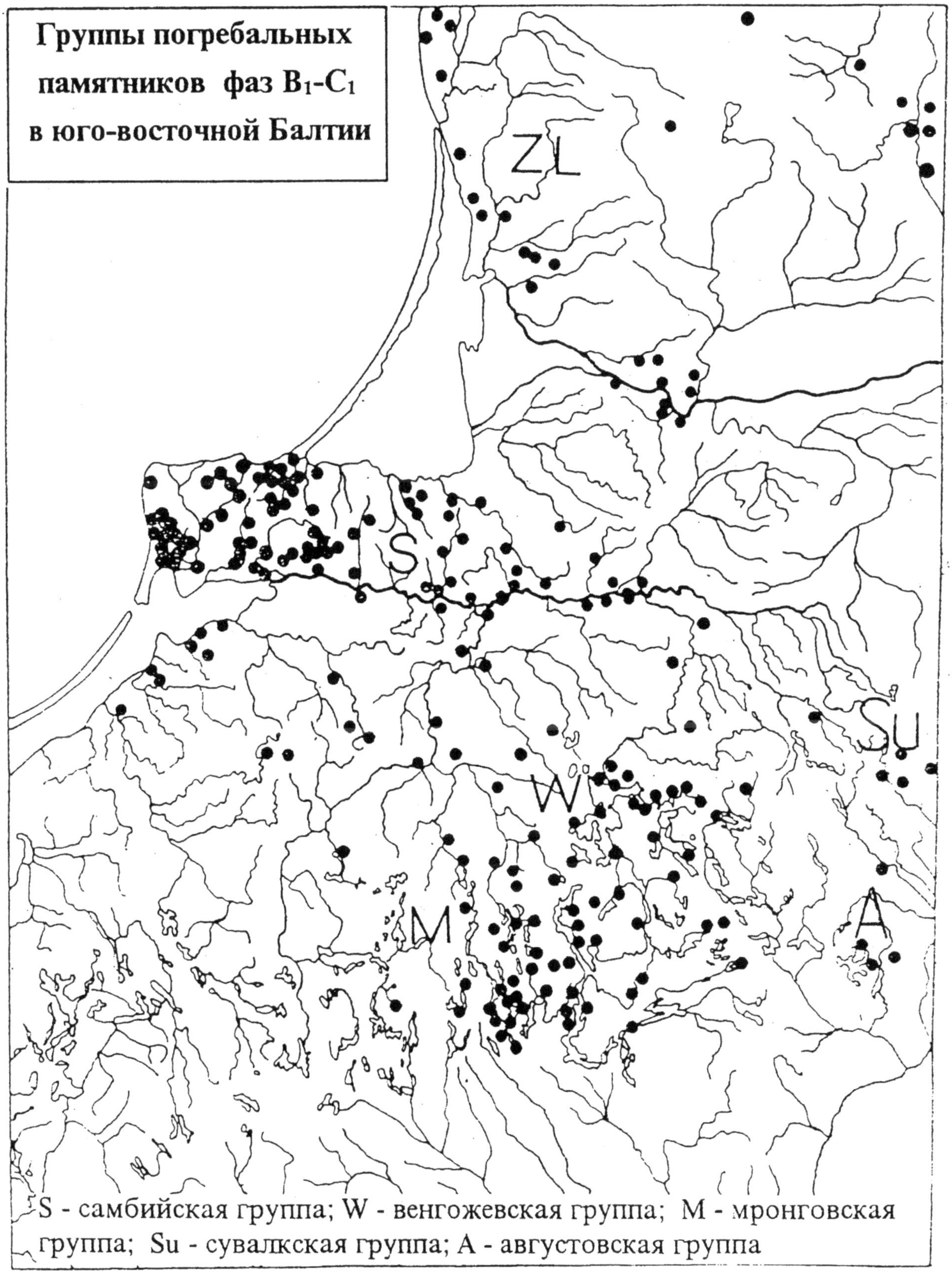

Fig. 4. Groups of phase B_1-C_1 burial sites in the southeast Baltic region.

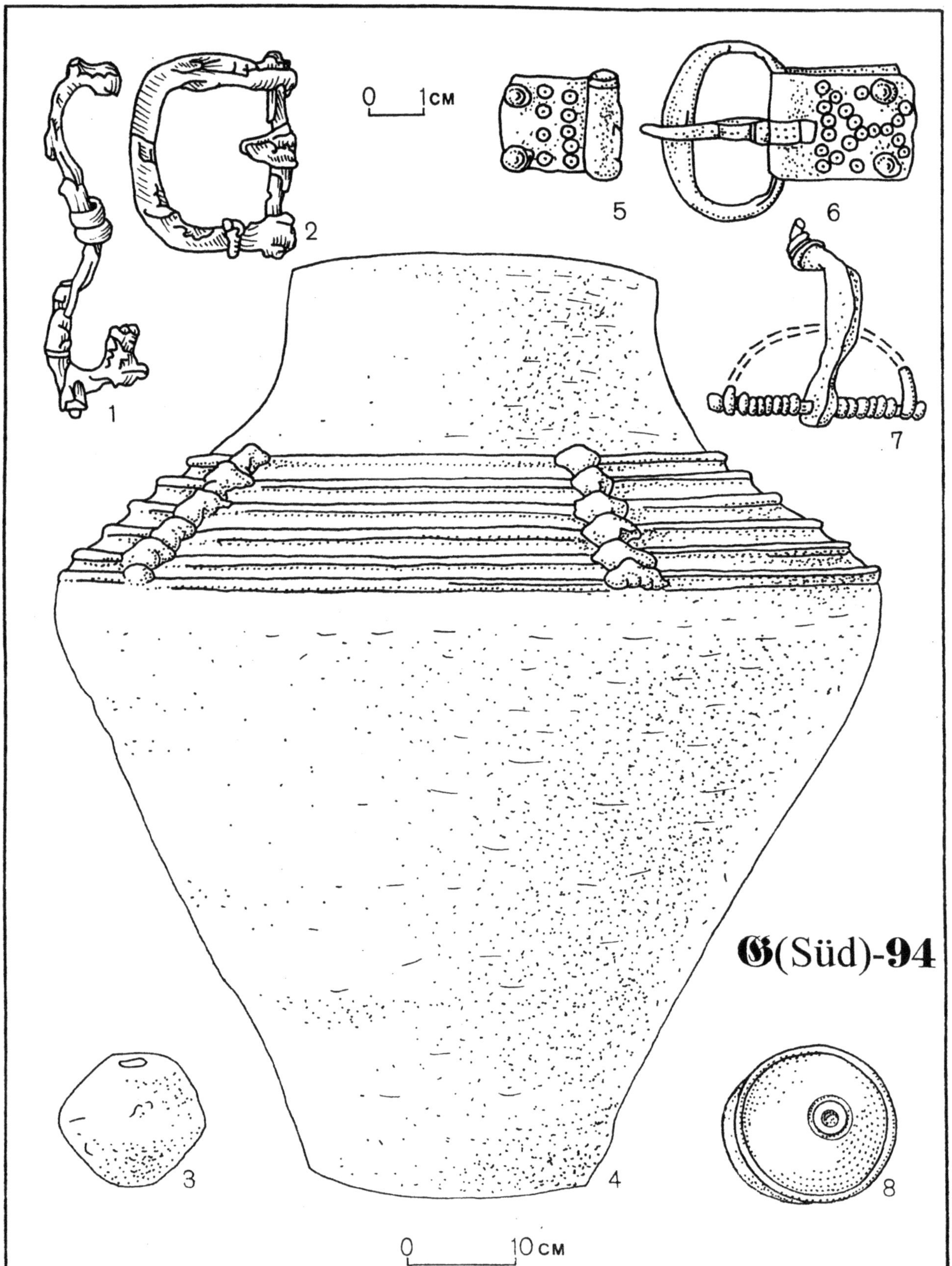

Fig. 5. Grave 94 assemblage from the Grebieten (Süd) cemetery.

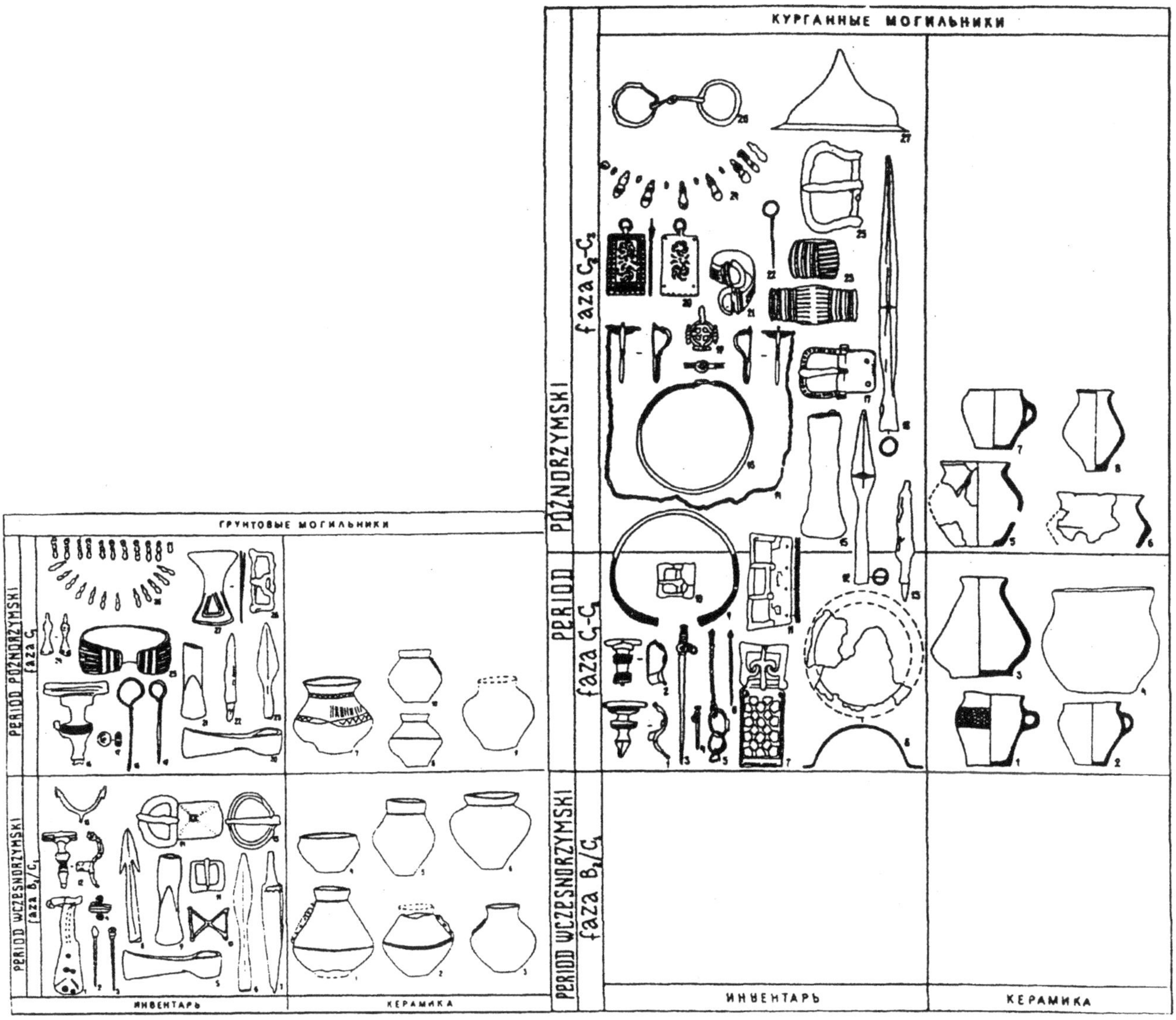

Fig. 6. 2nd-4th cent. AD antiquities from East Masuria.

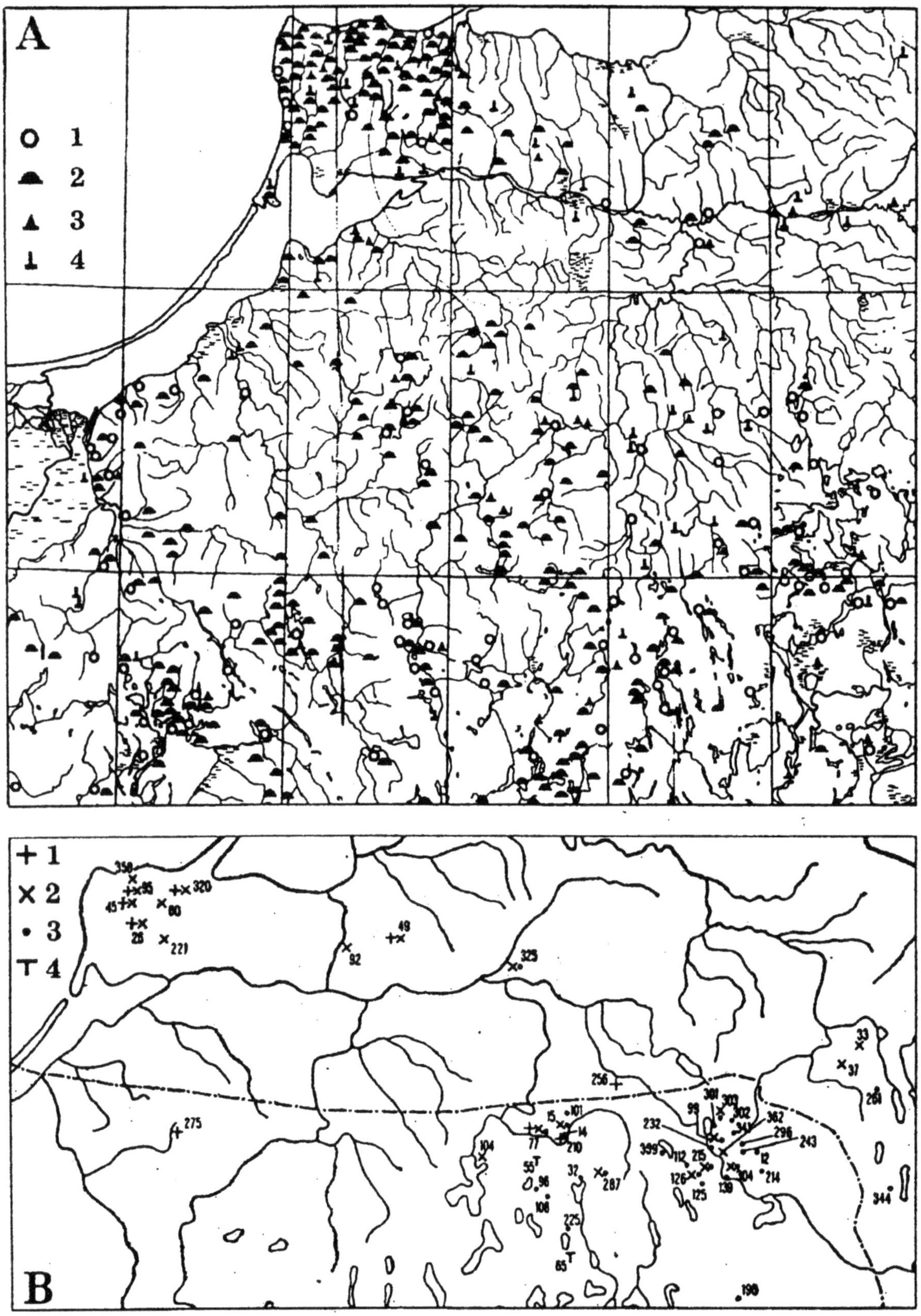

Fig. 7. The southeast Baltic region at the end of the 1st mill. BC/beginning of the 1st mill. AD: A - sites of western-Balt barrow culture (1 - defended sites, 2 - barrow cemeteries, 3 - hoards, casual finds - Okulicz J., 1970, map I); B - barrows and barrow-like mounds in the southeast Baltic region from the 1st-4th cent. AD (1 - early Roman era, 2 - late Roman era, 3 - era of the great migrations, 4 - burials from the 1st-4th cent. AD; site numbers after J. Jaskanis - 26 - Khrustal'noye, 45 - Dubrovka, 60 - Geroyskoye, 92 - Izobil'noye, 95 - Yaroslavskoye, 221 - Perkuicken, 256 - Raduzhnoye, 320 - Warengen, 325 - Wengerin, 350 - Zarechye, actually - "Gora Velikanov" - Jaskanis J., 1974, map II).

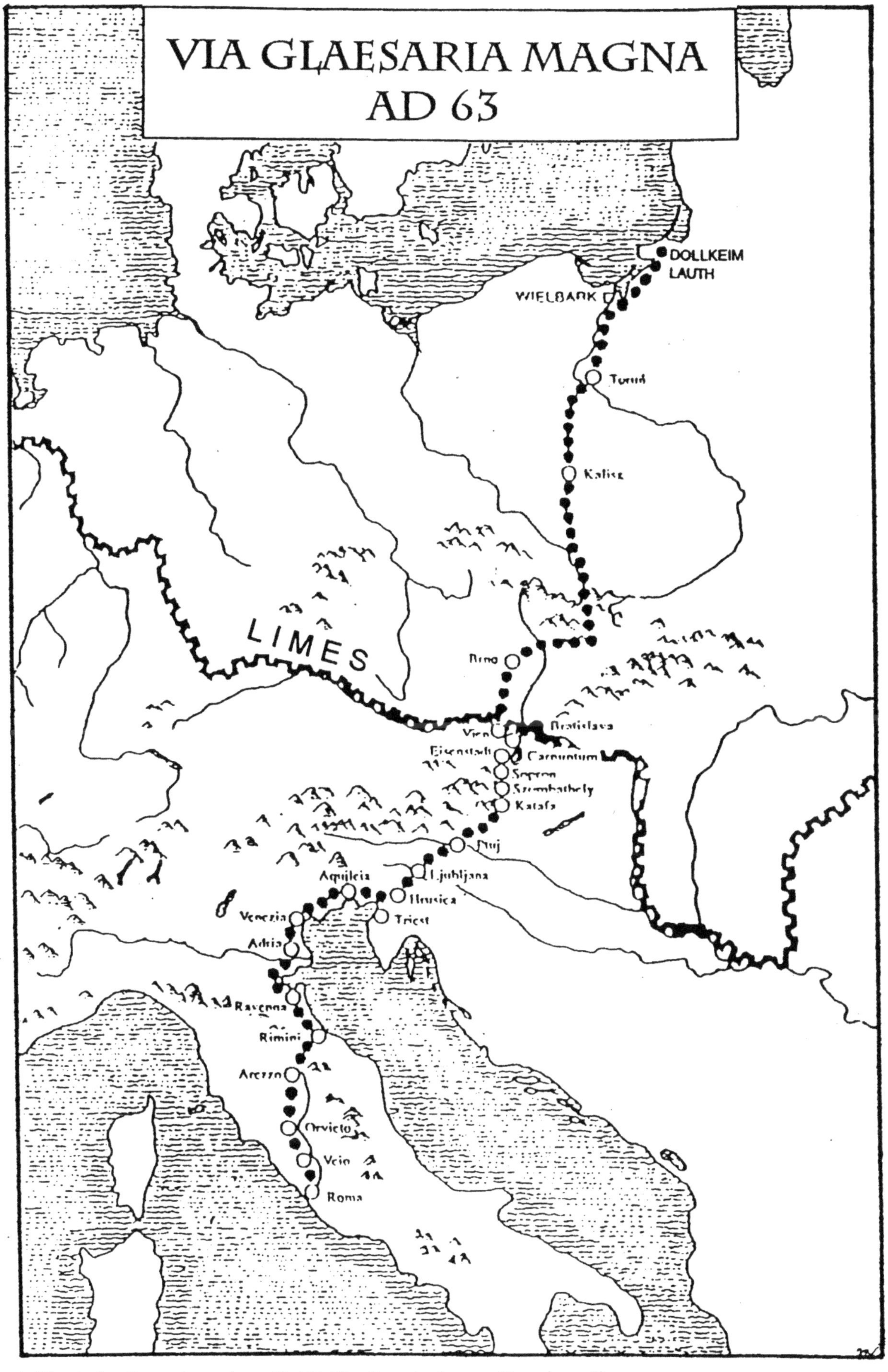

Fig. 8. The Amber Route from about 63 AD. The *limes* is shown with a zigzag line.

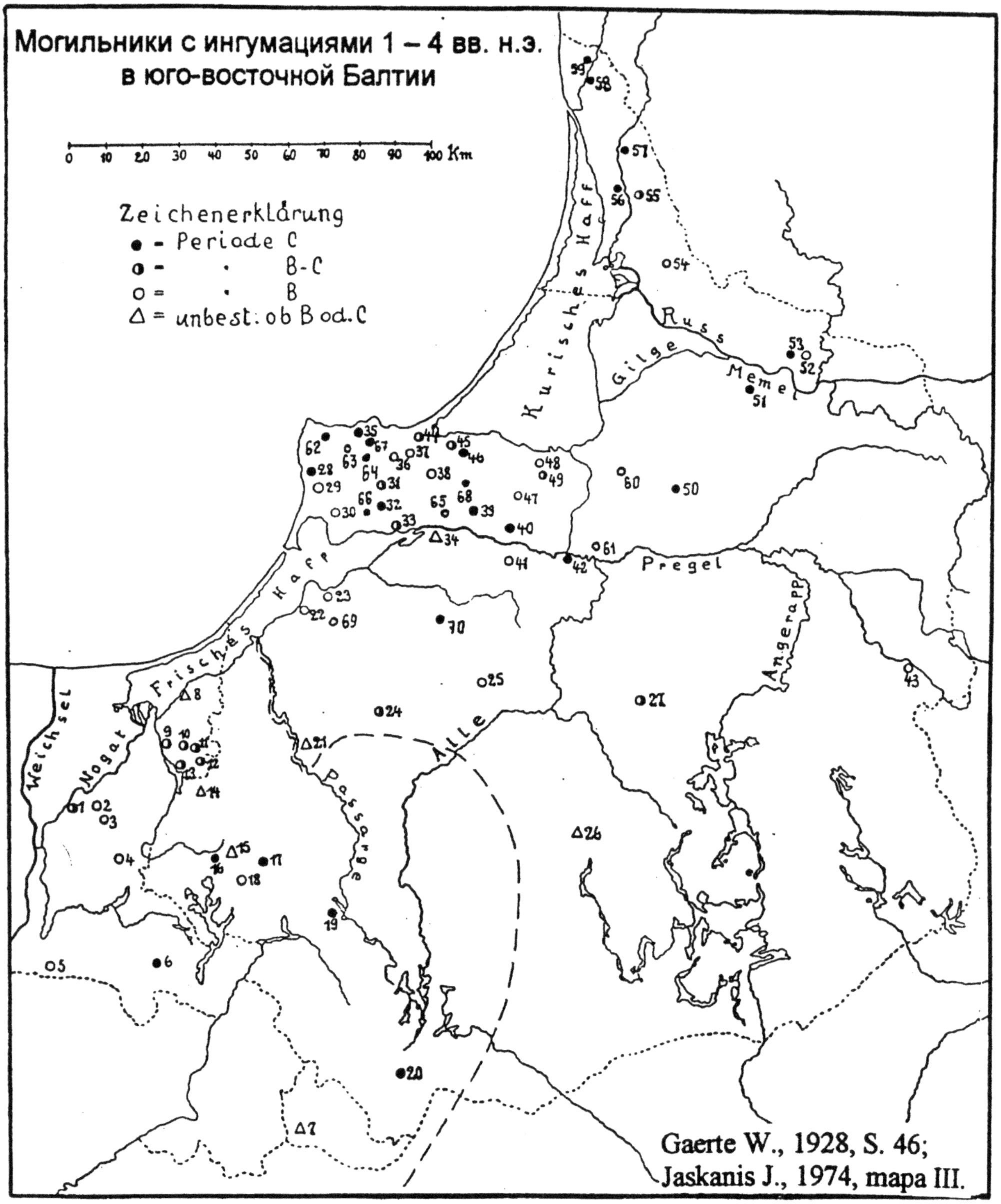

Fig. 9. Cemeteries with inhumations from the 1st-4th cent. in the southeast Baltic region.

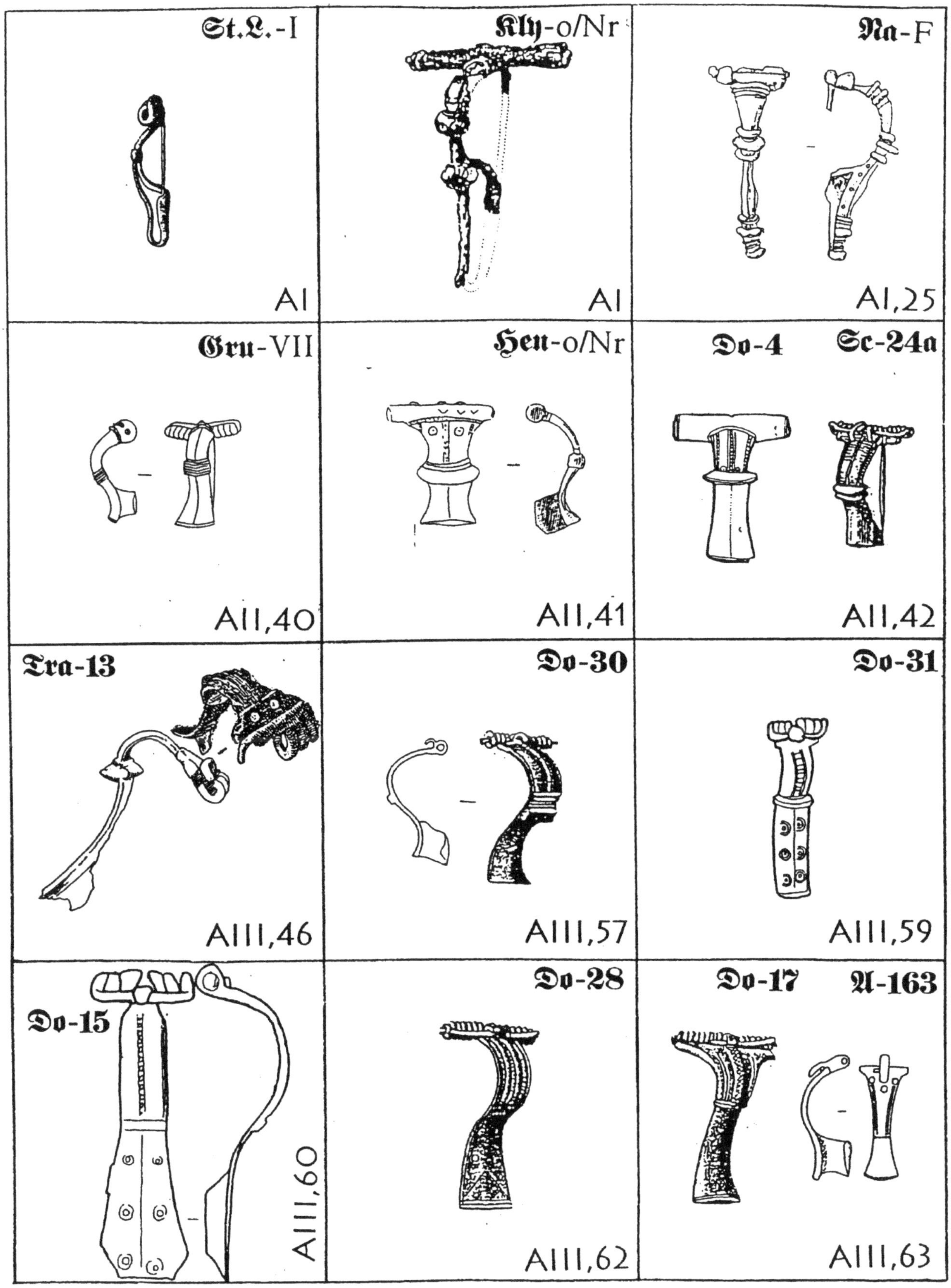

Fig. 10. Types of brooches of groups AI, AII, AIII from the 'SNG' area.

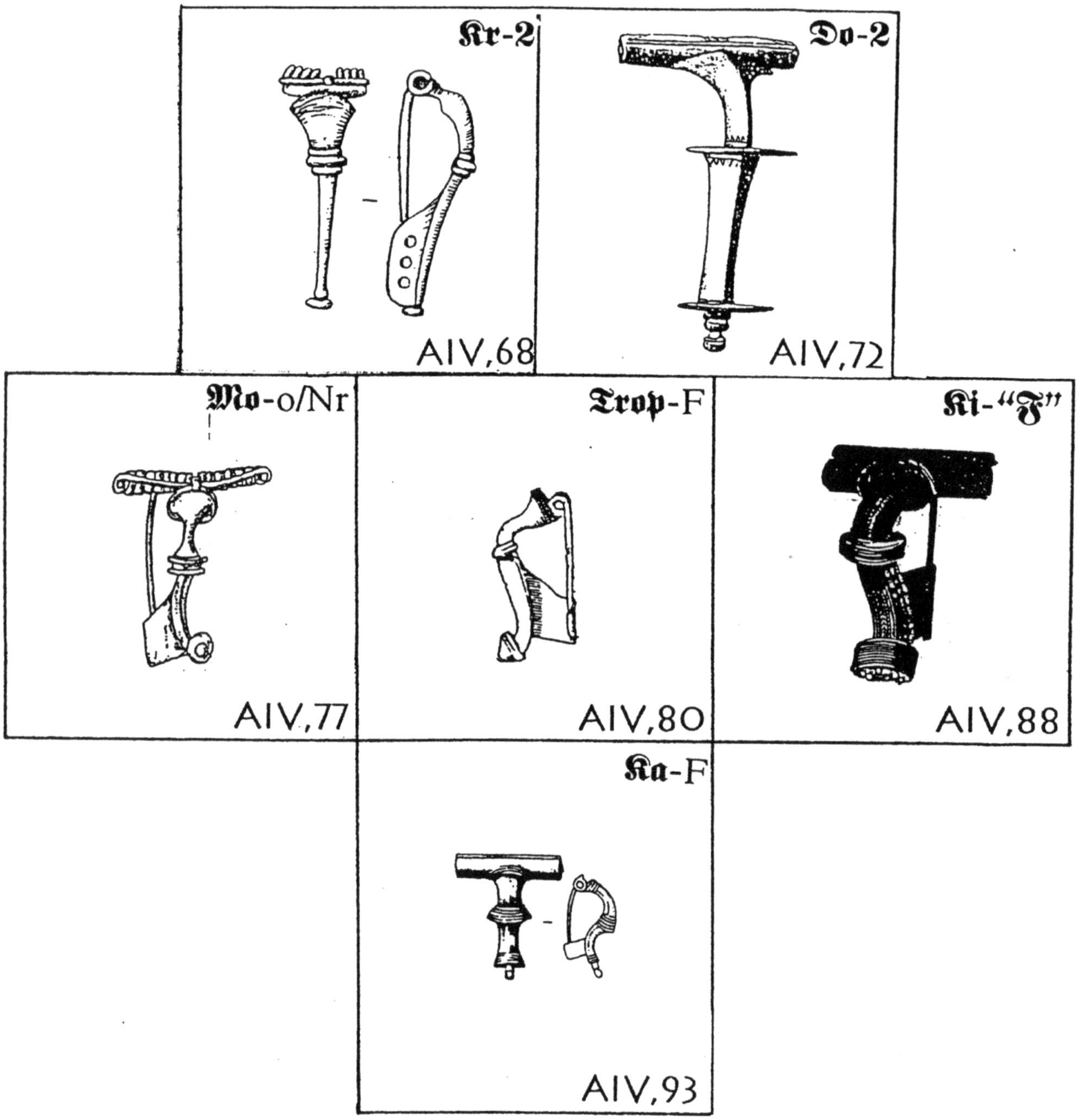

Fig. 11. Types of brooches of groups AIV.

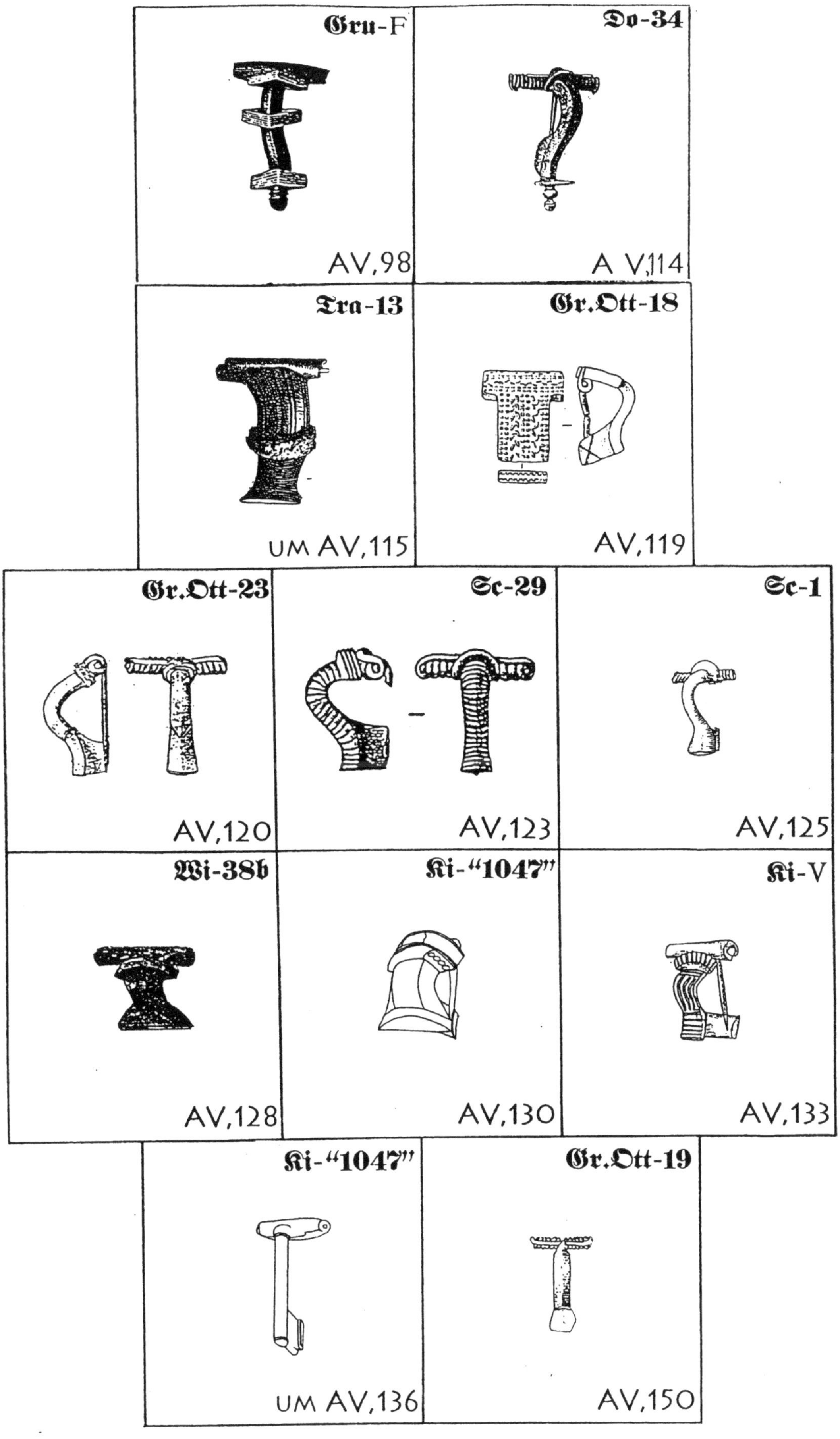

Fig. 12. Types of brooches of groups AV.

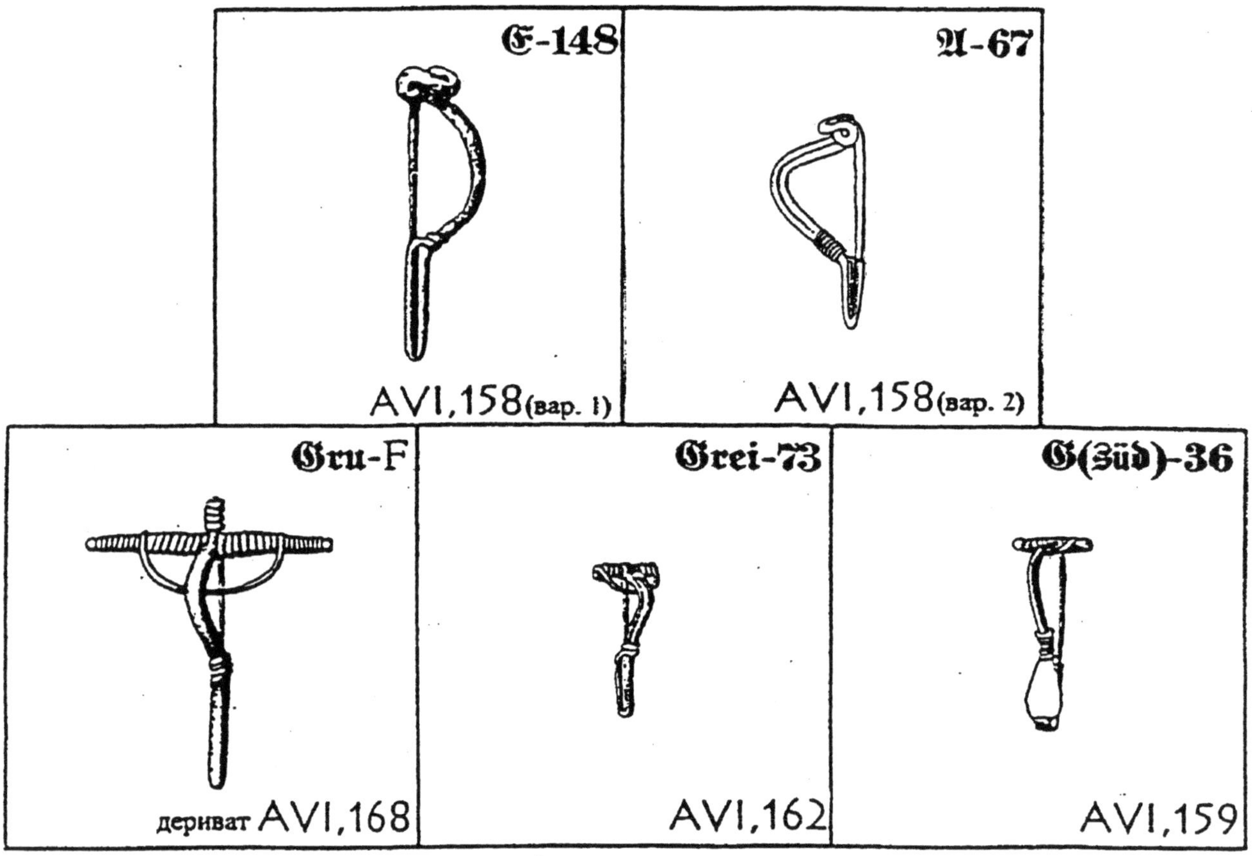

Fig. 13. Types of brooches of groups AVI.

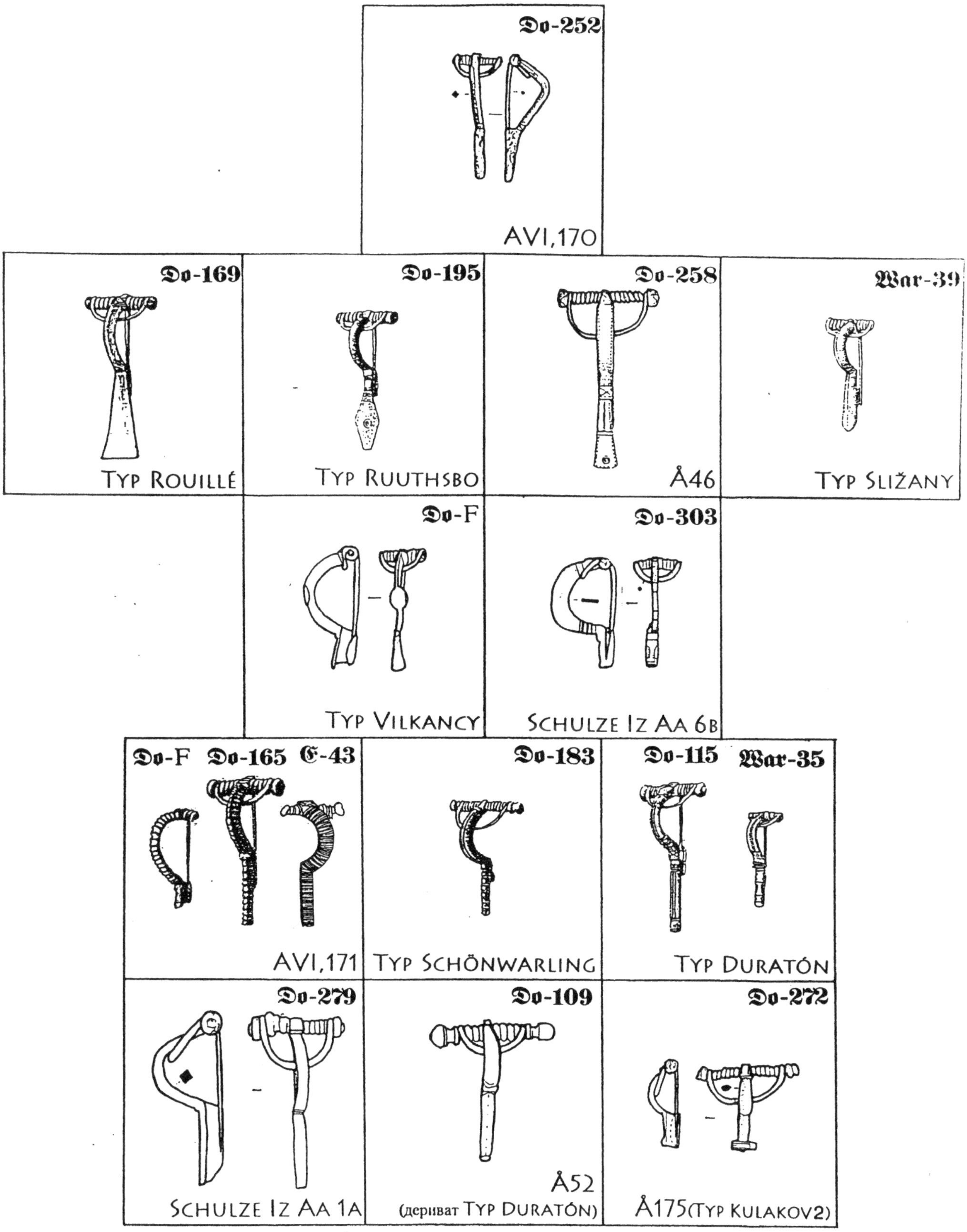

Fig. 14. Types of crossbow brooch with cast pin-receivers.

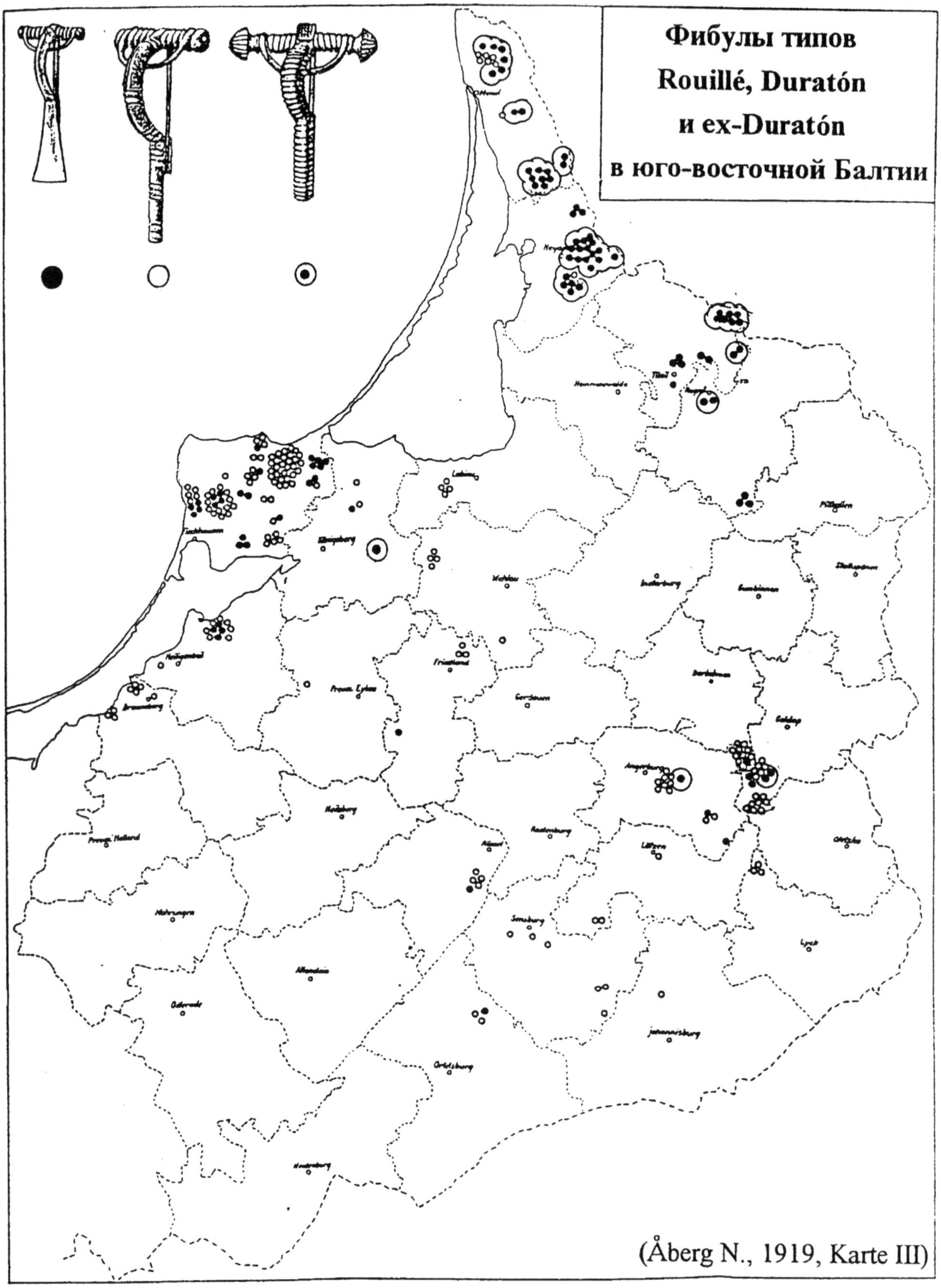

Fig. 15. Brooches of types Rouillé, Duratón and ex-Duratón in the southeast Baltic region.

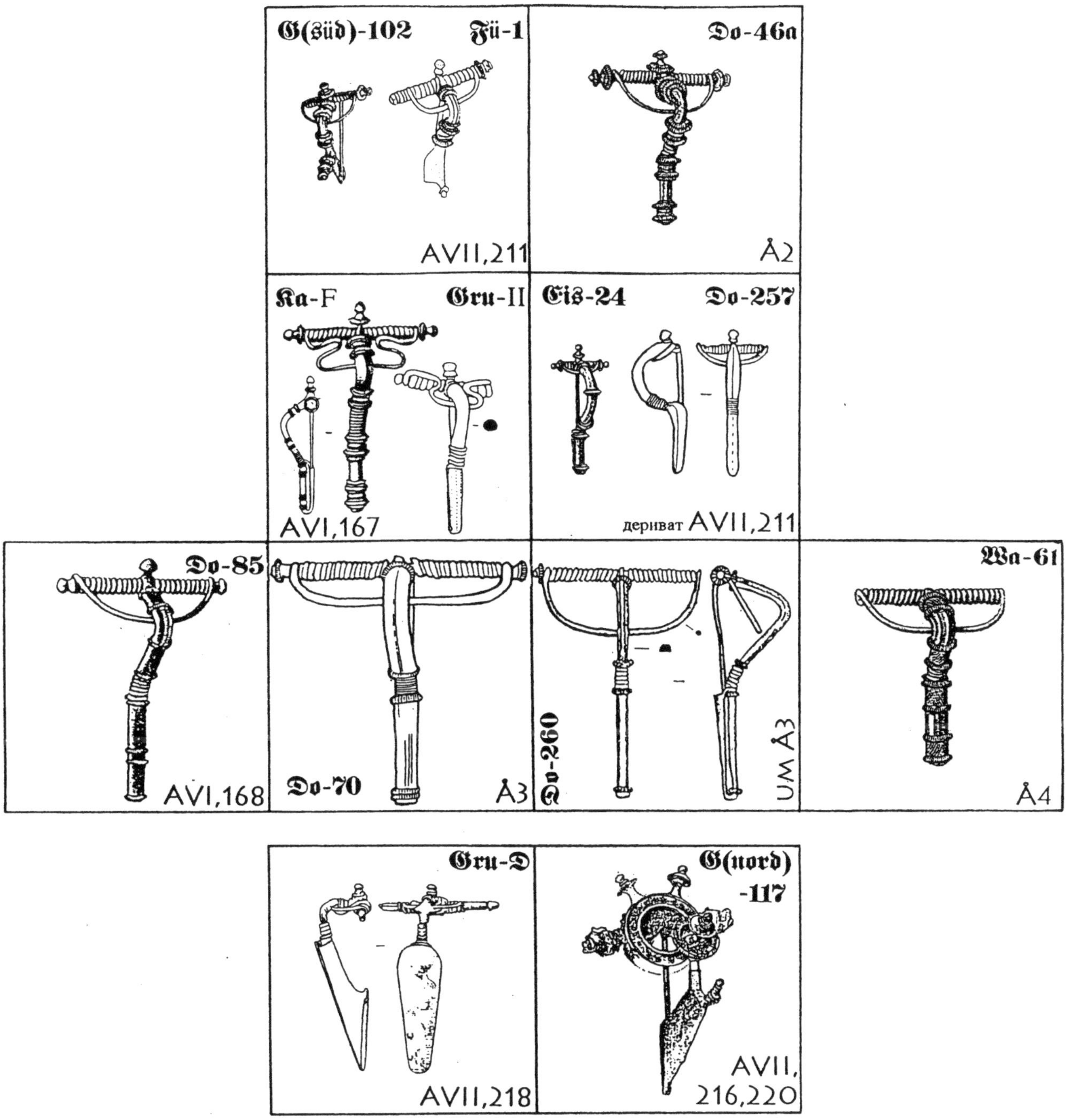

Fig. 16. The prototype AVII, 211 and types of crossbow brooch of groups AVII.

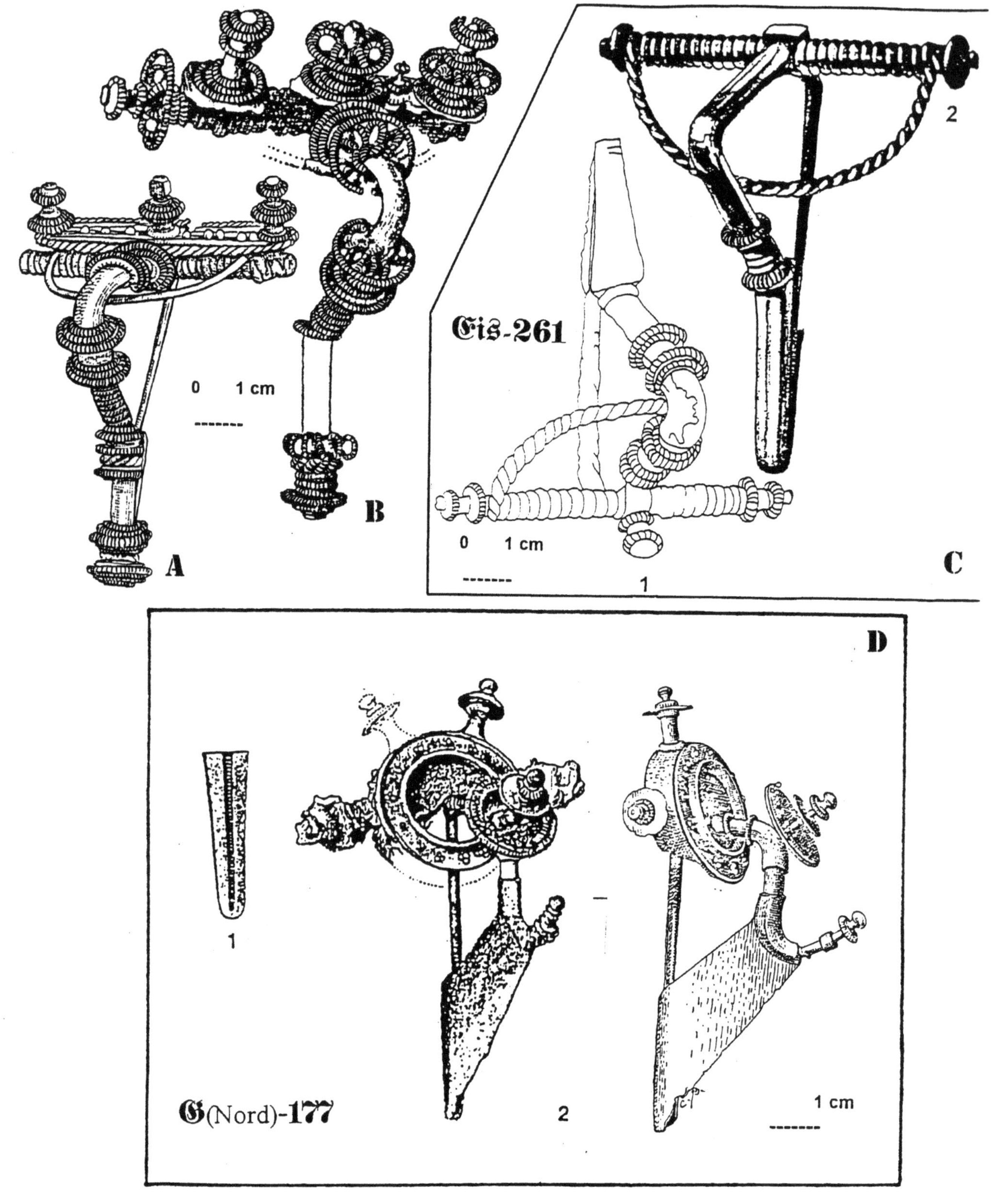

Fig. 17. 'Monstruosa' brooches from Sambia and analogies: A - Orekhovo (Zelenogradsk District, ehem. Schuditten, Kr. Samland); B - former Warengen, Kr. Samland; C - brooches of type grosse ABF (one with silver mount on pin-receiver) from Grave 261 at Geroyskoye-Roshchino (Zelenogradzk District, ehem. Eisliethen, Kr. Samland); D - brooch from Grave 177 ehem. Grebieten (Nord), Kr. Samland.

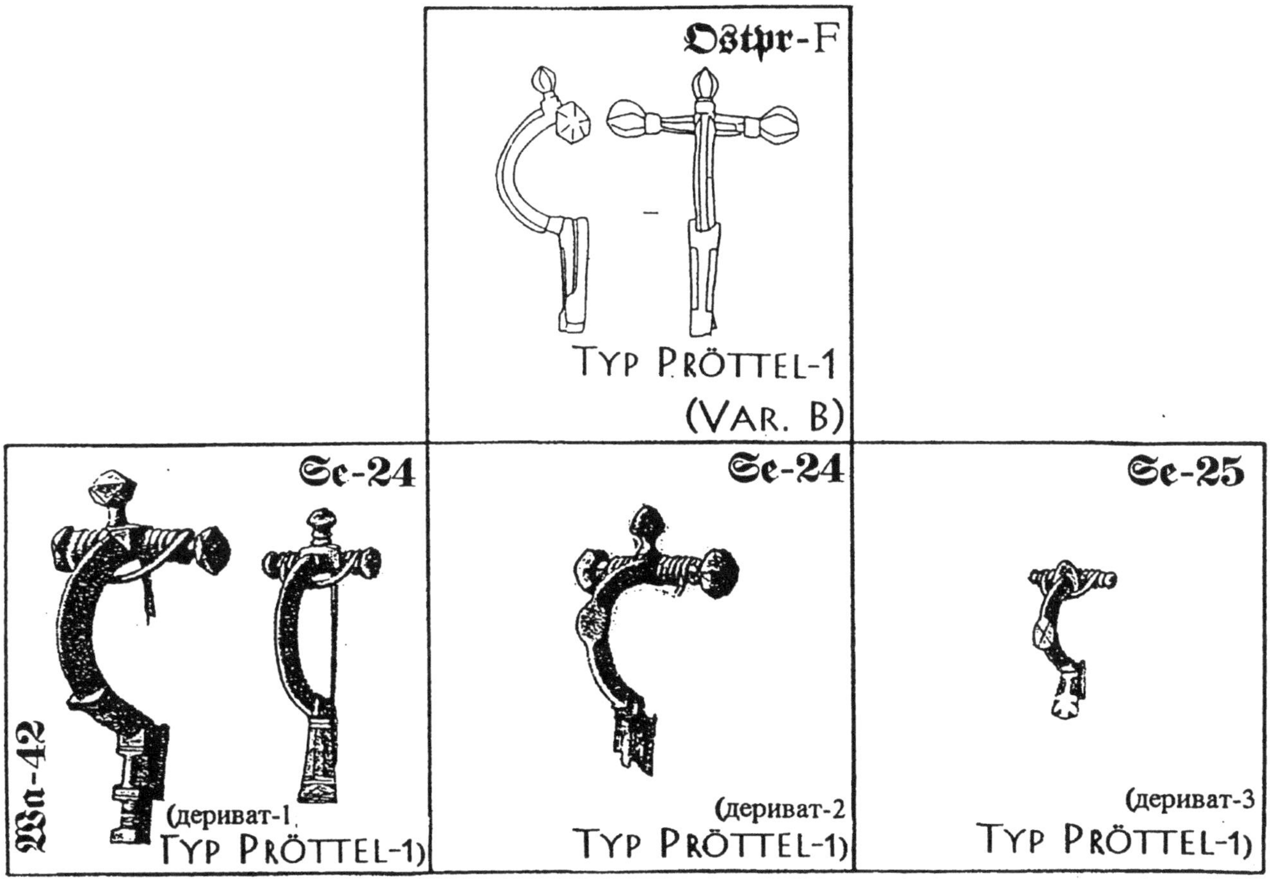

Fig. 18. Prototype of joint-pin brooch with bulbous head and crossbow derivatives.

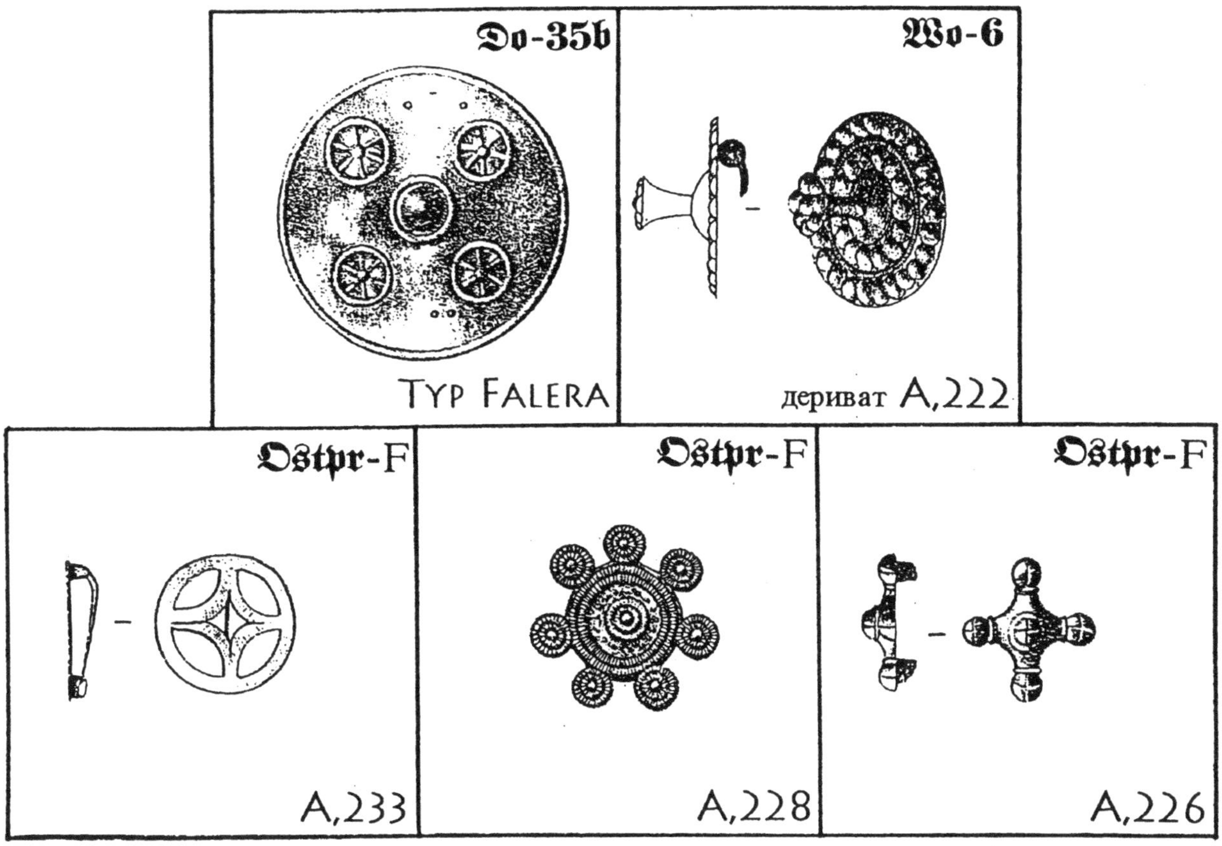

Fig. 19. Types of plate brooch.

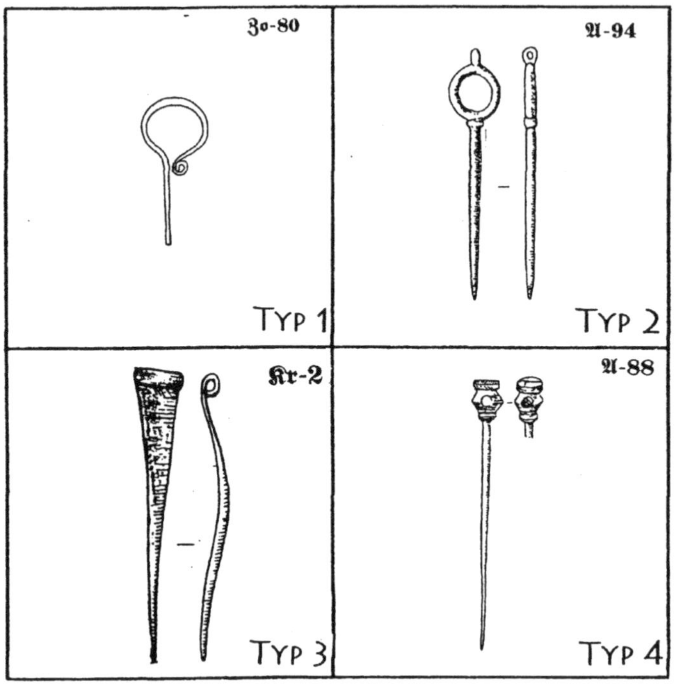

Fig. 20. Types of pin.

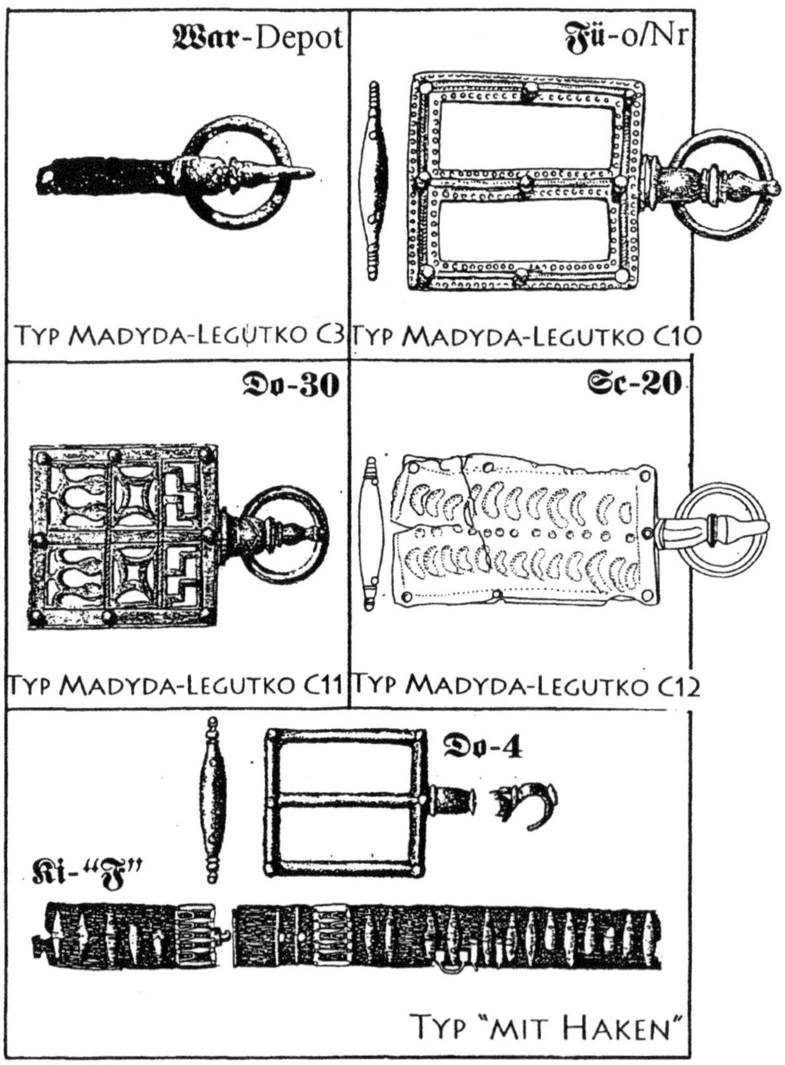

Fig. 21. Types of buckle with immovable tongue.

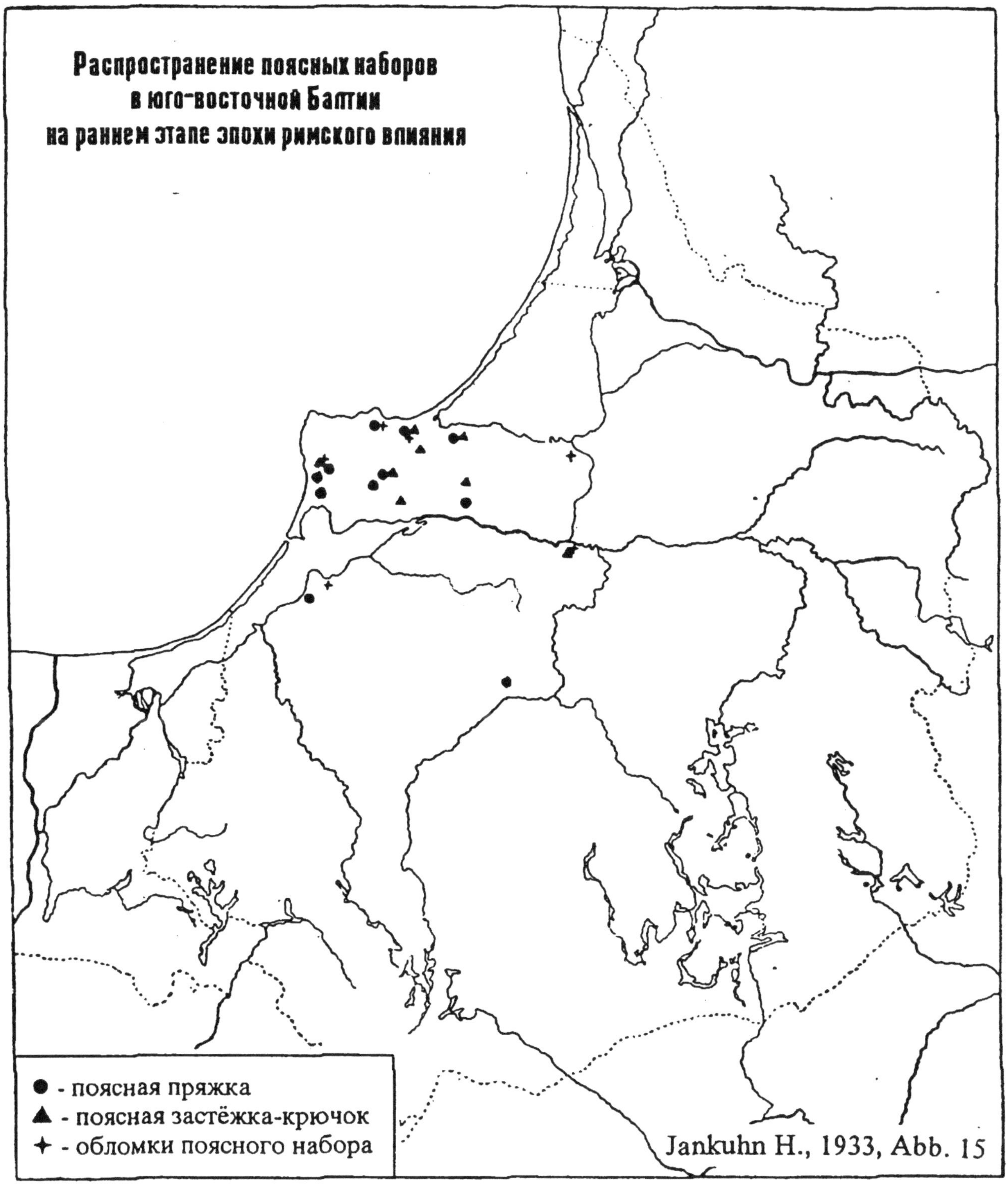

Fig. 22. Distribution of belt sets in the southeast Baltic region at the early stage of Roman influence.

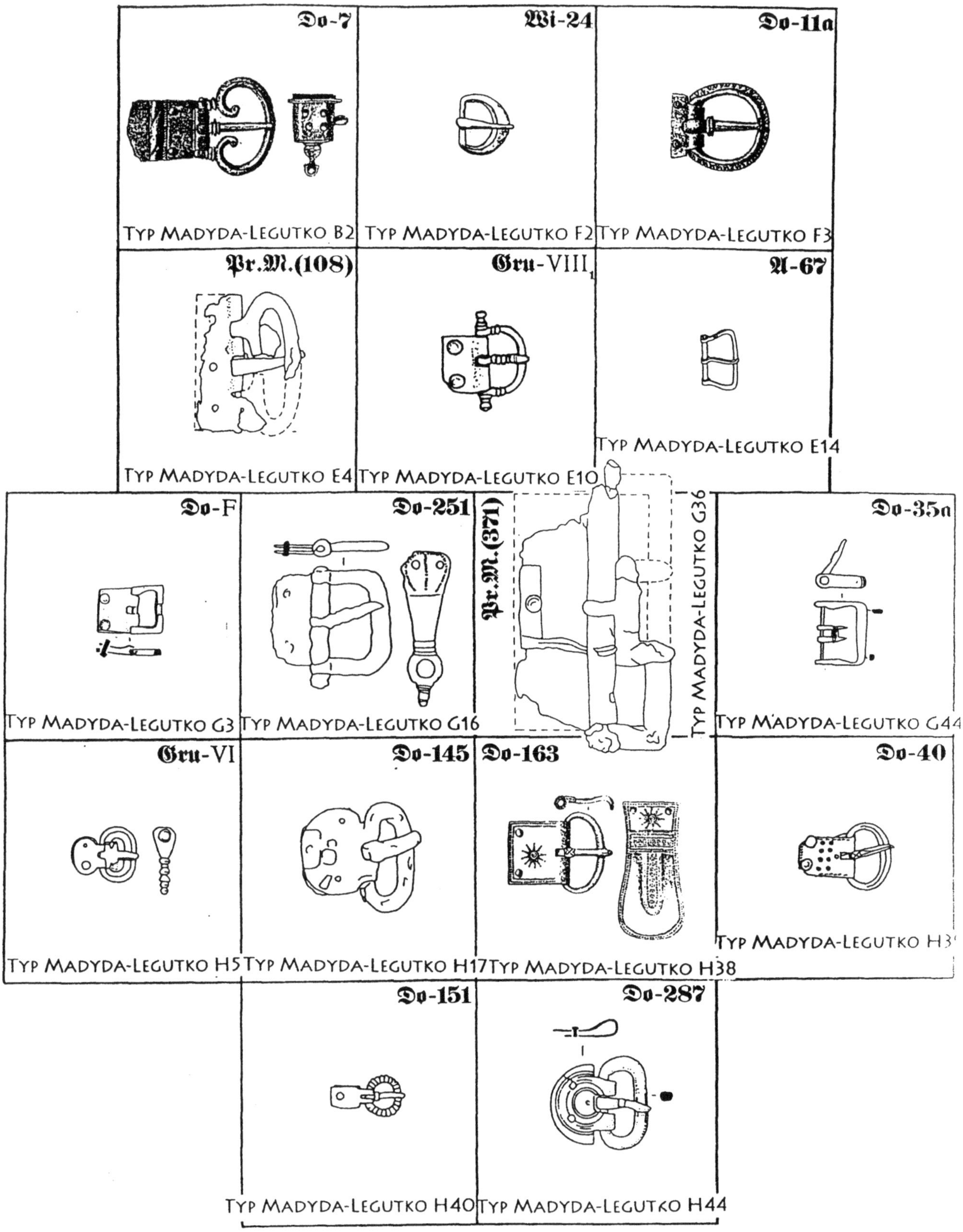

Fig. 23. Types of the buckle with movable tongue and appropriate strap-ends.

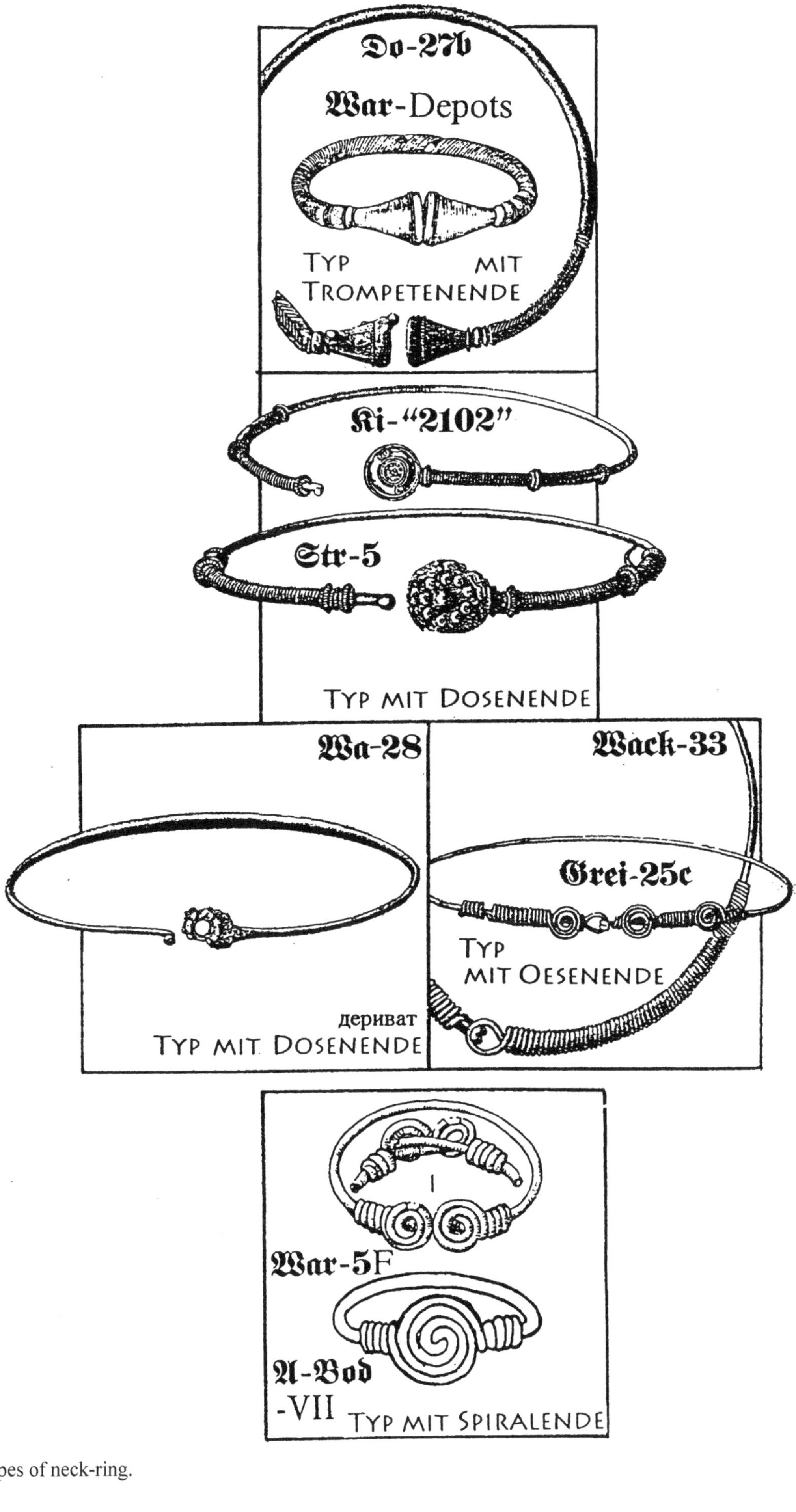

Fig. 24. Types of neck-ring.

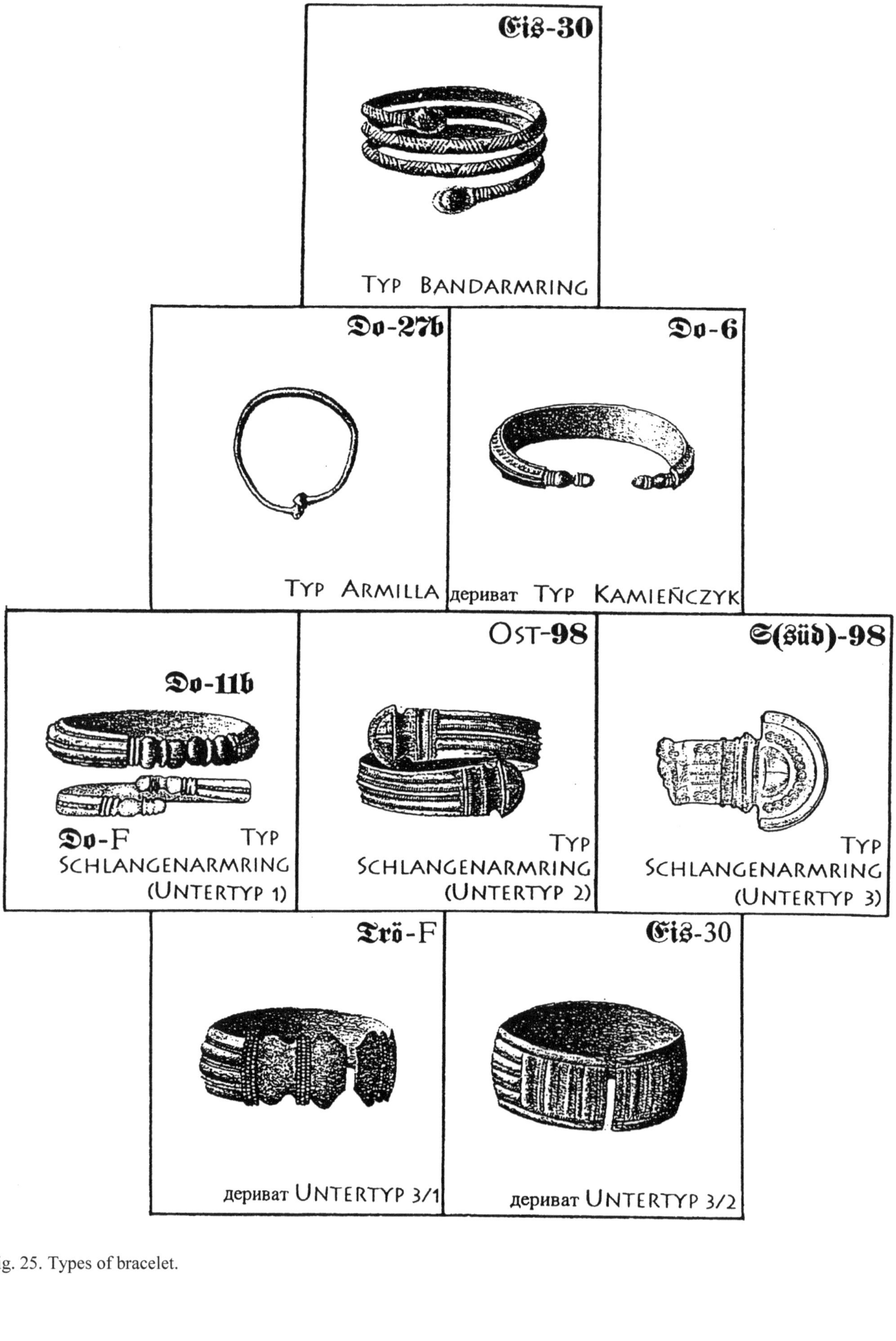

Fig. 25. Types of bracelet.

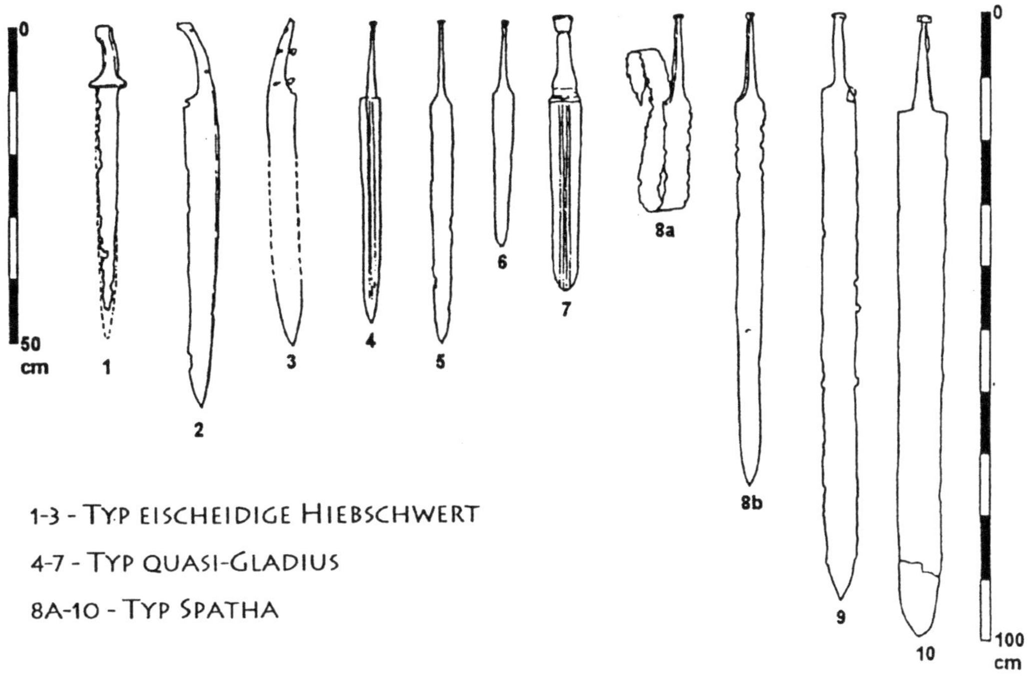

Fig. 26. Types of sword from the 2nd-5th cent. AD in the southeast Baltic region: 1 - Moritten, 2 - Moythienen, Grave 2, 3 - Kotzeck, Grave 122, 4 - Skomatzko, Grave 23, 5 - Grunajki, 7 - Wiekau, grave XXXIV, 8 - Gaffken, 9 - Szwajcaria, Barrow 25, 10 - Szwajcaria, Barrow 2.

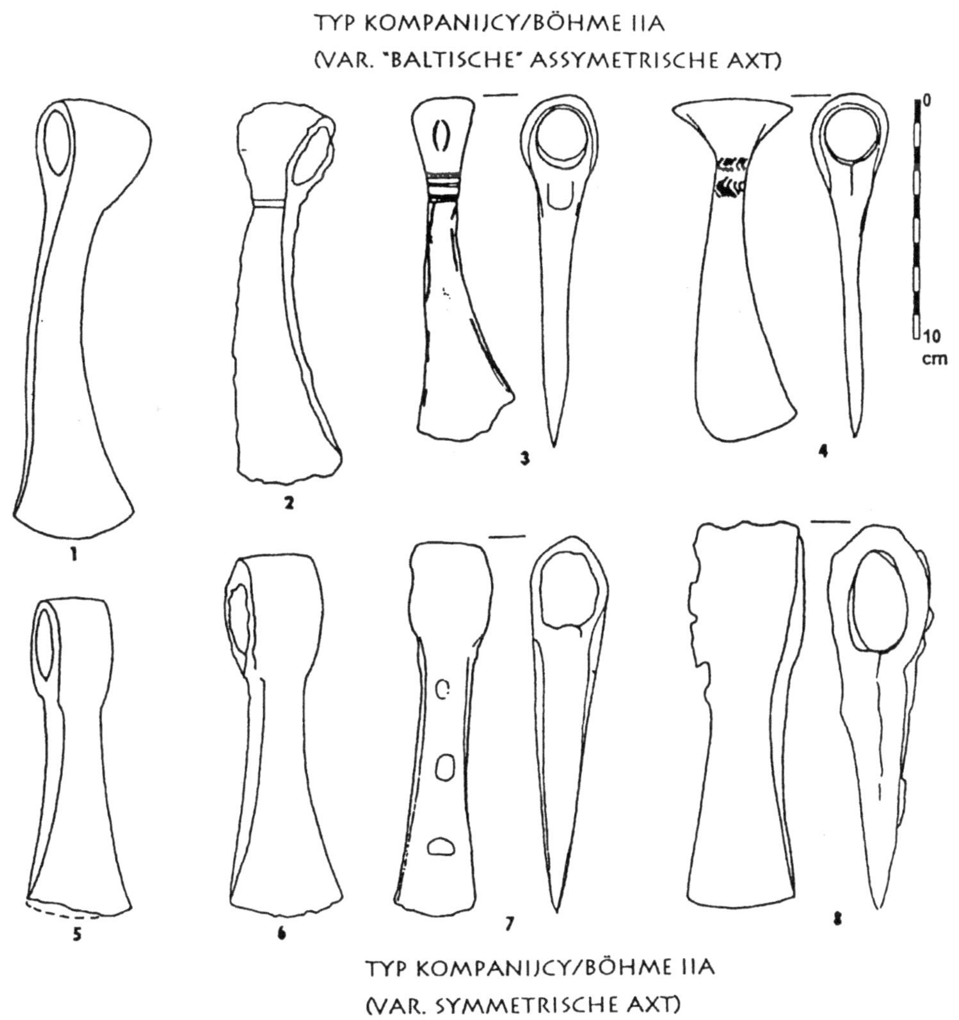

Fig. 27. Variants of axes with lugs from the 2nd-4th cent. AD in the southeast Baltic region.

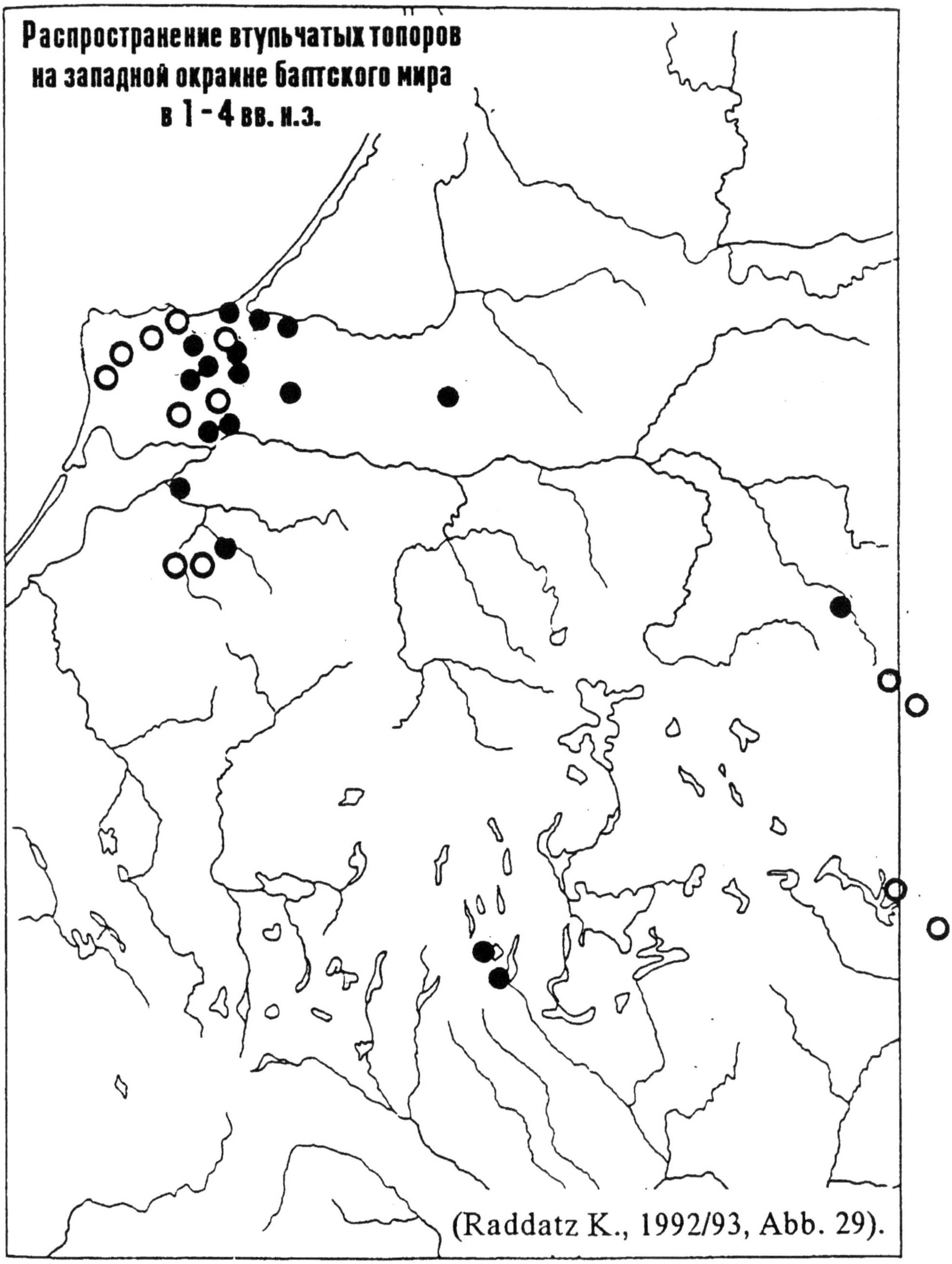

Fig. 28. Distribution of socketed axes on the western outlying district of the Balt world in the 1st cent. AD.

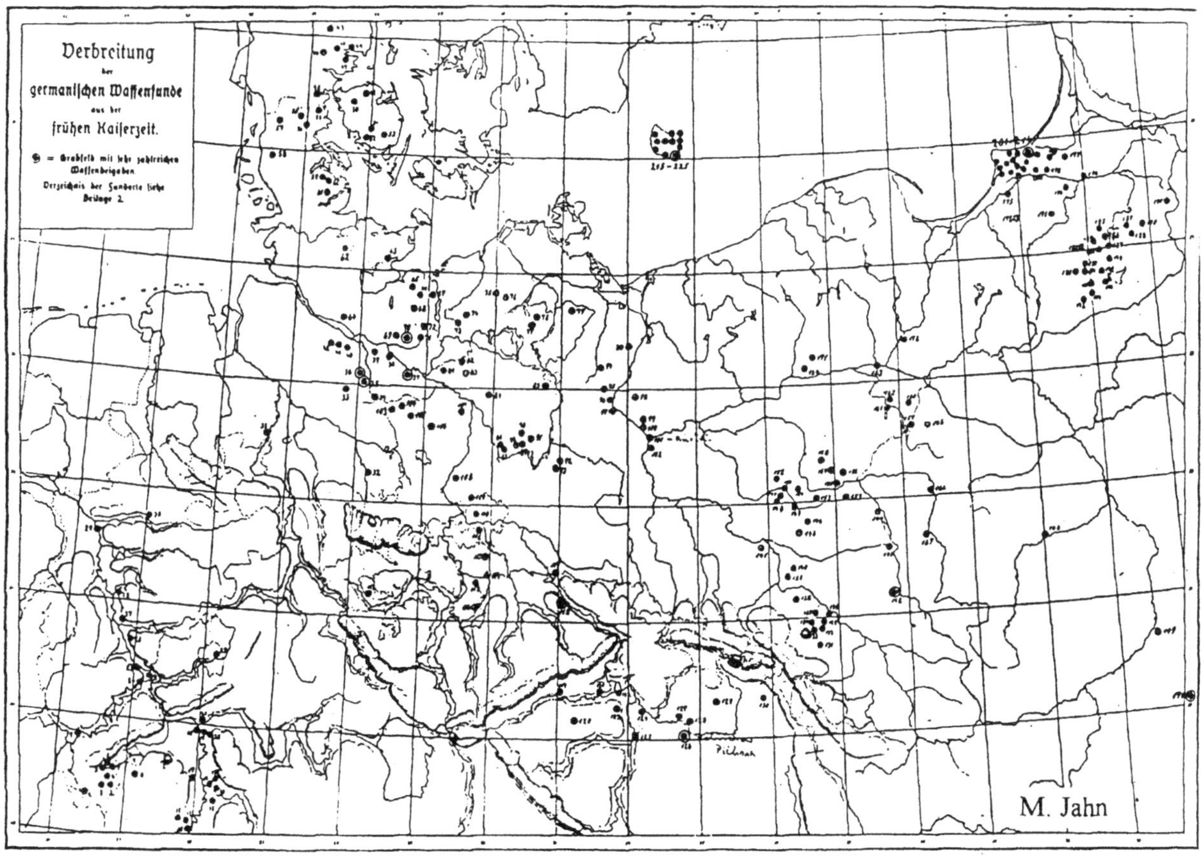

Fig. 29. Barbaricum graves with weapons from the early stage of Roman influence.

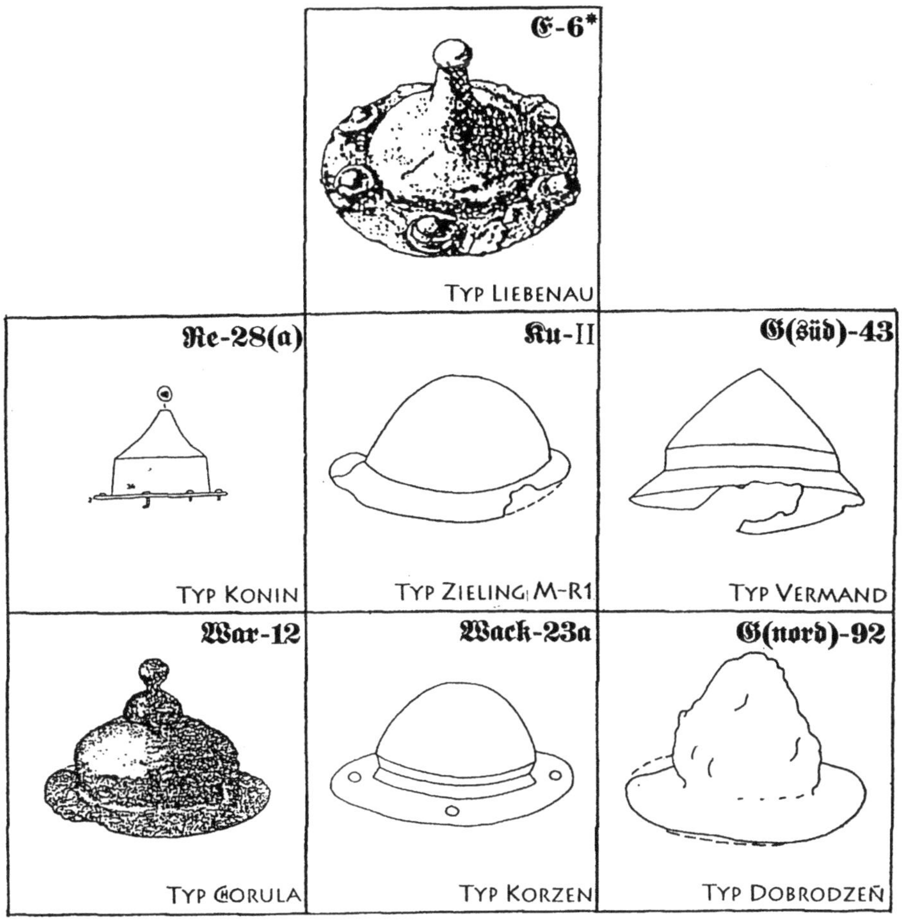

Fig. 30. Types of shield-boss.

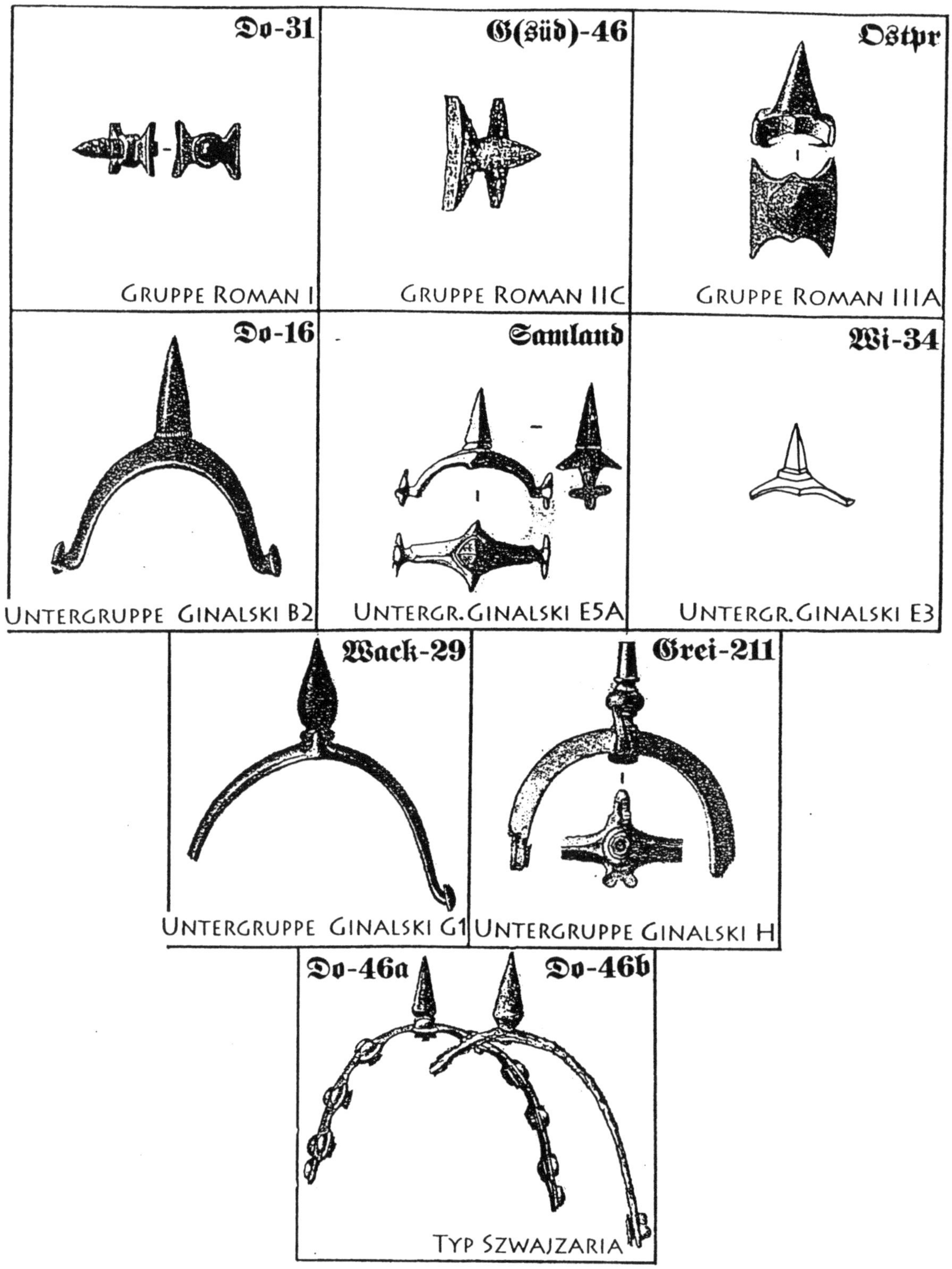

Fig. 31. Types of spur.

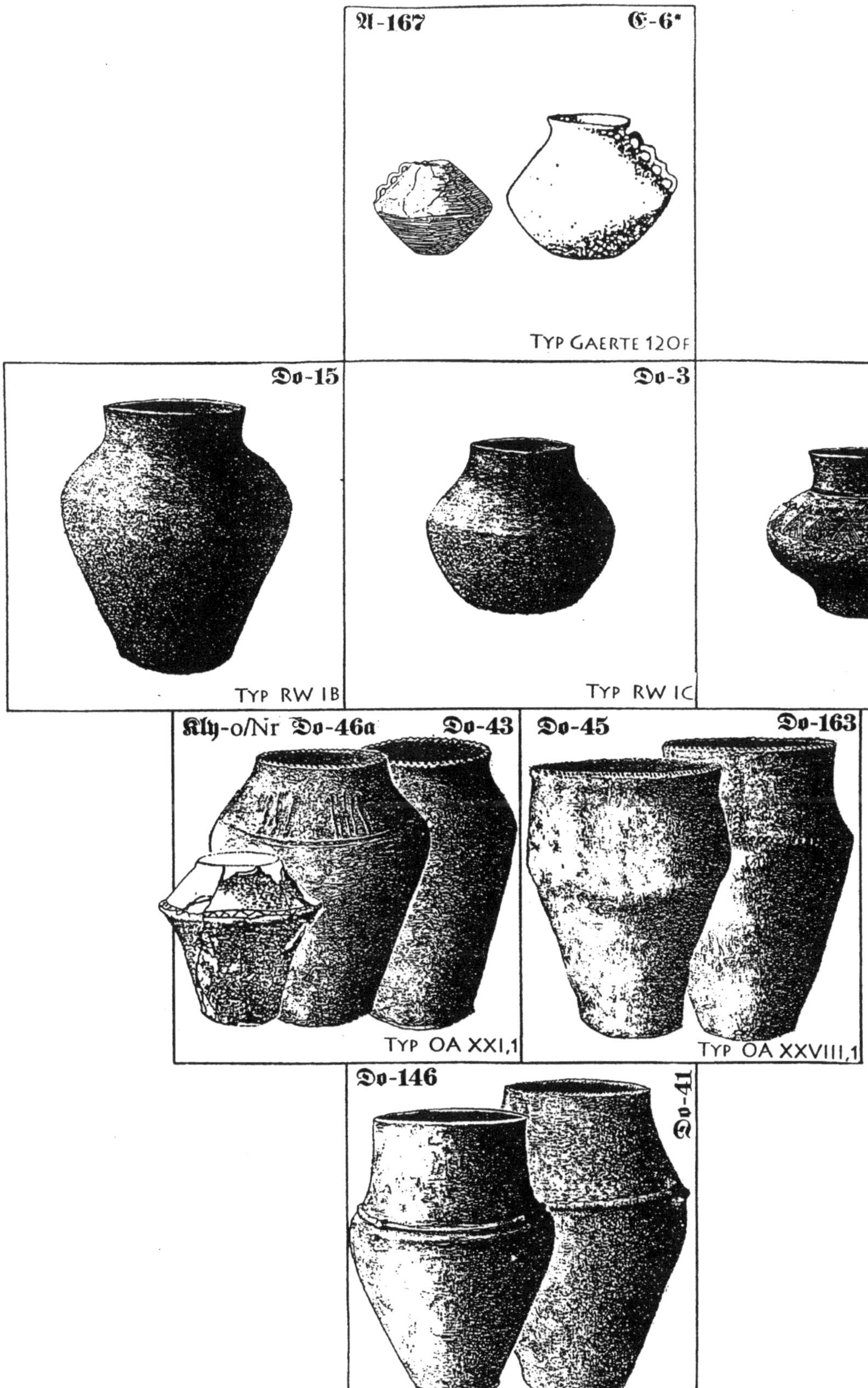

Fig. 32. Types of urn.

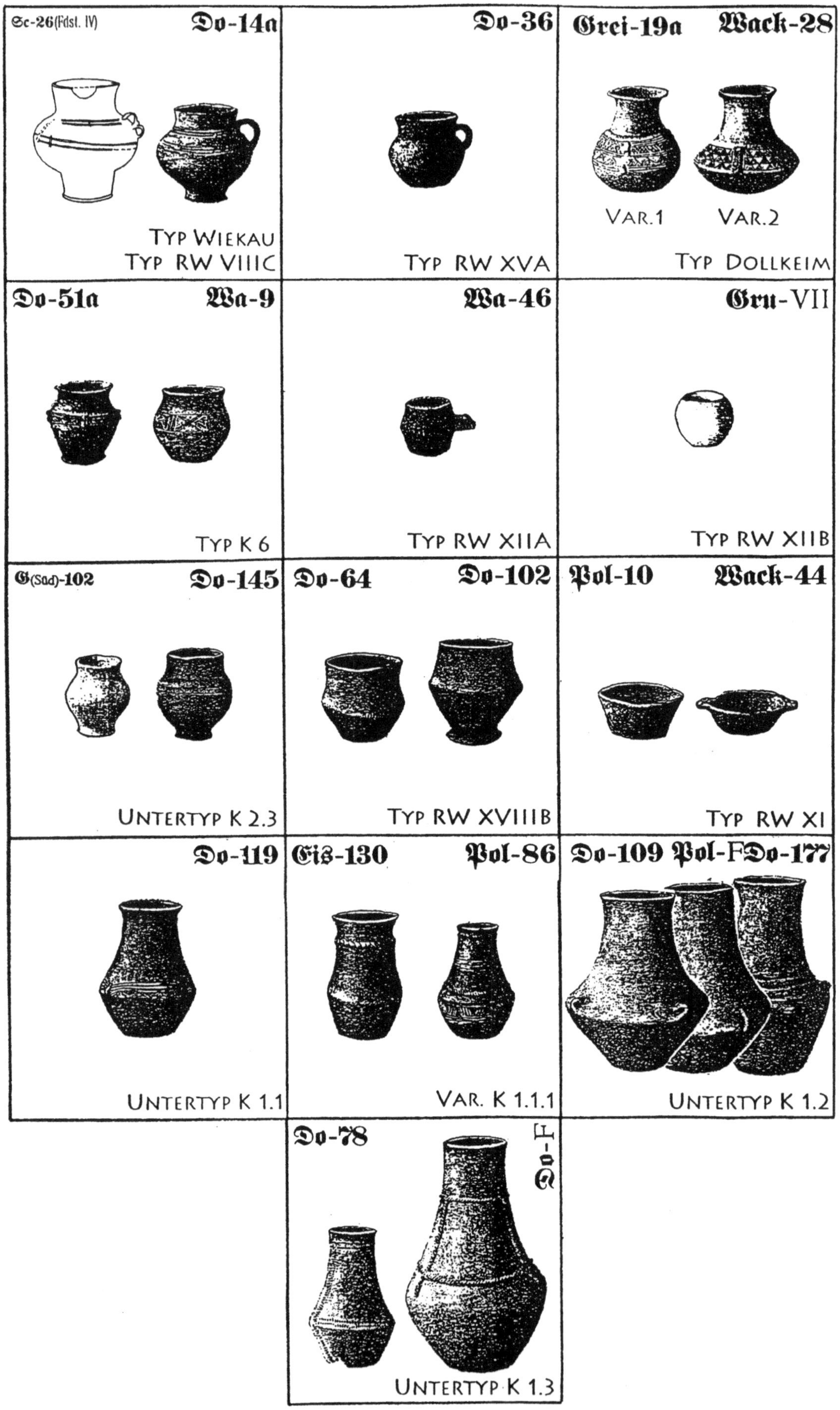

Fig. 33. Types of accessory vessel.

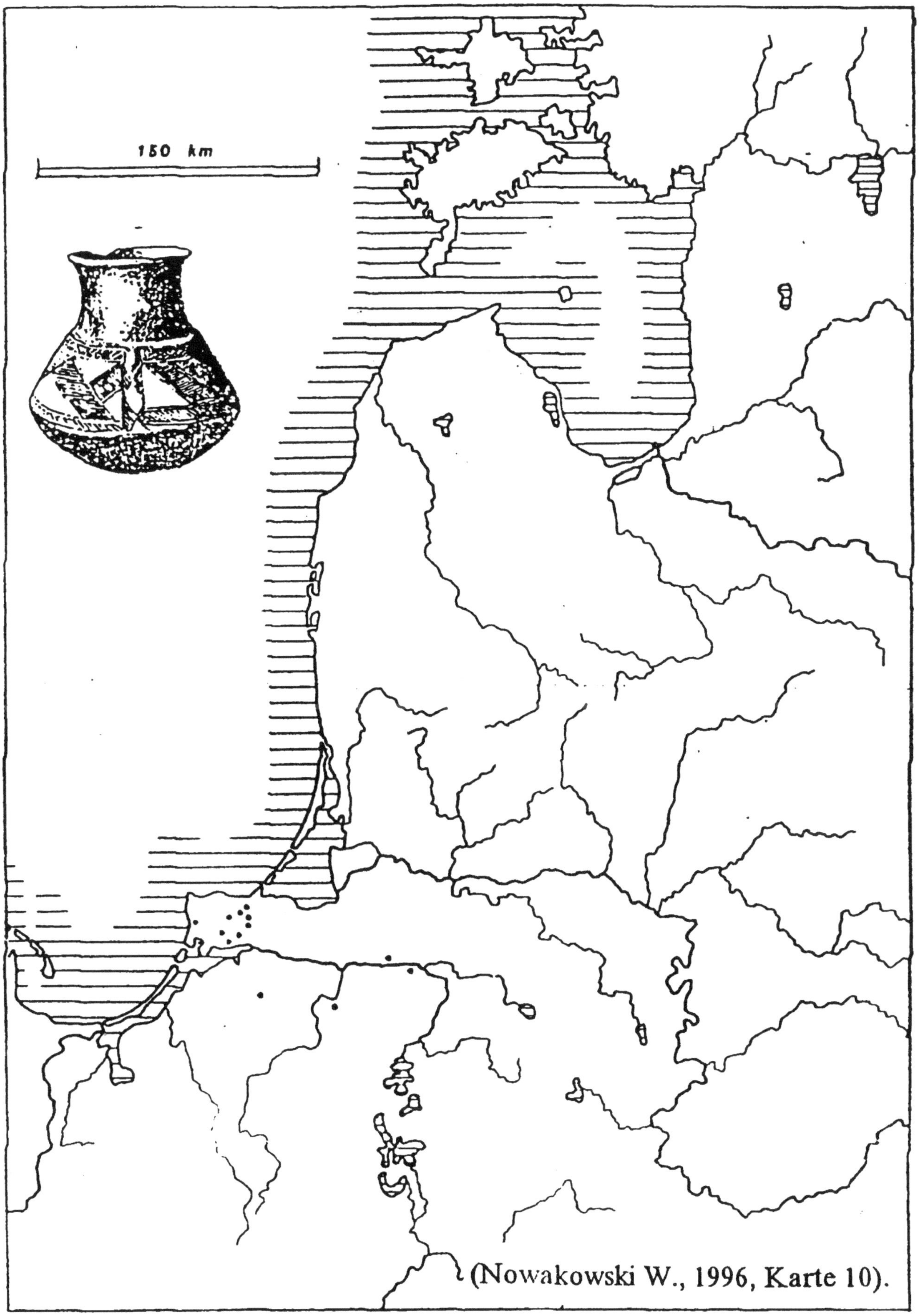

Fig. 34. Distribution of "Typ Dollkeim" vessels.

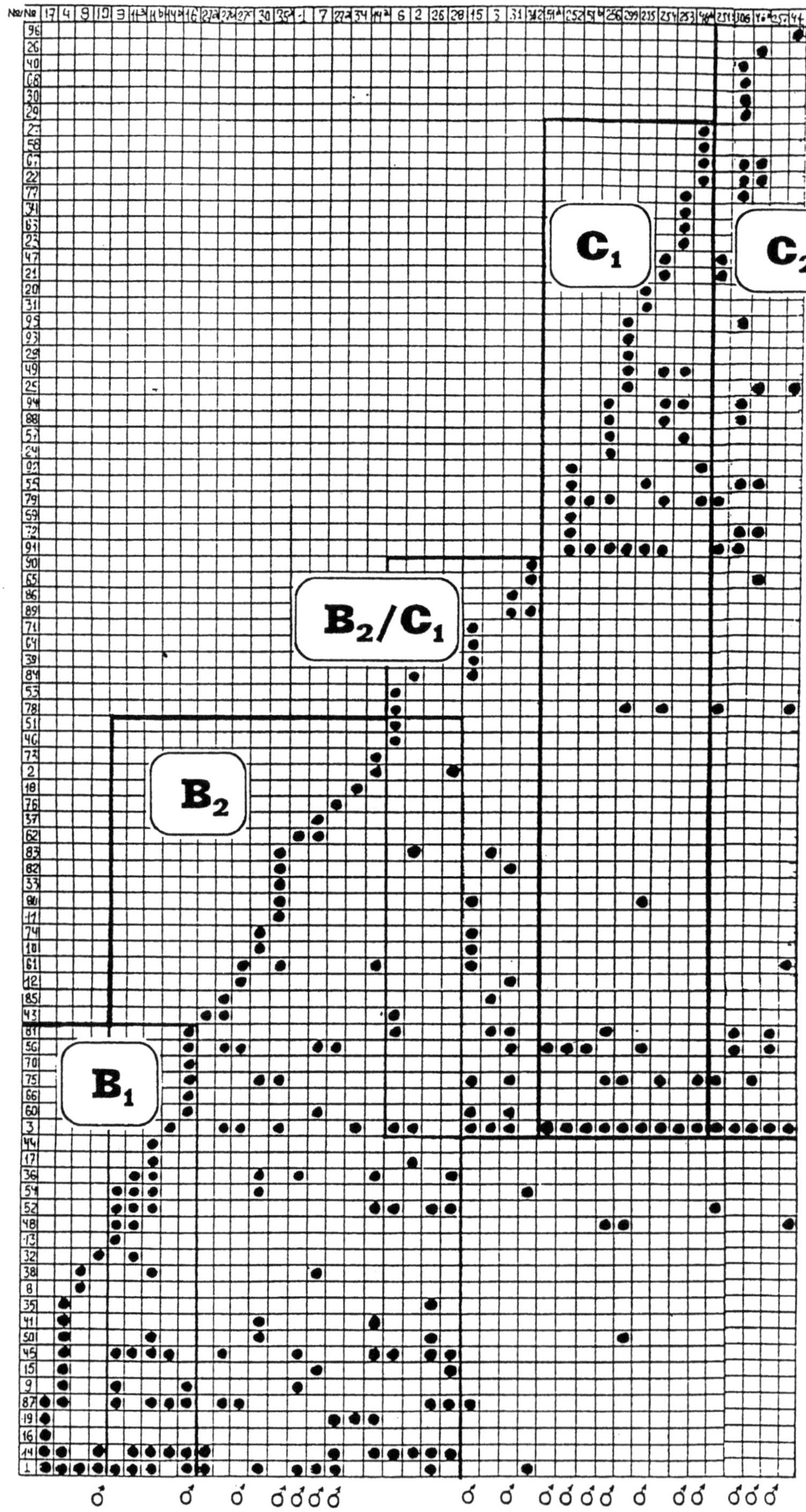

Fig. 35. Dollkeim. A database of the graves of phases B_1-C_2.

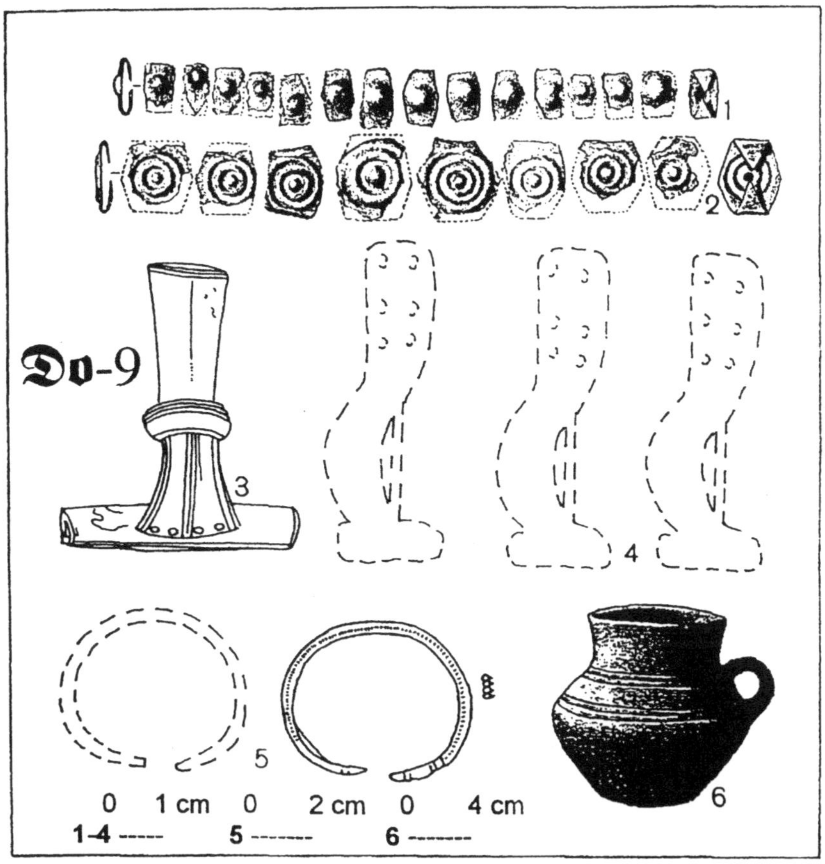

Fig. 36. Assemblage Do-9.

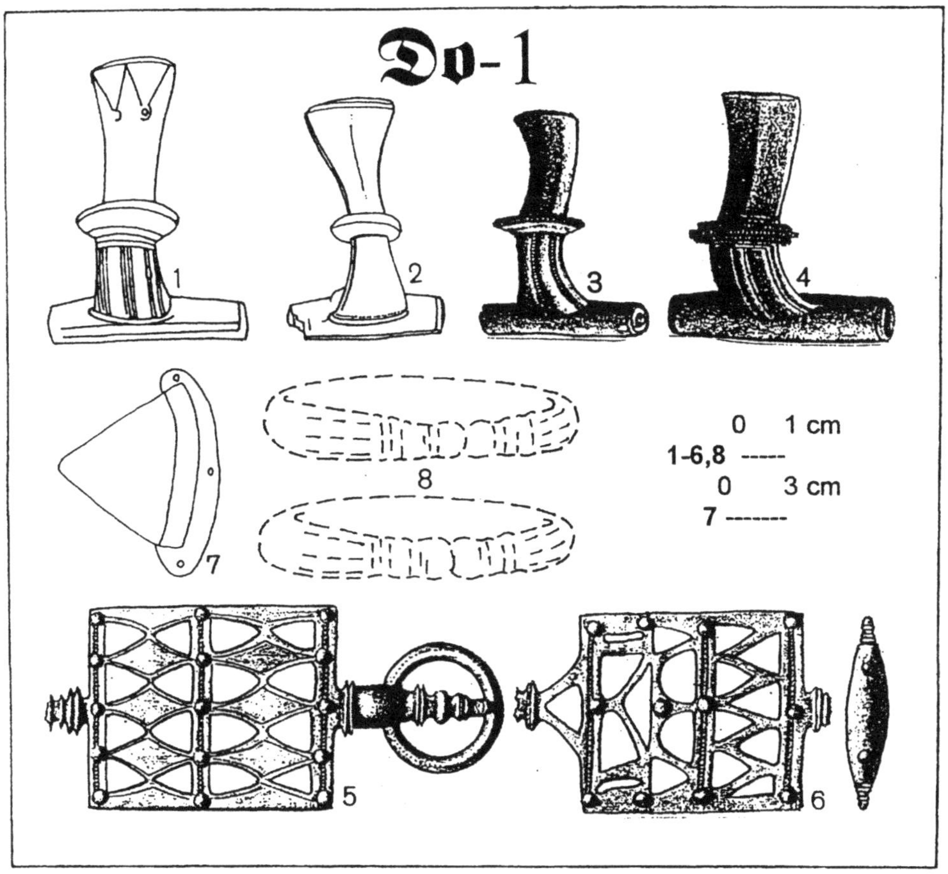

Fig. 37. Assemblage Do-1.

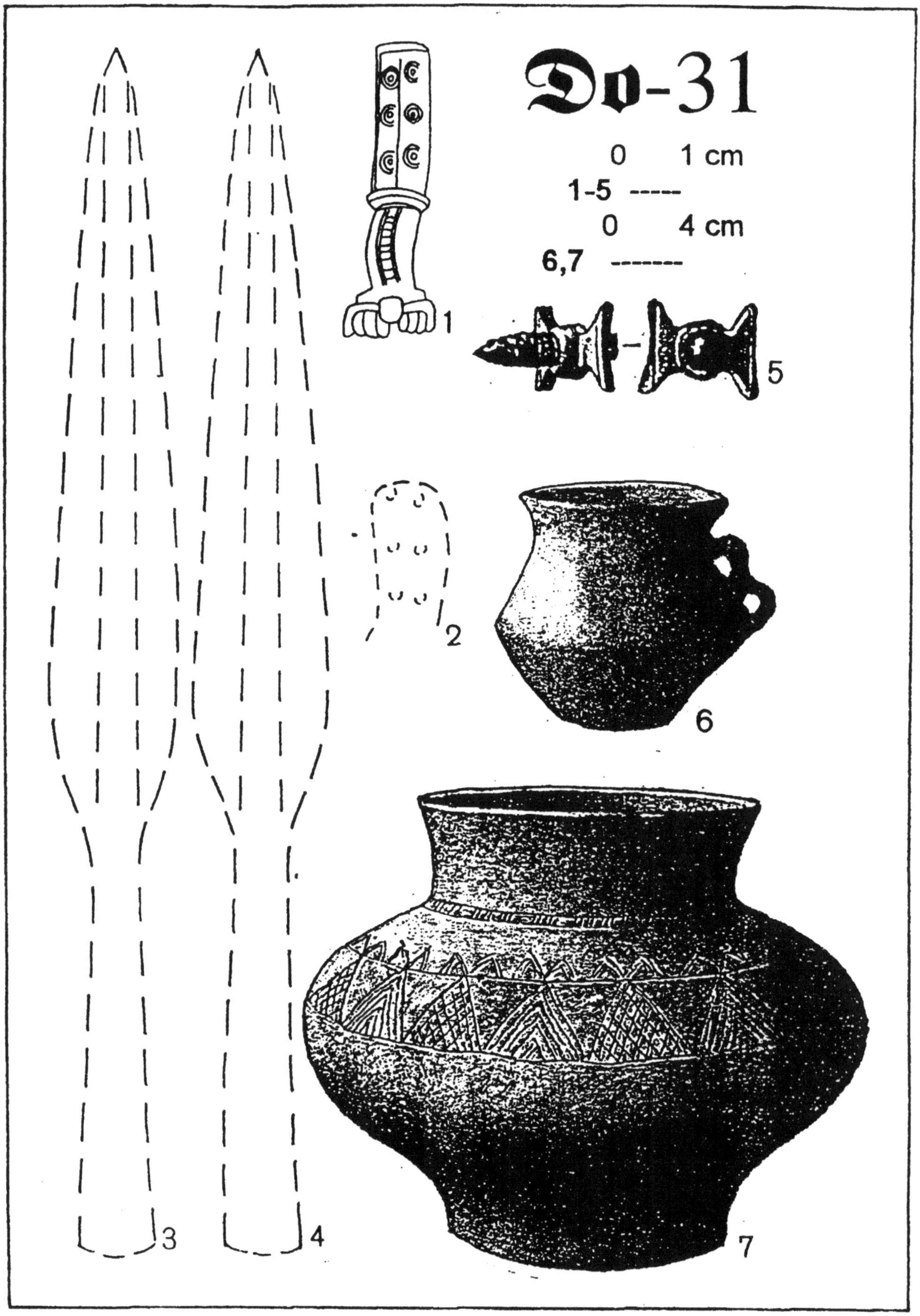

Fig. 38. Assemblage Do-31.

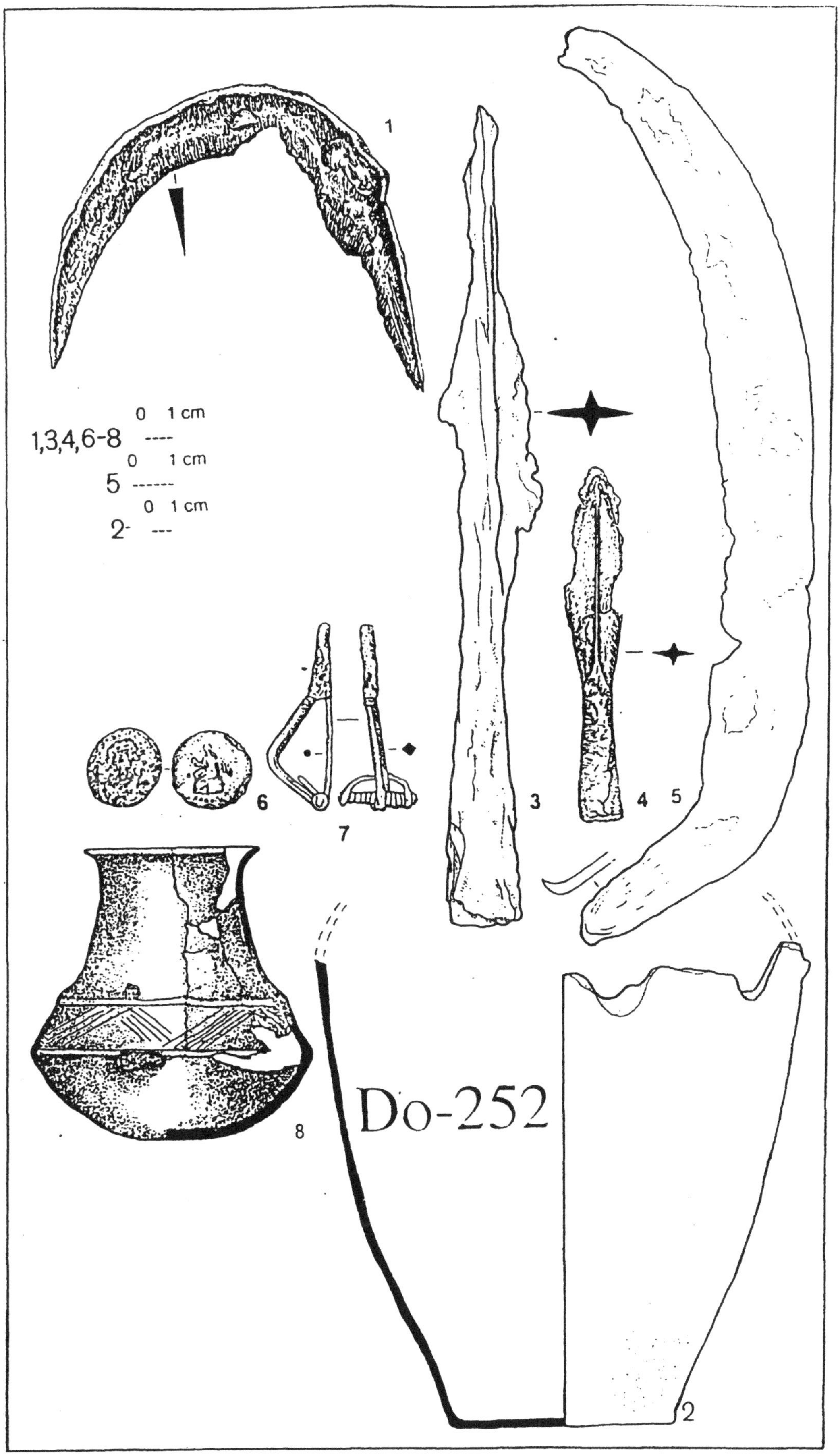

Fig. 39. Assemblage Do-252.

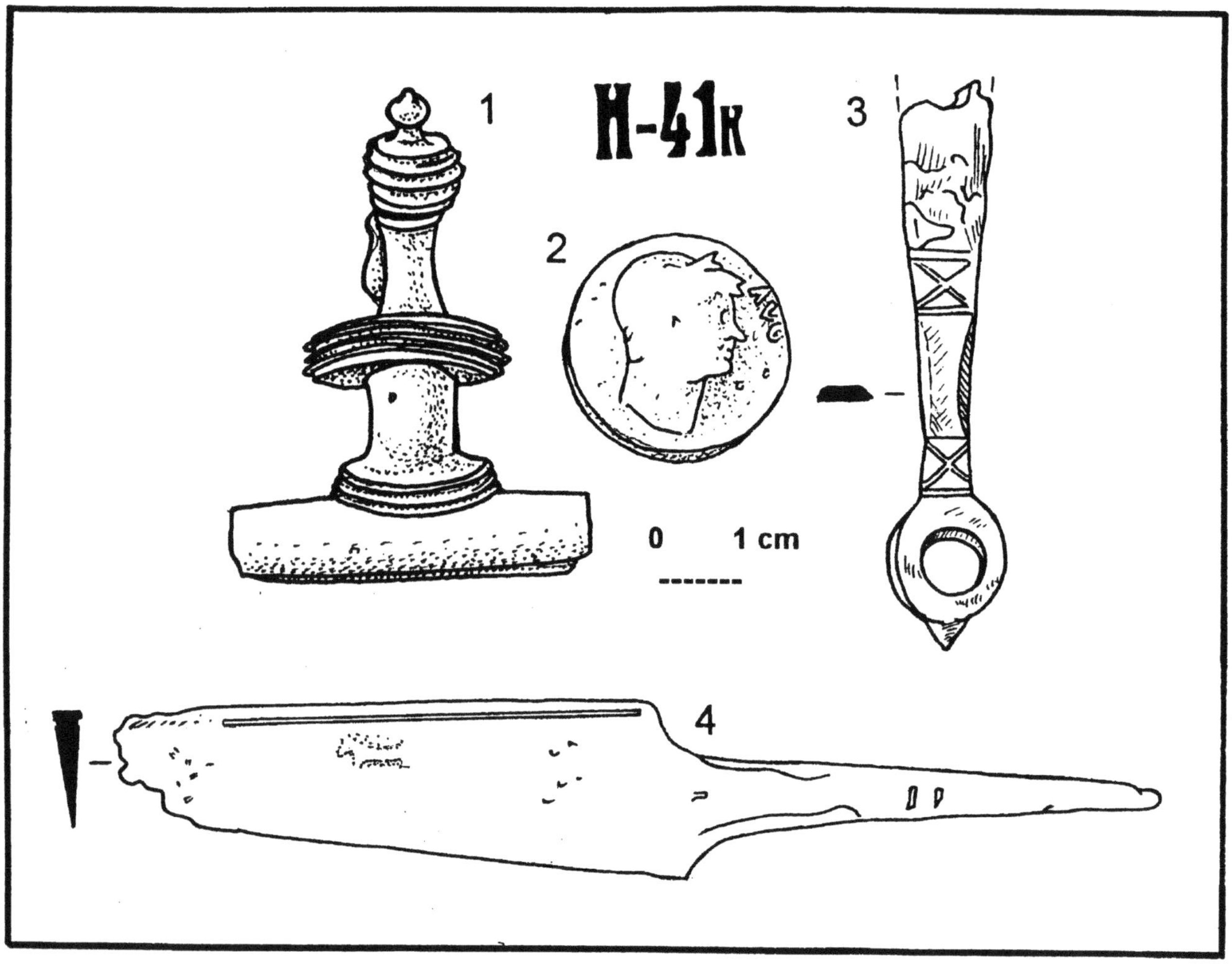

Fig. 40. Assemblage H-41k.

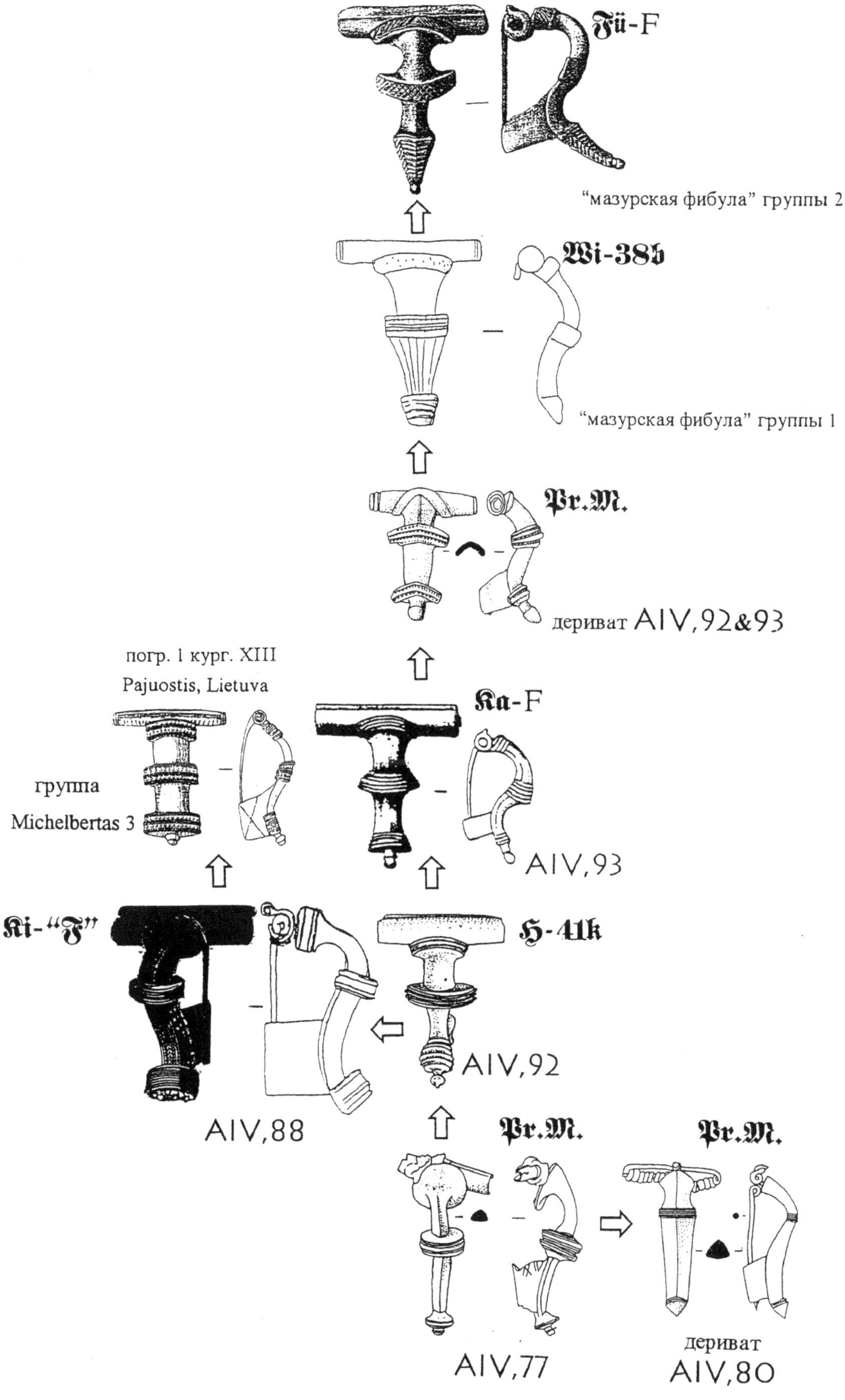

Fig. 41. Origination of "Mazurian type" brooch.

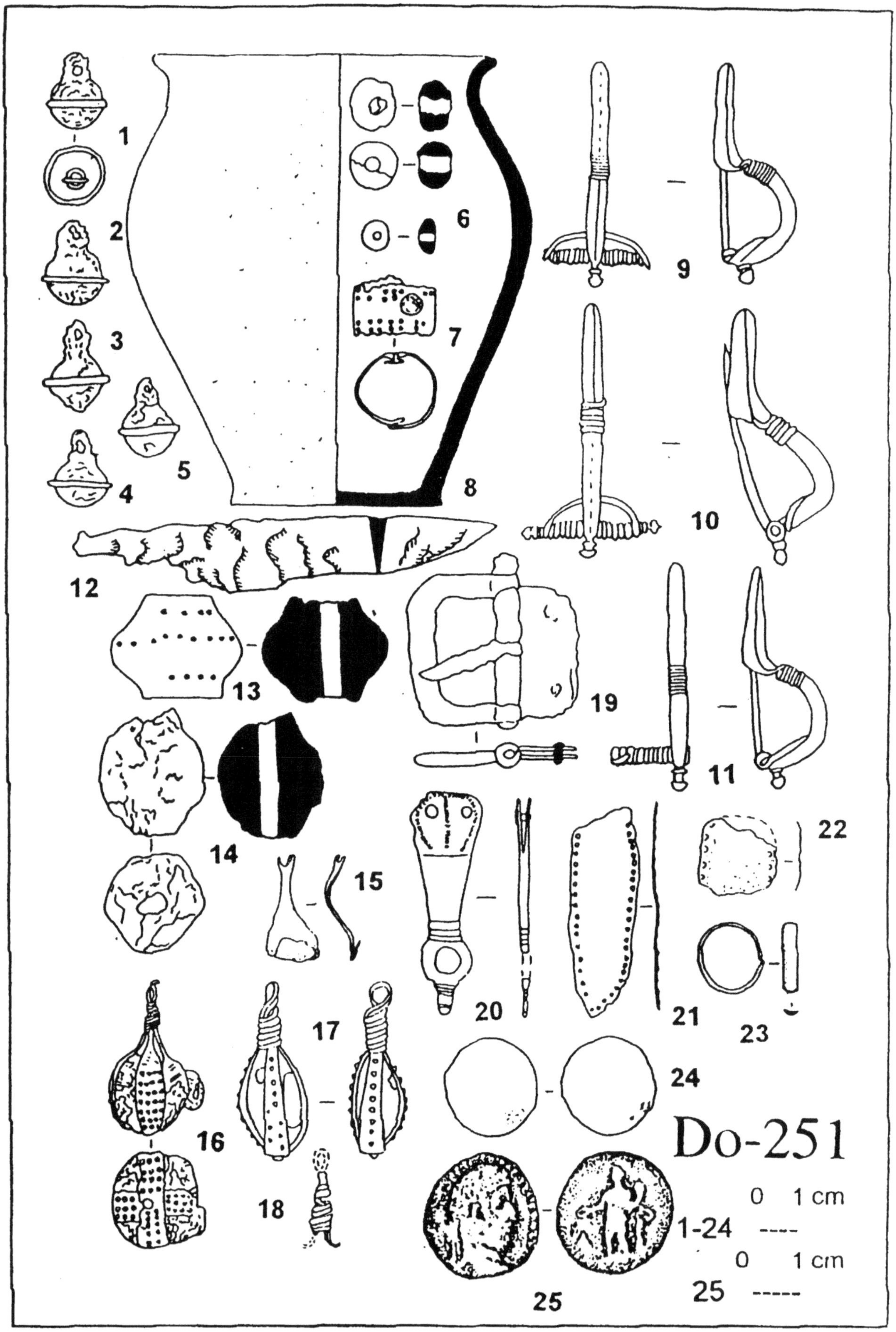

Fig. 42. Assemblage Do-251.

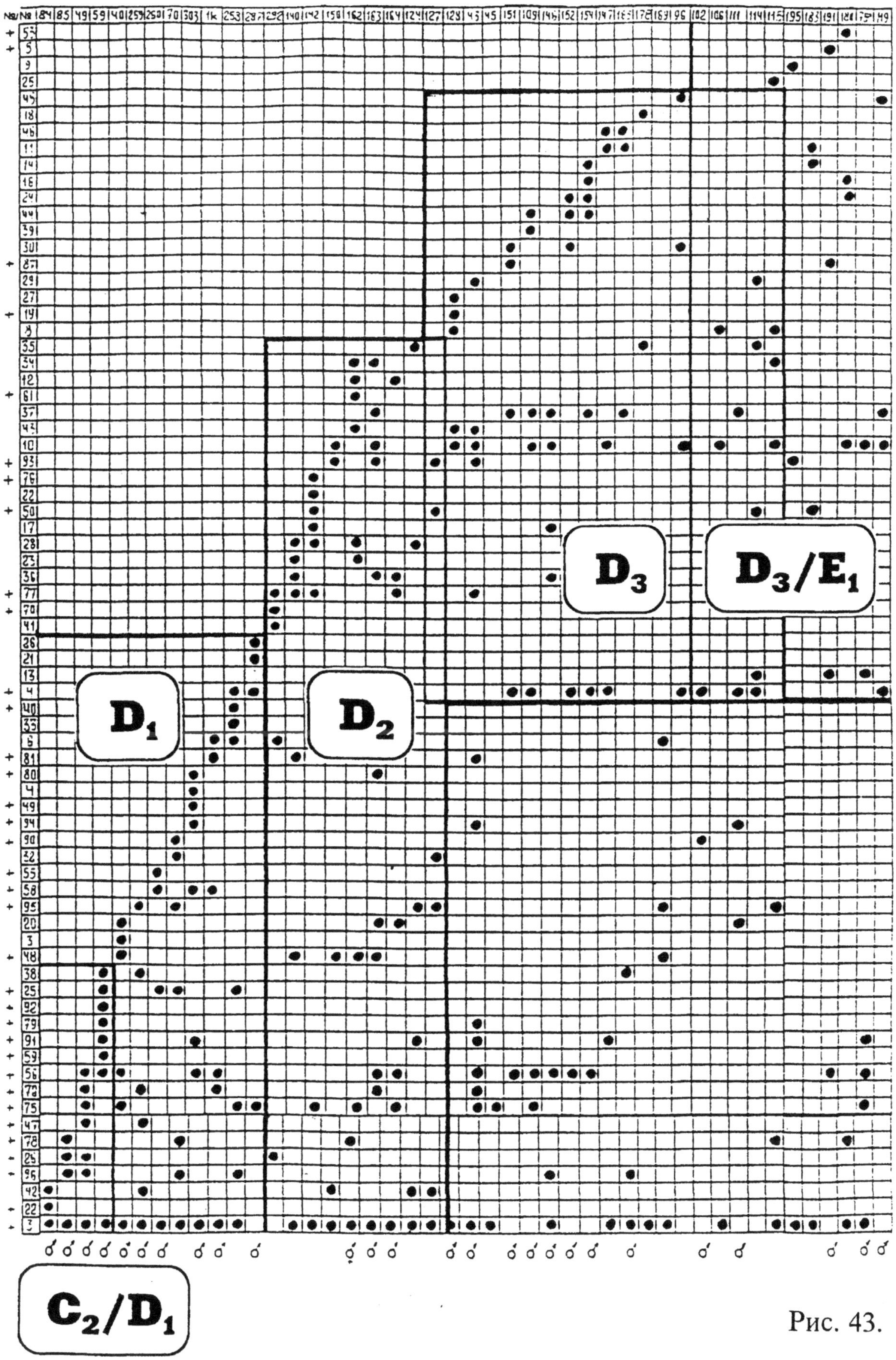

Рис. 43.

Fig. 43. Dollkeim. A database of the graves of phases C_2/D_1-D_3/E_1.

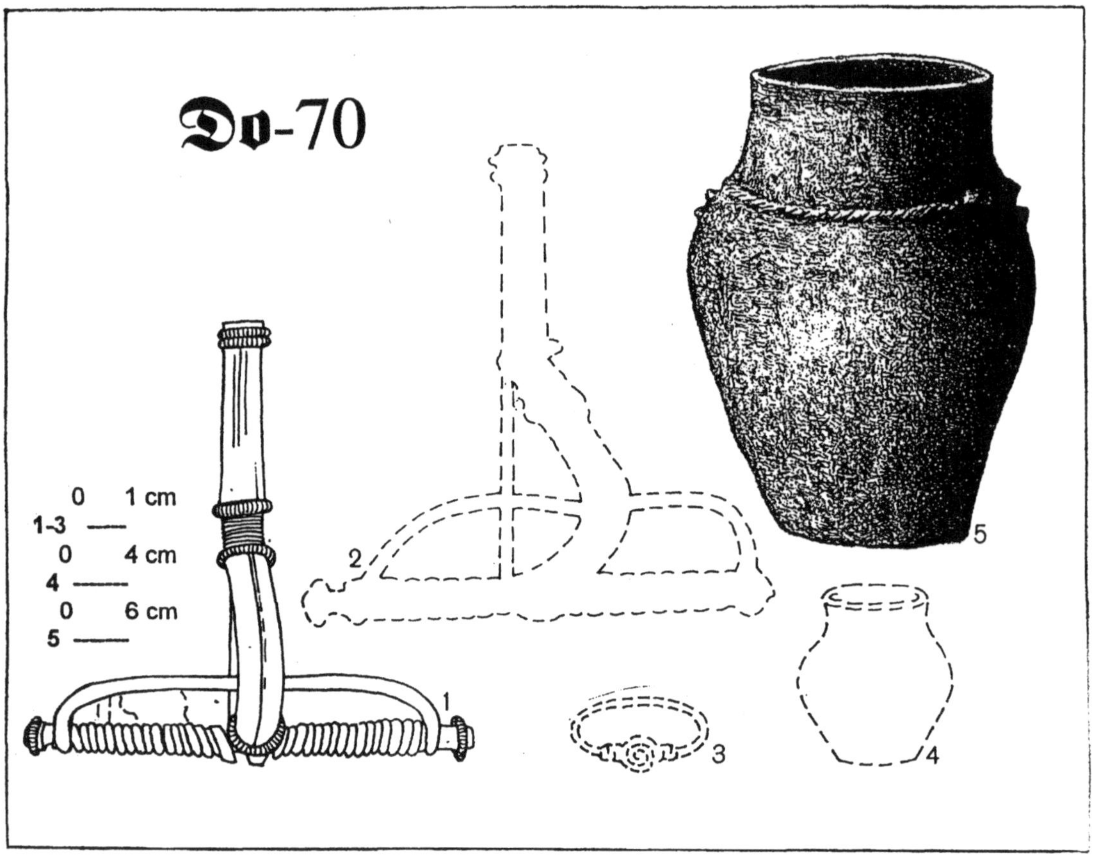

Fig. 44. Assemblage Do-70.

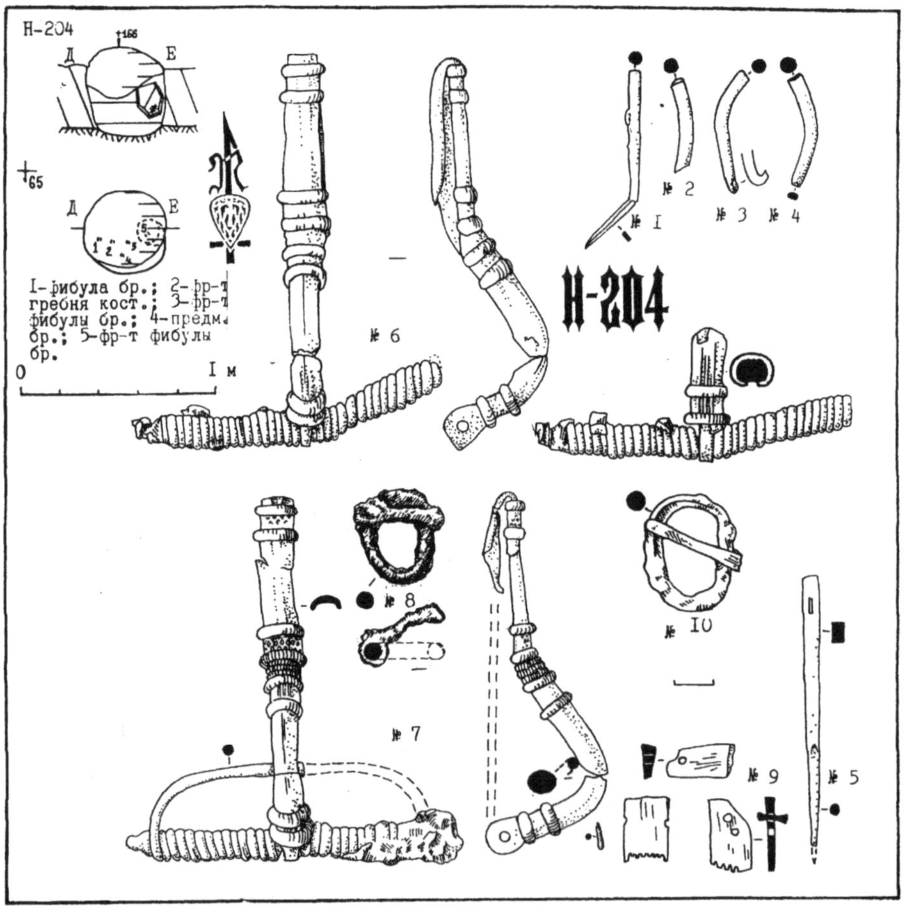

Fig. 45. Assemblage H-204.

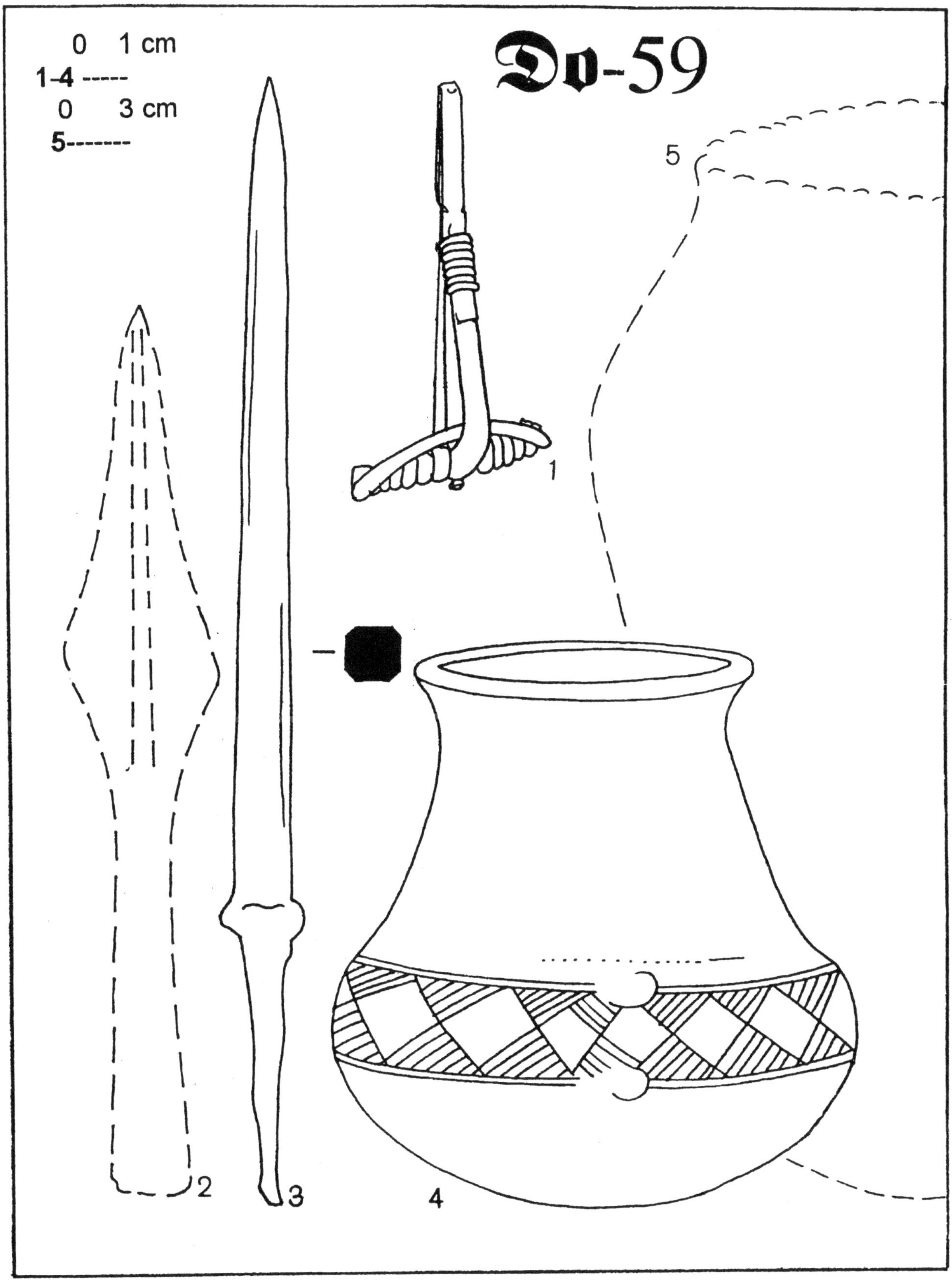

Fig. 46. Assemblage Do-59.

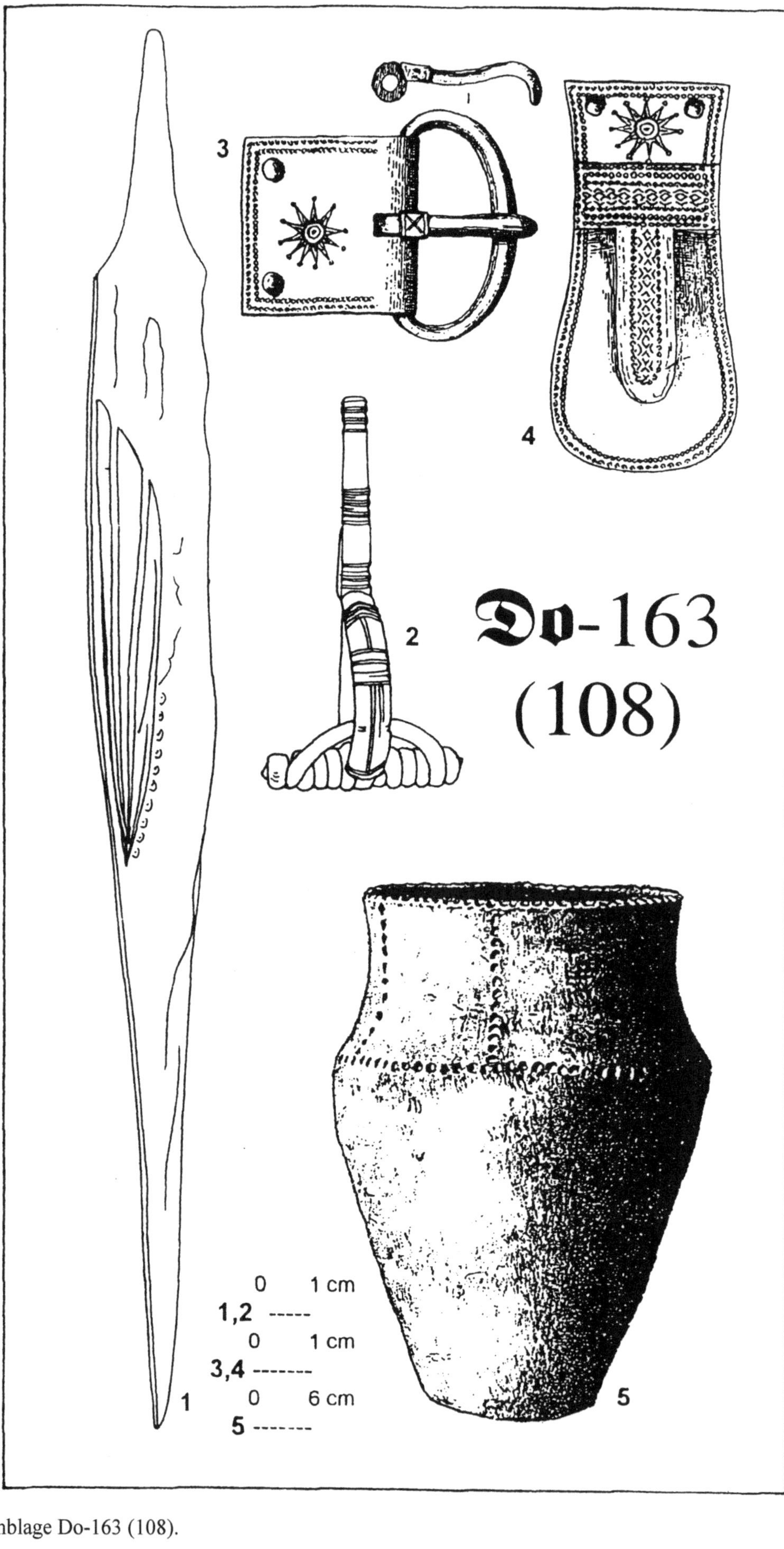

Fig. 47. Assemblage Do-163 (108).

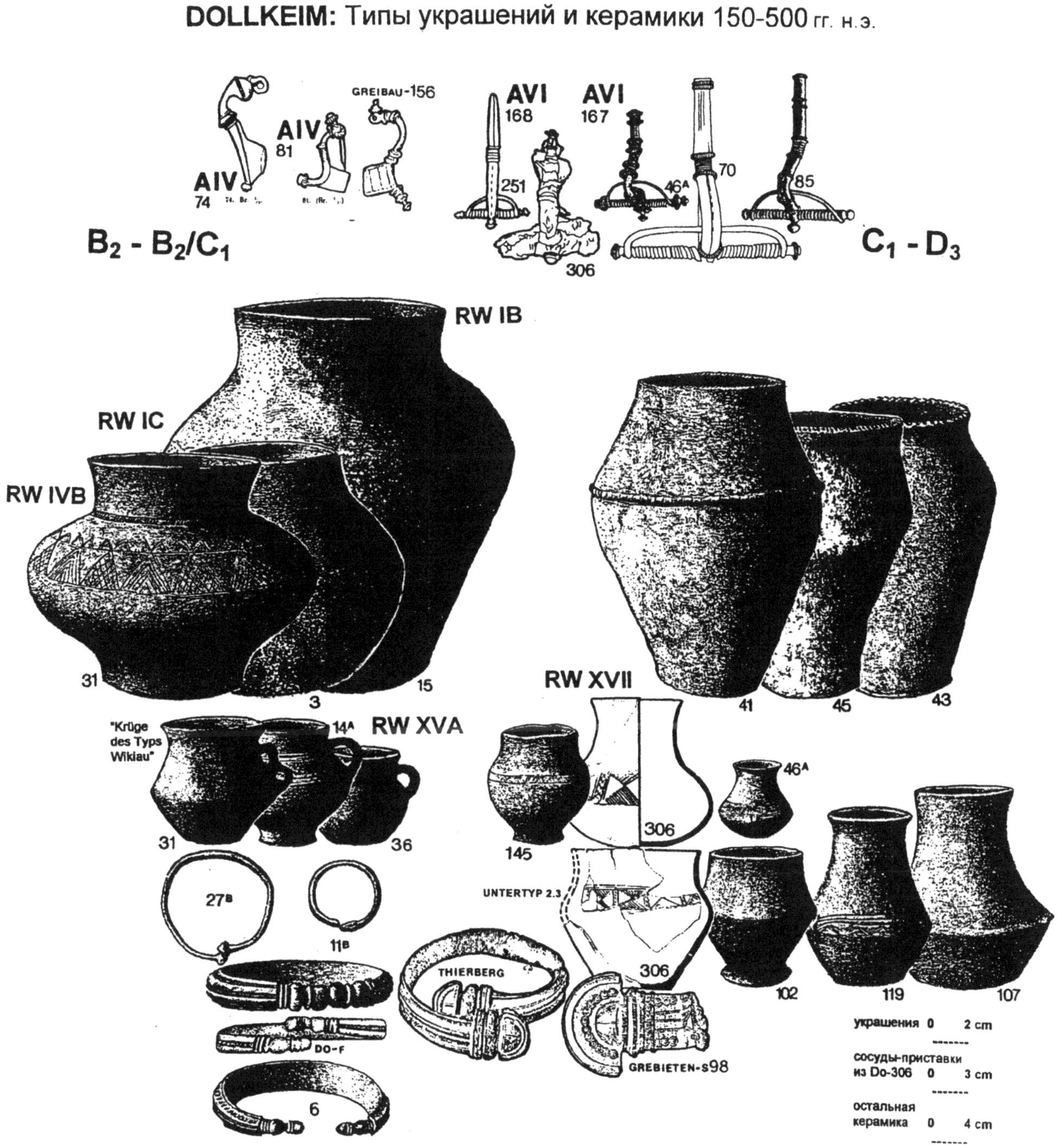

Fig. 48. Dollkeim. Types of jewellery and pottery from 150-500 AD.

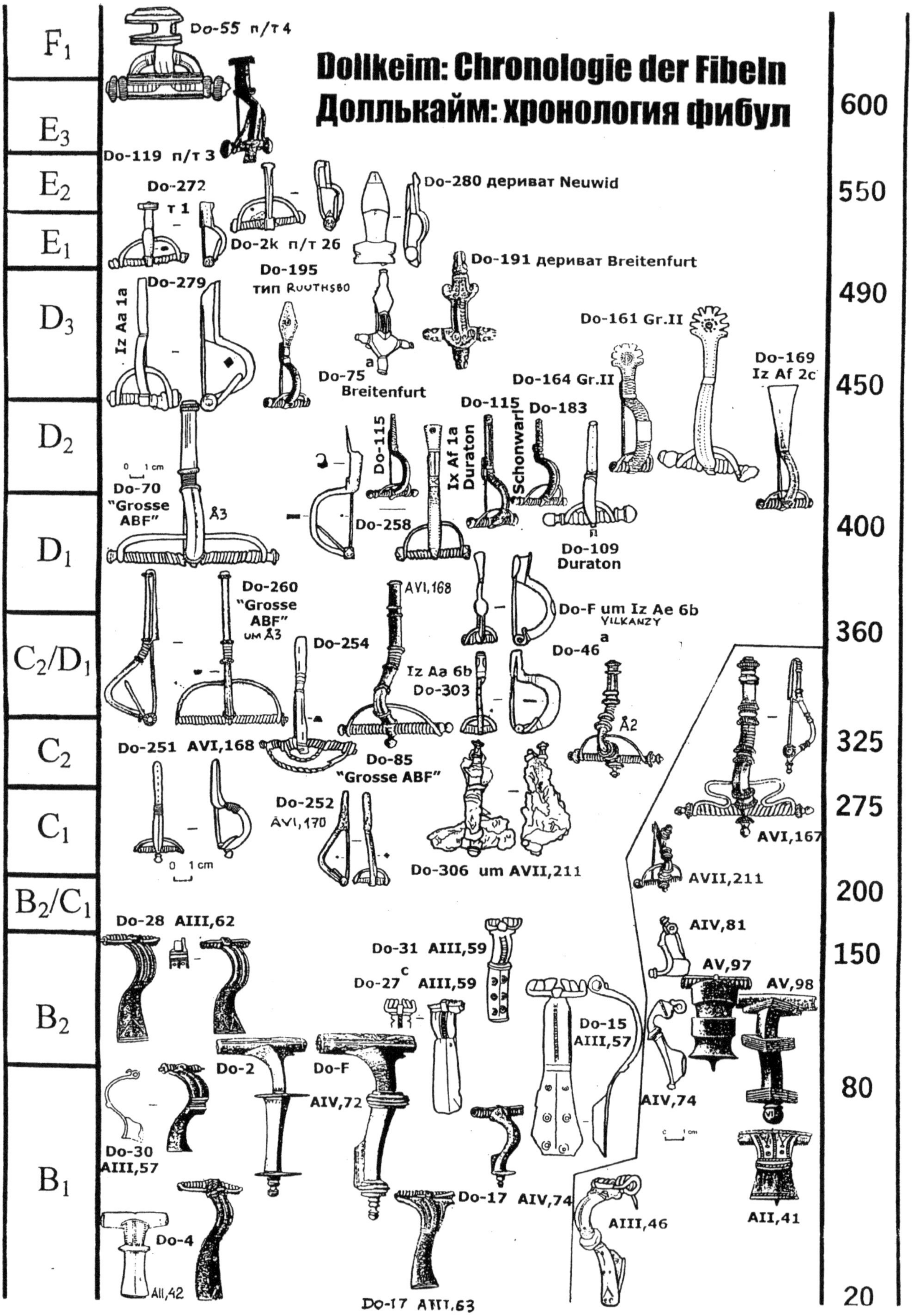

Fig. 49. Dollkeim. Chronology of brooches.

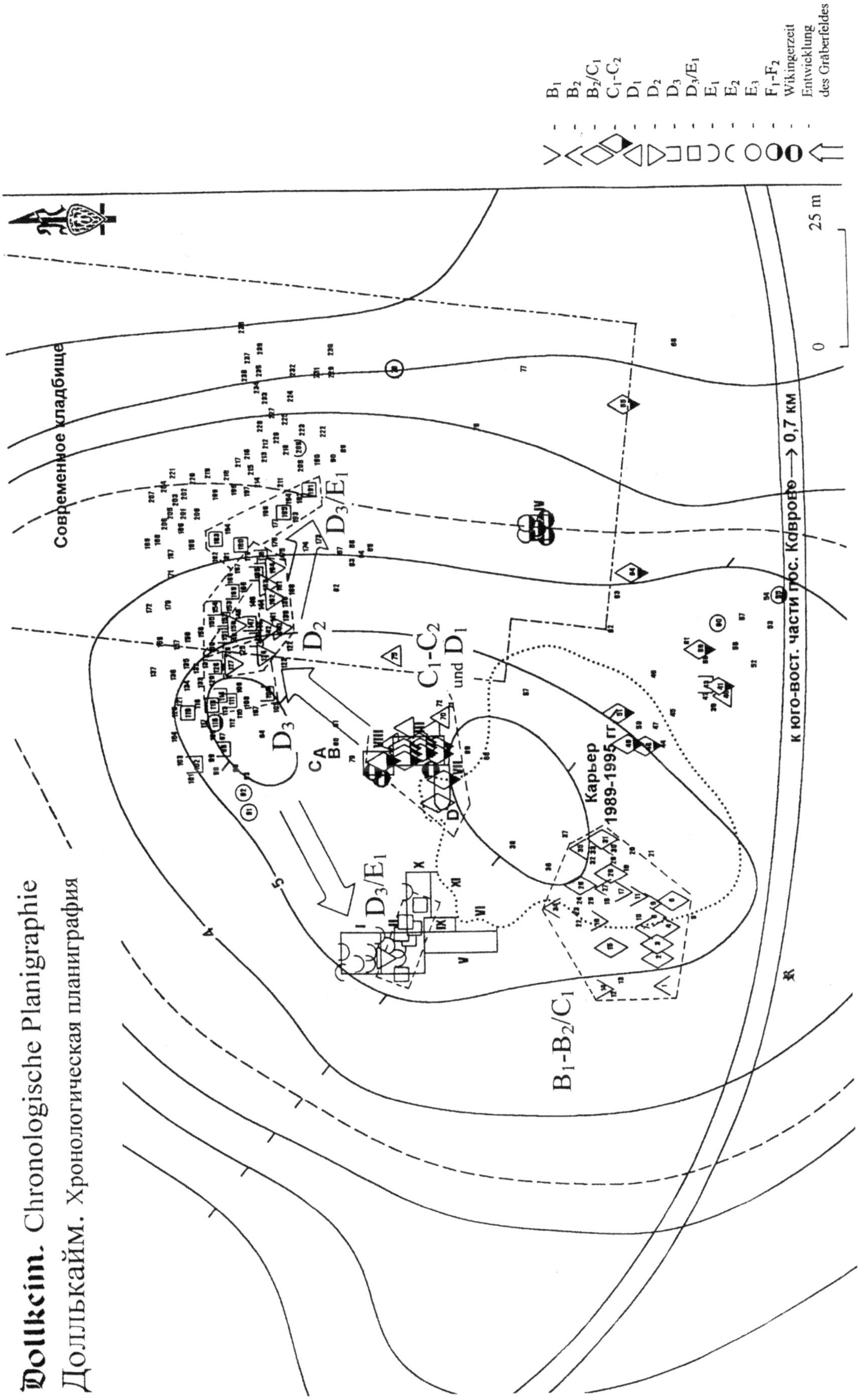

Fig. 50. Dollkeim. Chronological planigraphy.

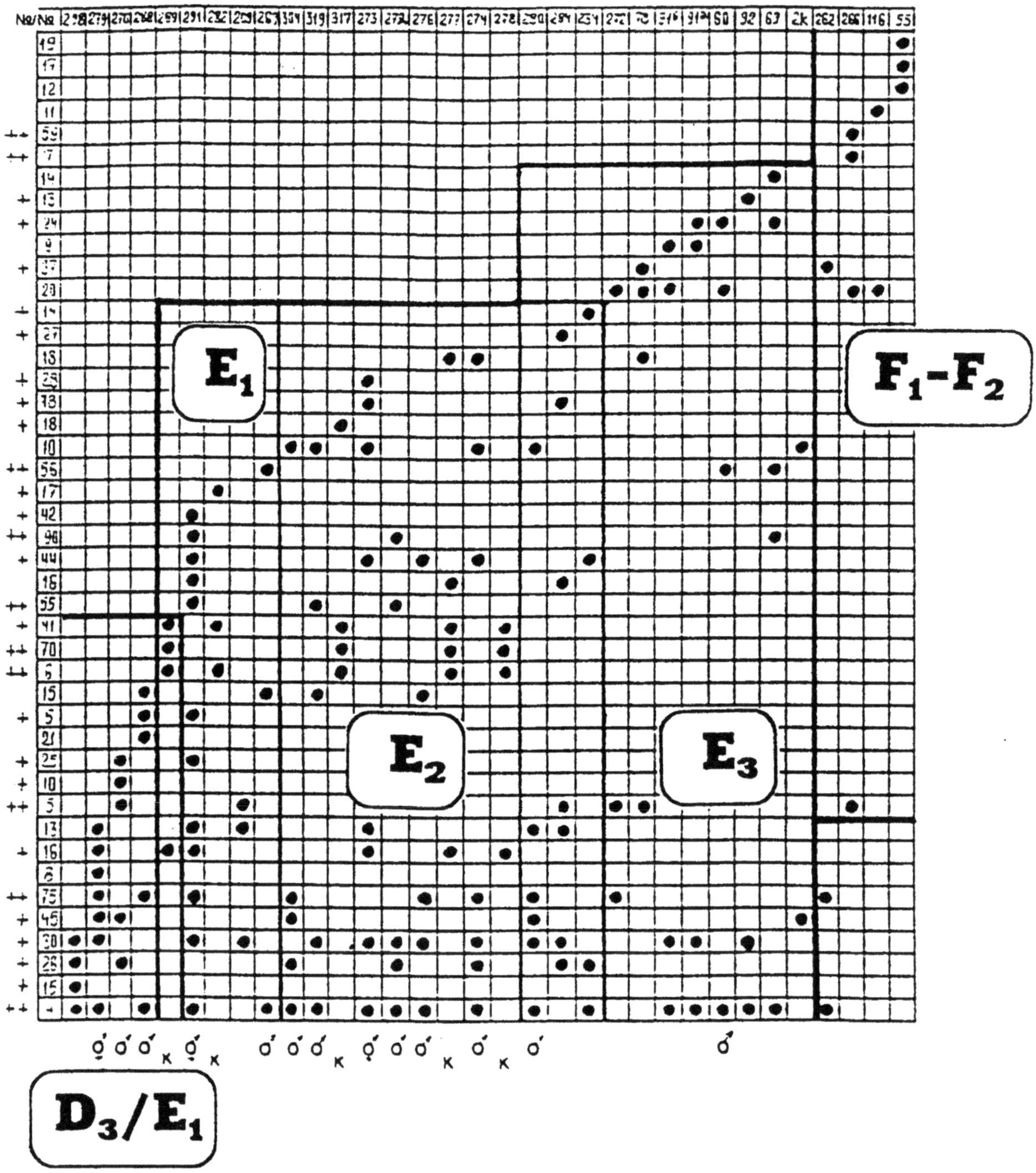

Fig. 51. Dollkeim. A database of the graves of phases D_3/E_1-F_1/F_2.

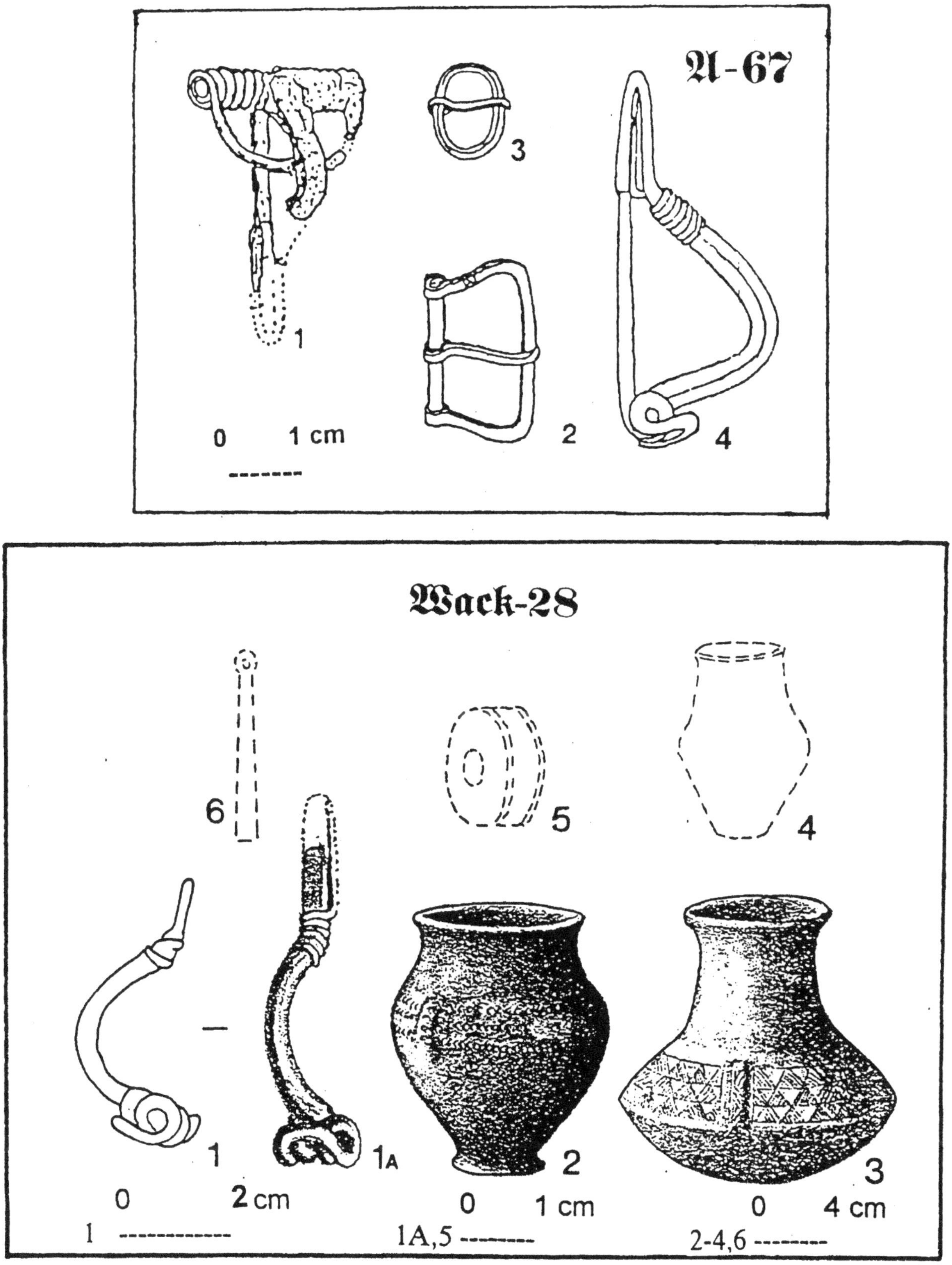

Fig. 52. The assemblage with tied, one-piece brooches: A-67 (CRM /cremation/ under a flat stone) - copper-alloy brooches and buckle (Grunert, 1935, S. 38); Wack-28: 1,1a - copper-alloy brooch, 2-4 - accessory vessels, 5 - amber bead, 6 - fire-steel (part of the object known only by description - Tischler, Kemke, 1902, S. 39).

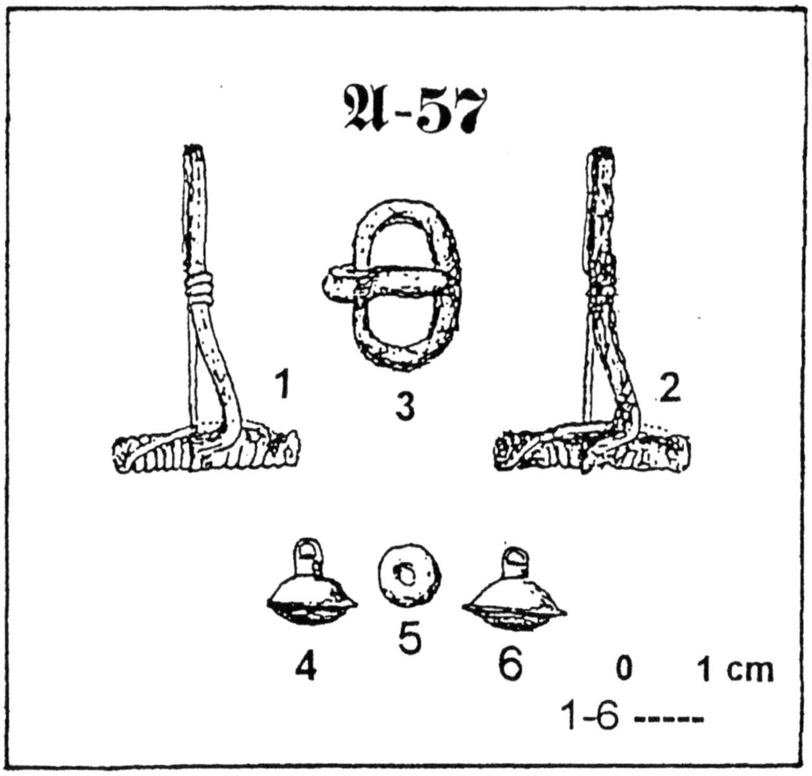

Fig. 53. Pair of iron, tied, two-piece brooches (1, 2) with an iron buckle (3), iron bell pendants (4, 6) and blue glass bead in without-urn CRM A-57 (Grunert, 1935, S. 39, Abb. 55).

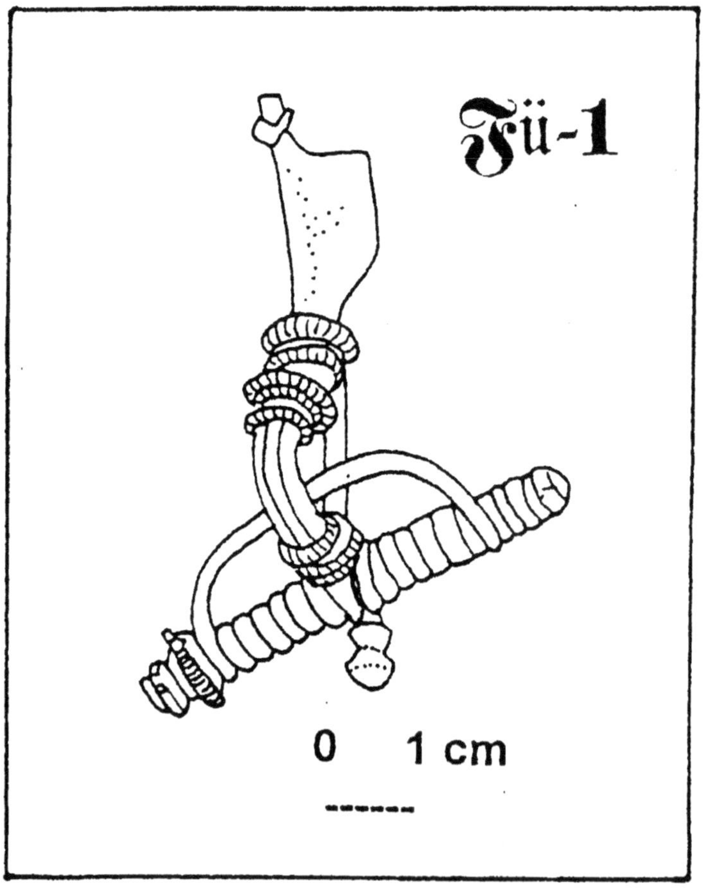

Fig. 55. Copper-alloy brooch of type AVII, 211 from Fü-1 (CRM as group of bones) with: copper-alloy needle, iron awl, clay spindle whorl, glass bead, Trajan coin (98-117 AD). Below CRM - skeleton of horse with three-piece bit (Kemke, 1901, S. 93).

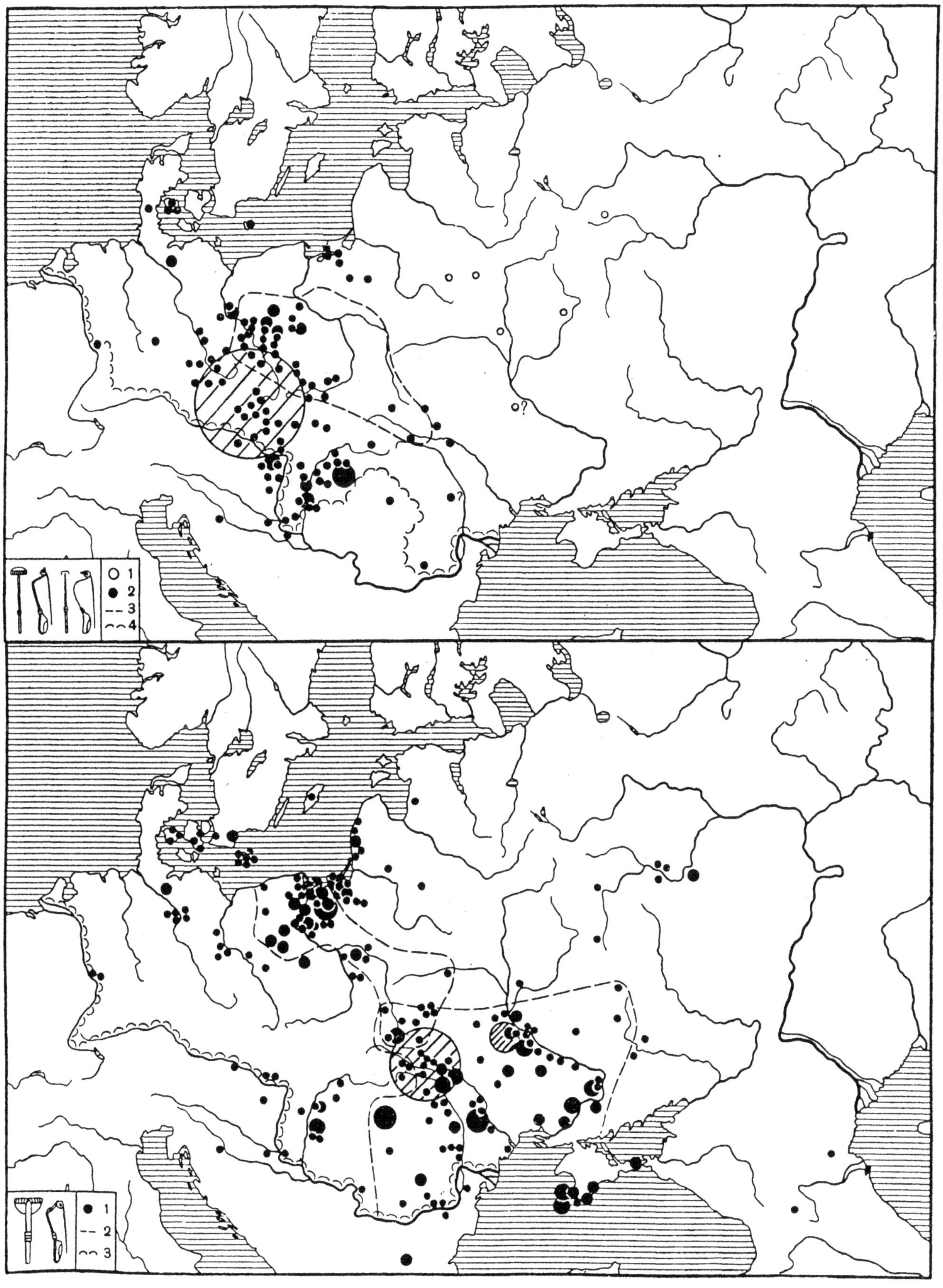

Fig. 54. Tied brooches in Barbaricum: 1 - finds of one-piece brooches with a narrow leg (1 - Upper Dnieper series, 2 - Middle Dnieper series, 3 - border of Przeworsk area, 4 - Roman *limes* in the 3rd cent. AD); 2 - finds of two-piece brooches of series AI (1) in Wielbark and Chernyakhov areas (2) to the northeast of the Roman *limes* of the 4th cent. (3) (Ambroz, 1966, tab. 23 with additions by the author).

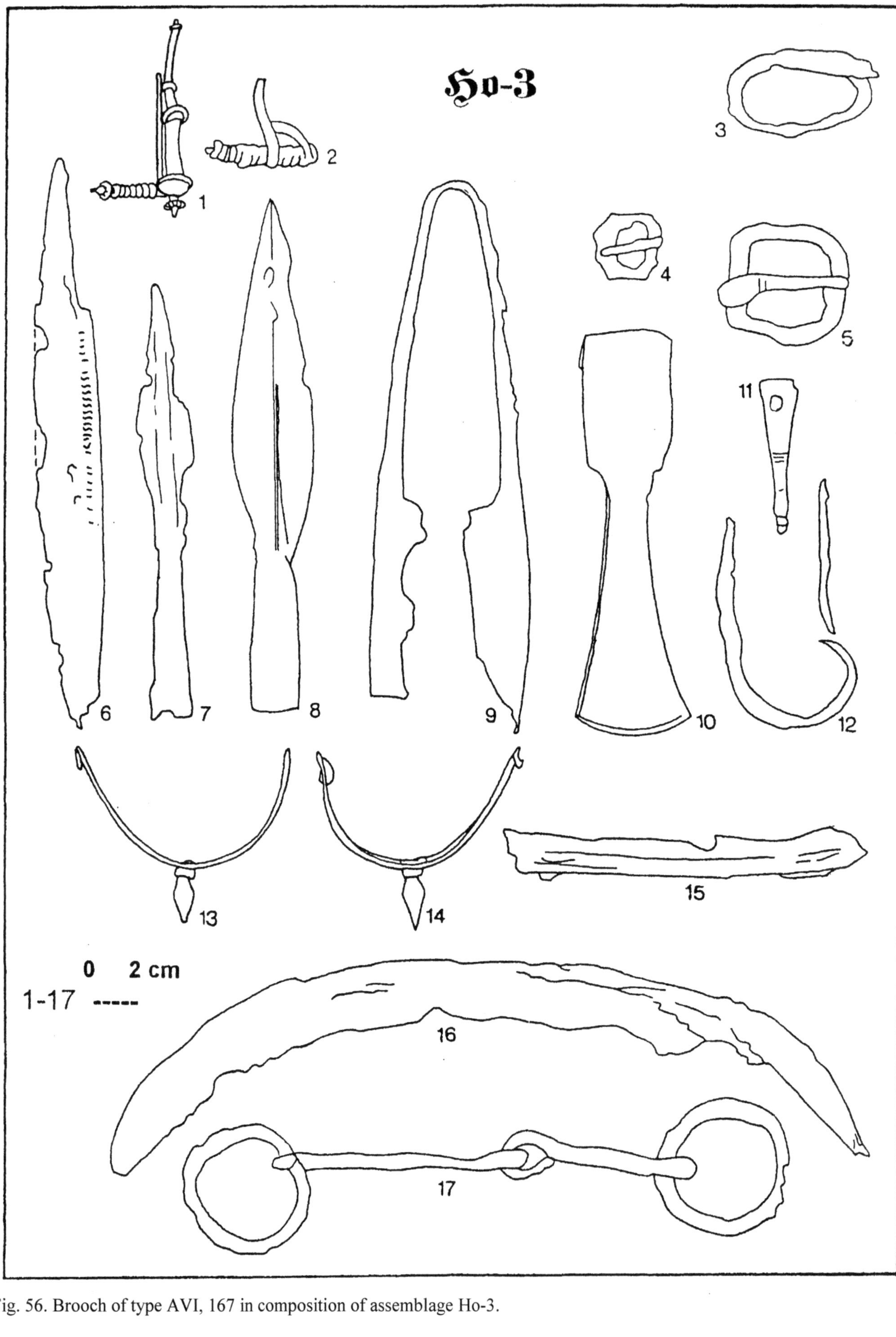

Fig. 56. Brooch of type AVI, 167 in composition of assemblage Ho-3.

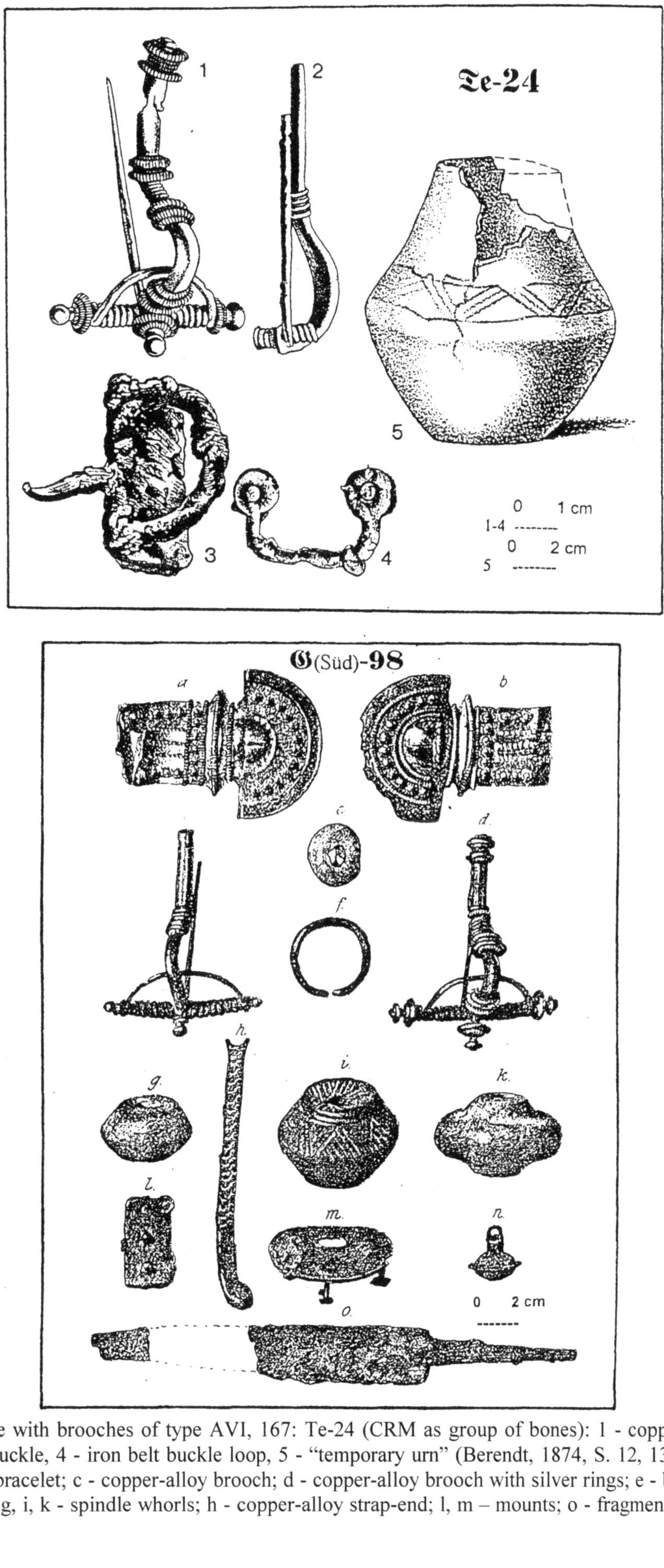

Fig. 57. Assemblage with brooches of type AVI, 167: Te-24 (CRM as group of bones): 1 - copper-alloy and 2 - iron brooches, 3 - iron buckle, 4 - iron belt buckle loop, 5 - "temporary urn" (Berendt, 1874, S. 12, 13); G (Süd)-98: a, b - fragments of silver bracelet; c - copper-alloy brooch; d - copper-alloy brooch with silver rings; e - bead of green glass; f - copper-alloy ring; g, i, k - spindle whorls; h - copper-alloy strap-end; l, m – mounts; o - fragments of knife (Heydeck, 1886/87, Taf. VII).

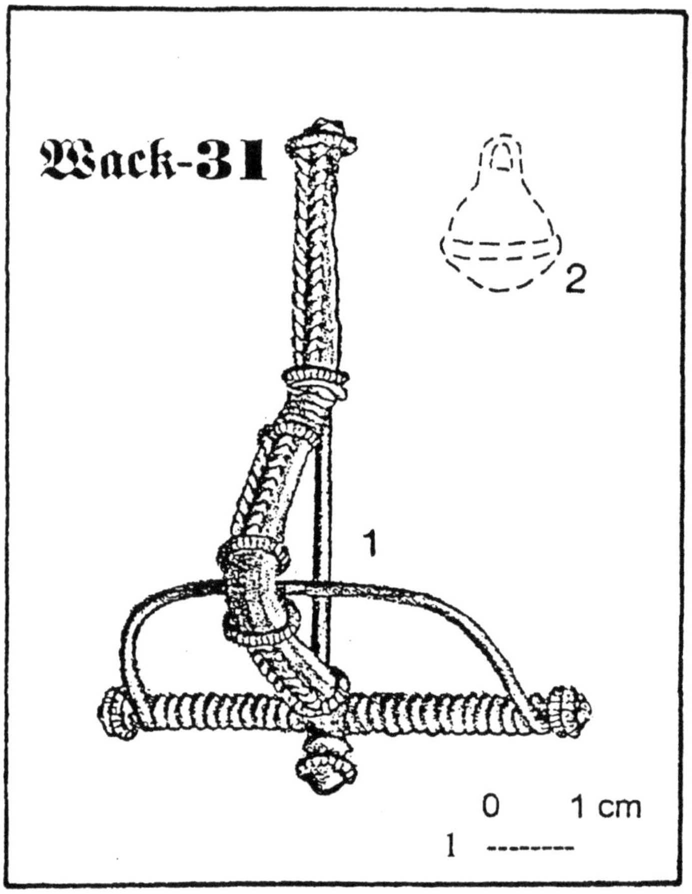

Fig. 58. Copper-alloy with silver ring brooch of type Å3 in Wack-31 (urn CRM). Beside it an iron bell pendant and iron buckle (Tischler, Kemke, 1902, S. 40).

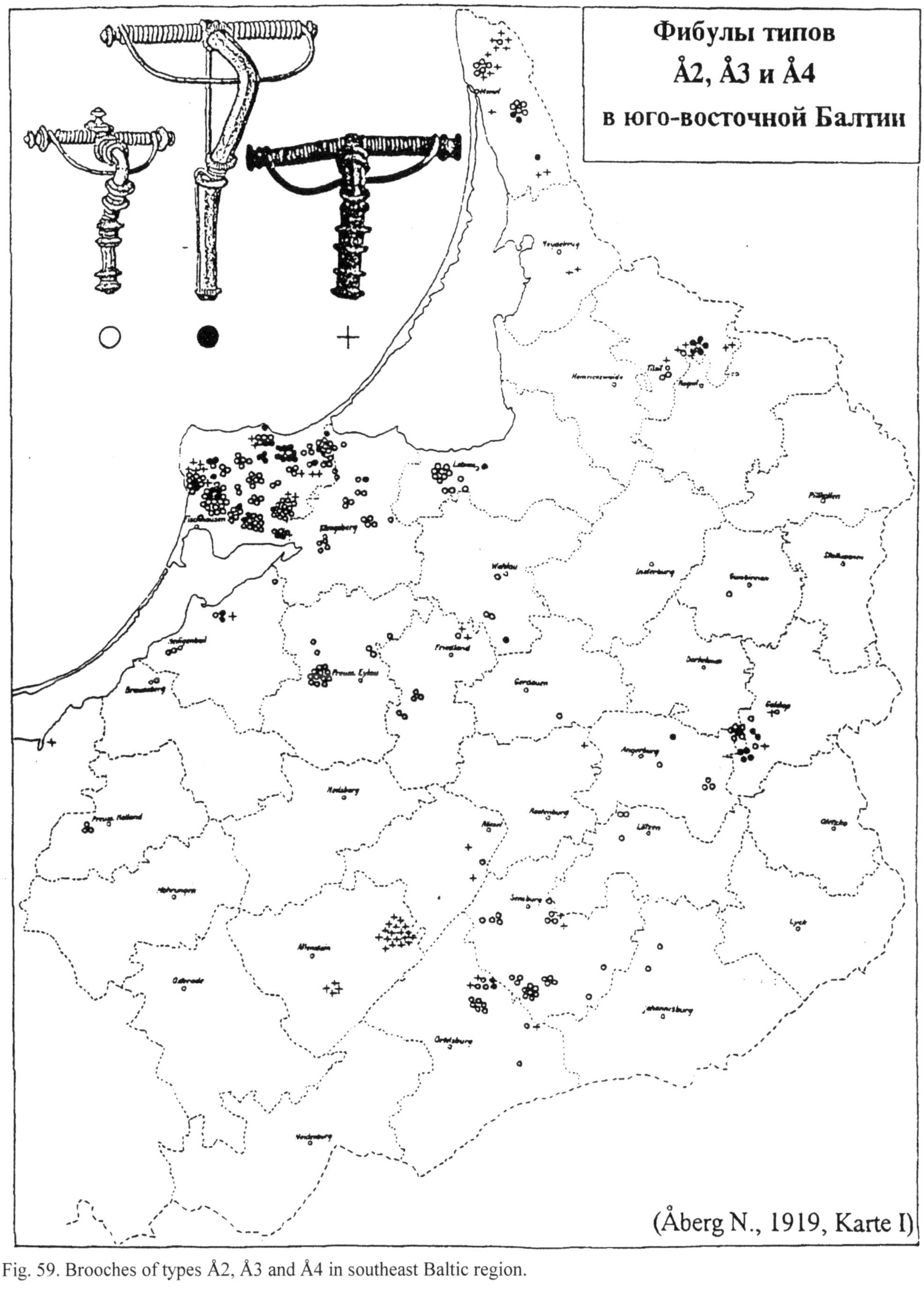

Fig. 59. Brooches of types Å2, Å3 and Å4 in southeast Baltic region.

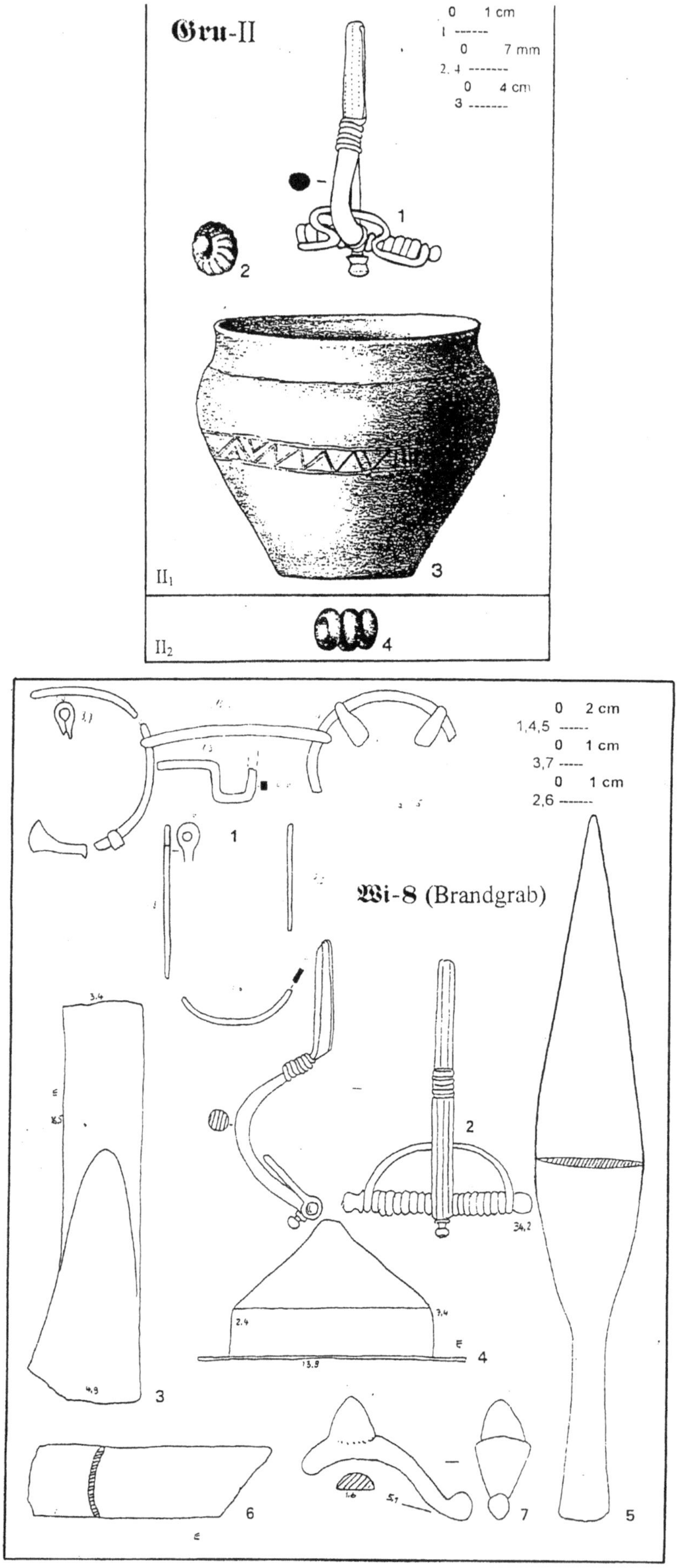

Fig. 60. Assemblage Gru-II and Wi-8 (Brandgrab).

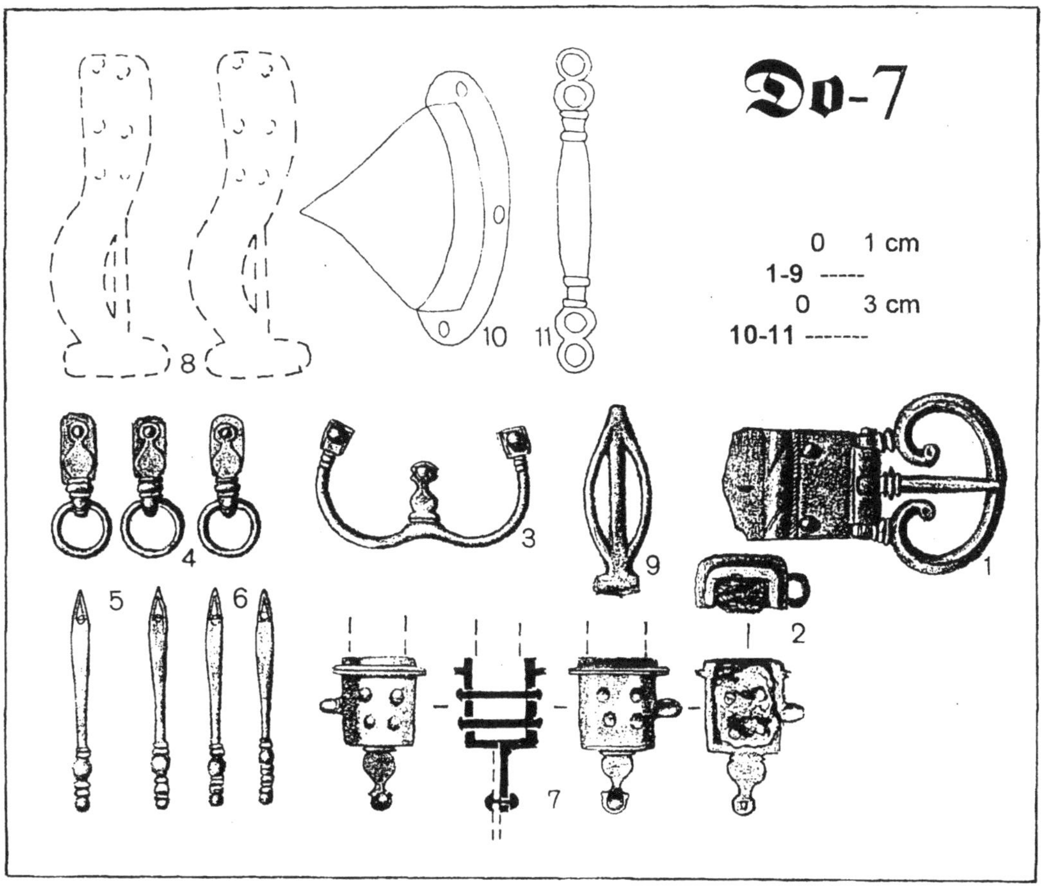

Fig. 61. Assemblage Do-7.

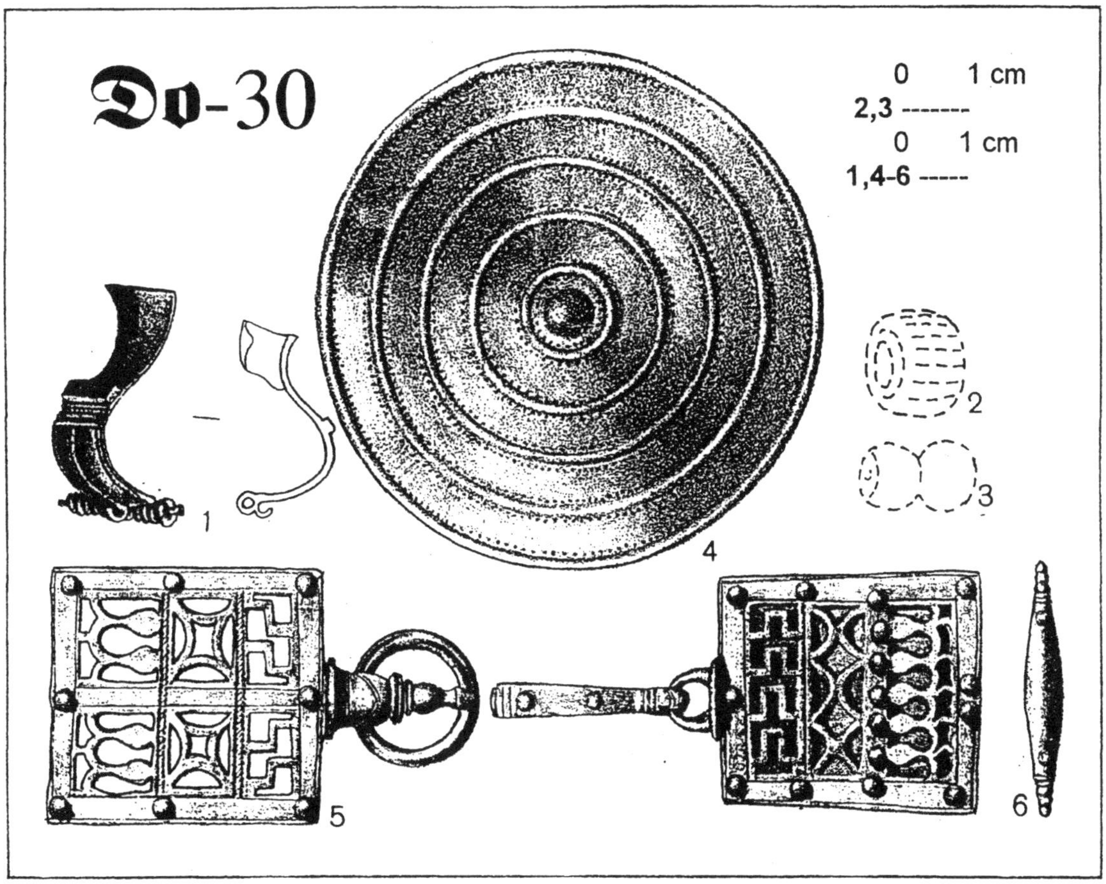

Fig. 62. Assemblage Do-30.

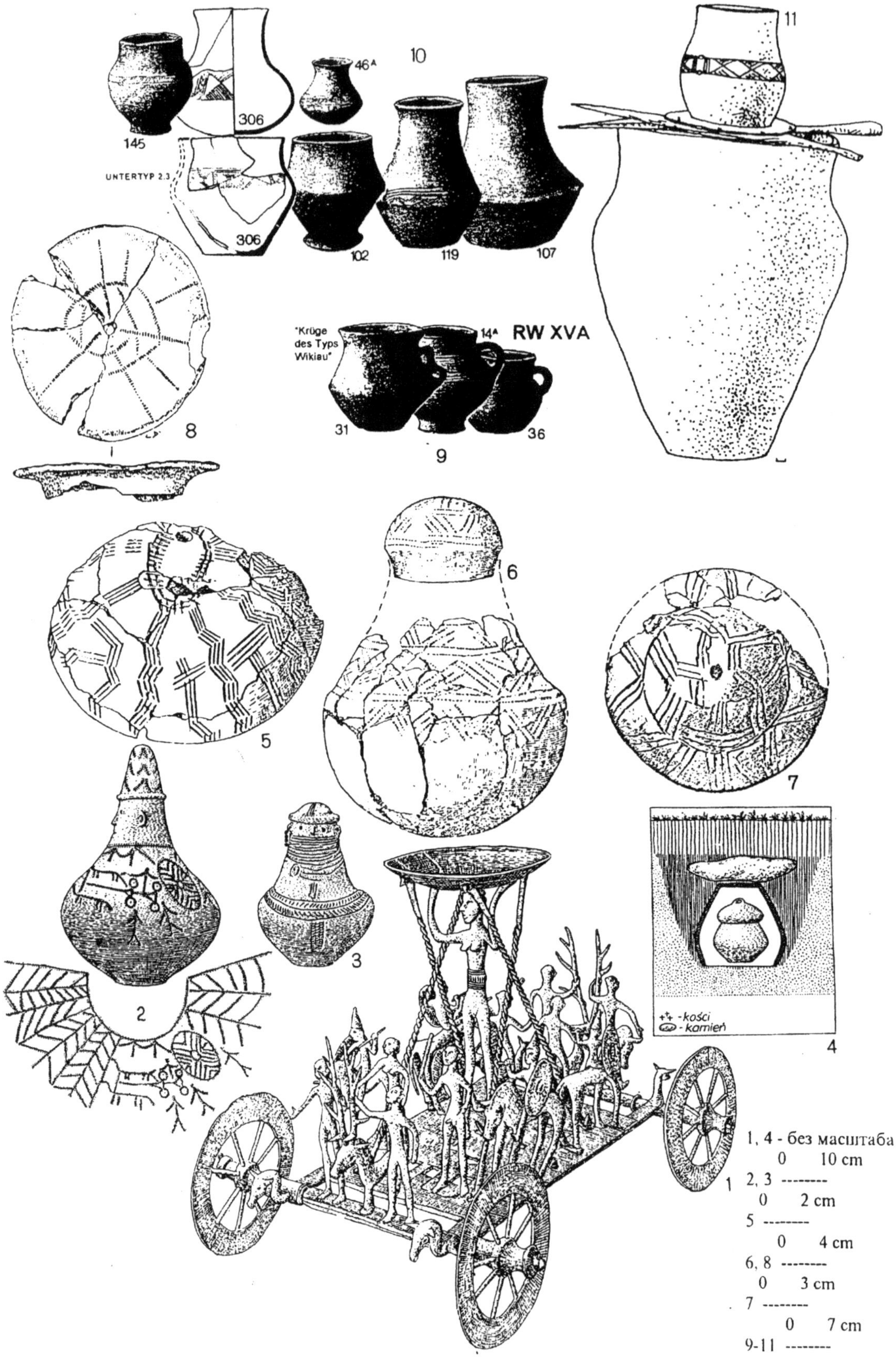

Fig. 63. Sacrificial bowls and vessels of small forms used in archaic rituals of Celts, Germans and Balts from 7th cent. BC - 6th cent. AD: 1 - copper-alloy cult vehicle from Strettung (Steiermark, Austria), 7th cent. BC; 2 - male urn with cover (+ unrolled picture of urn) from Grabowo Bobowsk (near Starogard Gdański), 6th-5th cent. BC; 3 - female urn with cover from Przegowo (vicinities of Gdańsk), the 6th-5th cent. BC; 4 - section of "under-cloched" burial (urn covered by inverted vessel) from Kazimierz (vicinities of Łódź), 4th cent. BC; 5 - cover of urn from Piórkowo, woj. Warmińsko-Mazurskie (ehem. Födersdorf, Kr. Brausberg; 6 - Żardeniki, woj. Warmińsko-Mazurskie (ehem. Scharnigk, Kr. Rößel); 7 - ehem. Linkau, Kr. Samland, nowadays - Zelenogradsk District of Kaliningrad Region; 8 - vicinities of Mamonovo, Bagrationovsk District of Kaliningrad Region (ehem. Wermten, Kr. Heiligenbeil) /positions 5-8 - the 5th-4th cent. BC/; 9 - accessory vessels from cemetery at Kovrovo, Zelenogradsk District of Kaliningrad Region (ehem. Dollkeim, Kr. Samland), phases B_1 - B_2/C_1; 10 - accessory vessels from cemetery at Kovrovo, phases C_1 - D_3; 11 - reconstruction of position of urn in Do-306 (C_{1b}/C_2 = 260-300).

Fig. 64. Wurshait before the beginning of the sacrifice, represented in the book of H. Maletius (1561-1562).

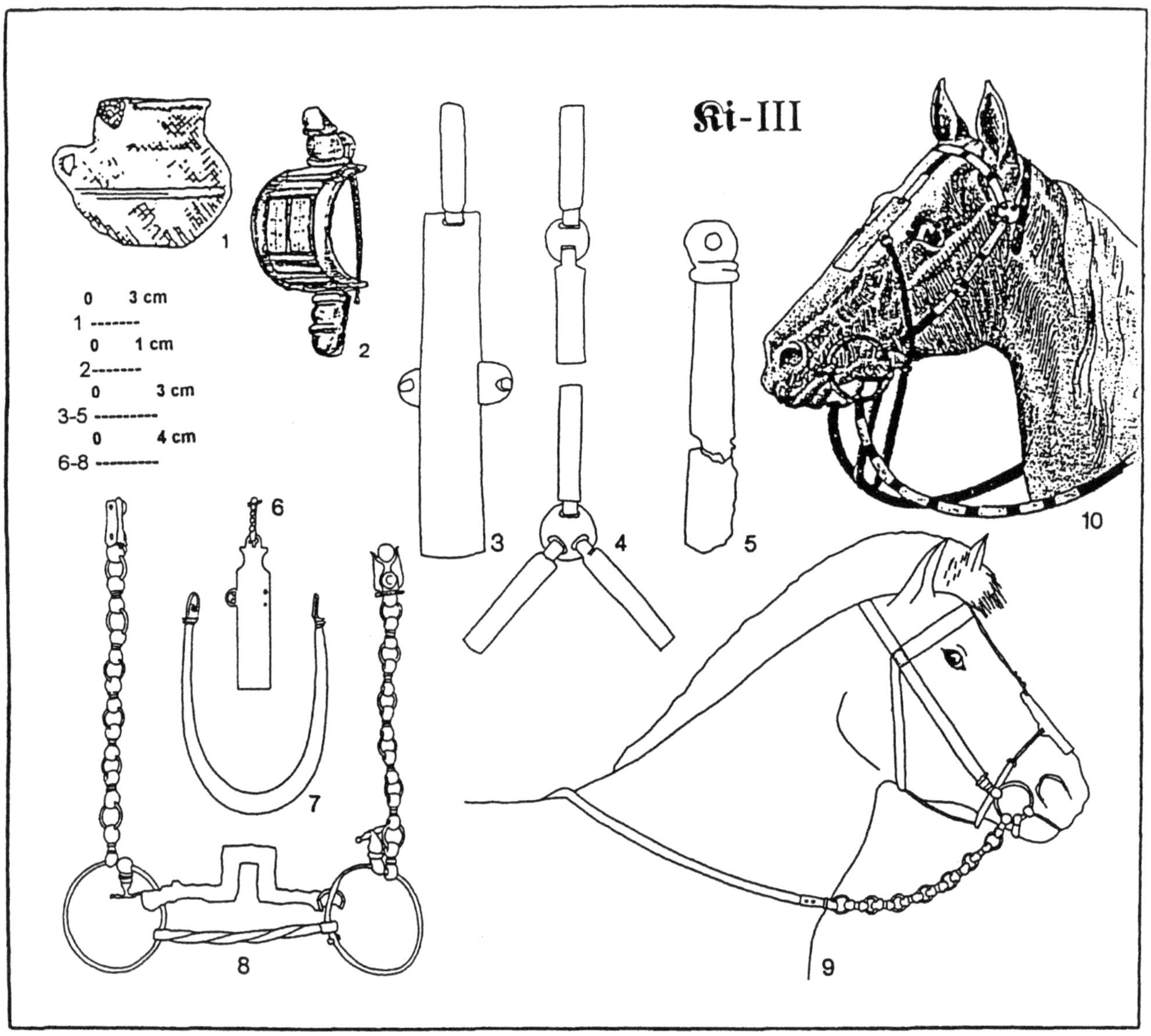

Fig. 65. Povarovka, Zelenogradsk District - Grave III, in the urn - two headpieces, in composition of one of them ("a": 6-9) - links of type Z1, strap holders of type Rh1 and straight snaffle of type Ks2b accompanied by the copper-alloy "eyed" brooch of type AIII, 61 (phase B_{1a}-?); also in the urn: 1 - accessory vessel, 2 - copper-alloy covered with silver joint-pin provincial-Roman brooch, close to scheme A24, 3 - copper-alloy covered with silver brow mount from headpiece "b"; 4 - copper-alloy covered with silver mounts from headpiece "b"; thin copper-alloy neck-ring (?), ornamented bone comb, knife, small copper-alloy spoon, fire-steel, three curry-combs, 5 - fragment of a copper-alloy arch of headpiece "b"; 9 - reconstruction of headpiece "a", 10 - reconstruction of headpiece "b" (1, 2 - Nowakowski W., 1996, Taf. 81, 1, 2; 3-5 - La Baume W., 1944, Abb. 7b, 8a; 6-9 - Wilbers-Rost S., 1994, S. 206, 207, Taf. 9, a, b).

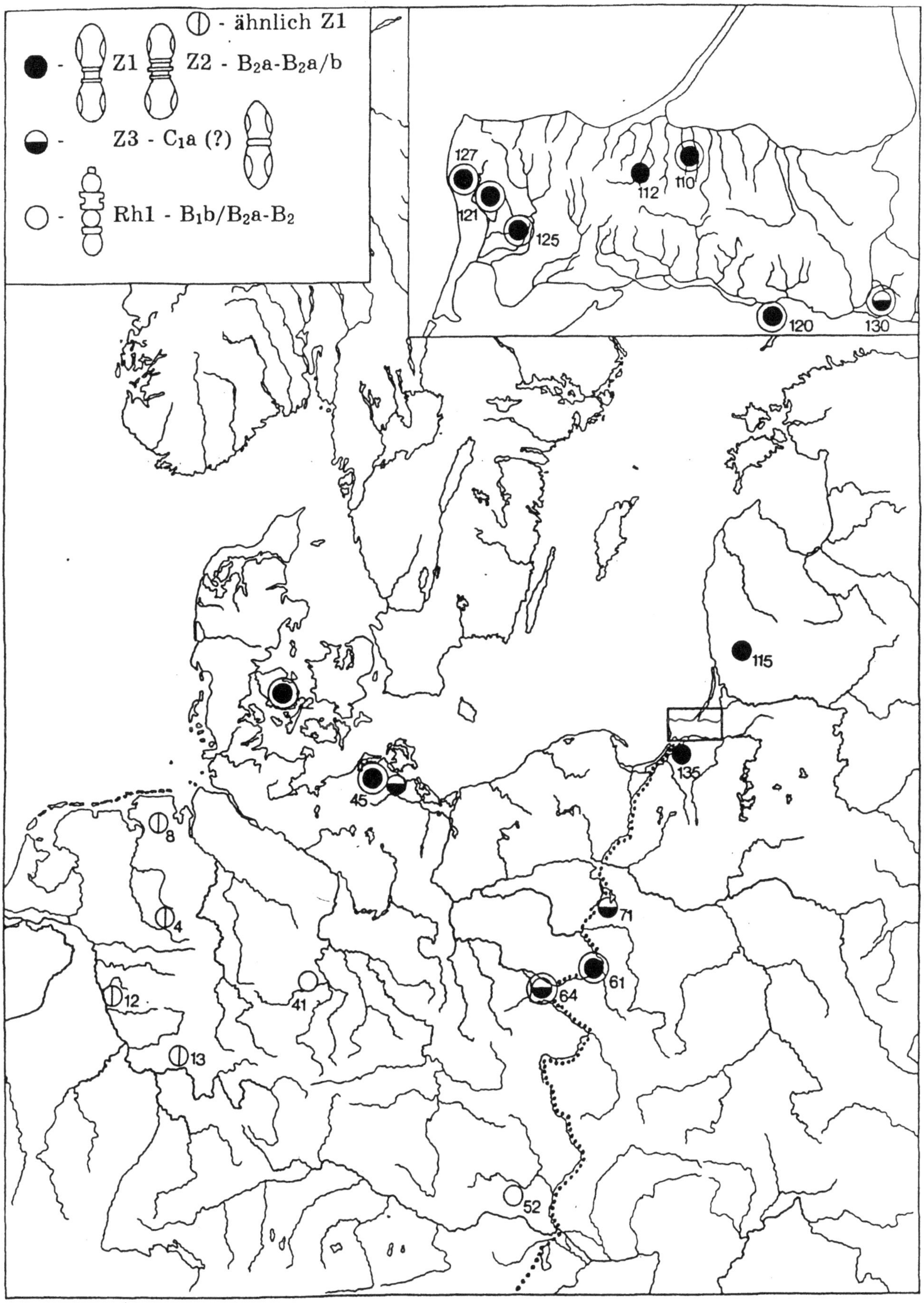

Fig. 66. Distribution of copper-alloy chain links of types Z1, Z2, Z2 and strap holders of type Rh1 in Barbaricum (Wilbers-Rost S., 1994, Karten 3, 3a, 7, 7a). Inserted map above - finds of these artefacts in Sambia.

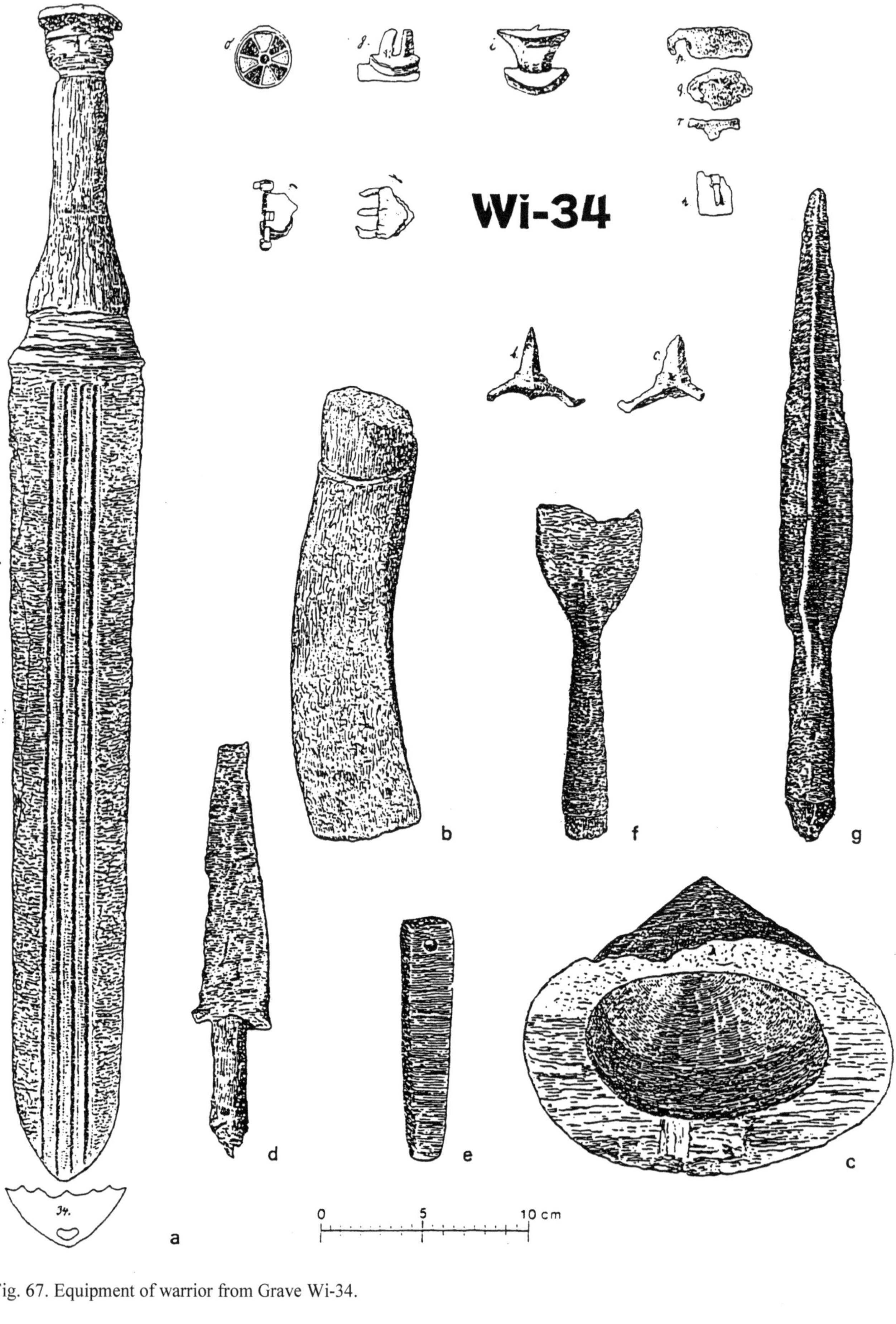

Fig. 67. Equipment of warrior from Grave Wi-34.

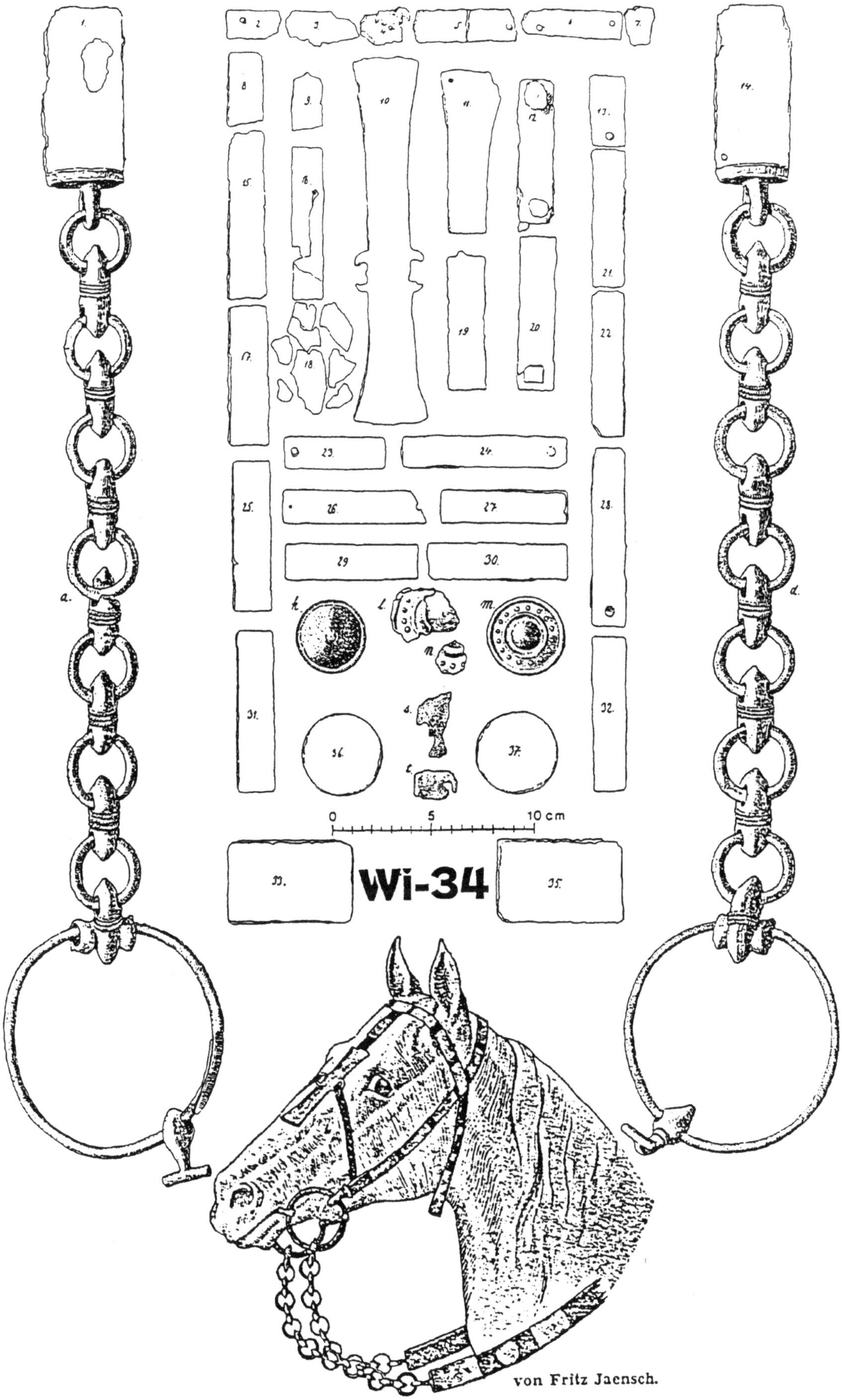

Fig. 68. Links of type Z4 and strap holders of type Rh2a in composition of headpiece of Vimose type in grave goods of Grave Wi-34, appertaining to horse equipment. In the lower section of the figure - reconstruction by Fritz Jenzsch (Raddatz K., 1992/93, Abb. 4; reconstruction of horse equipment - La Baume W., 1944, Abb. 10).

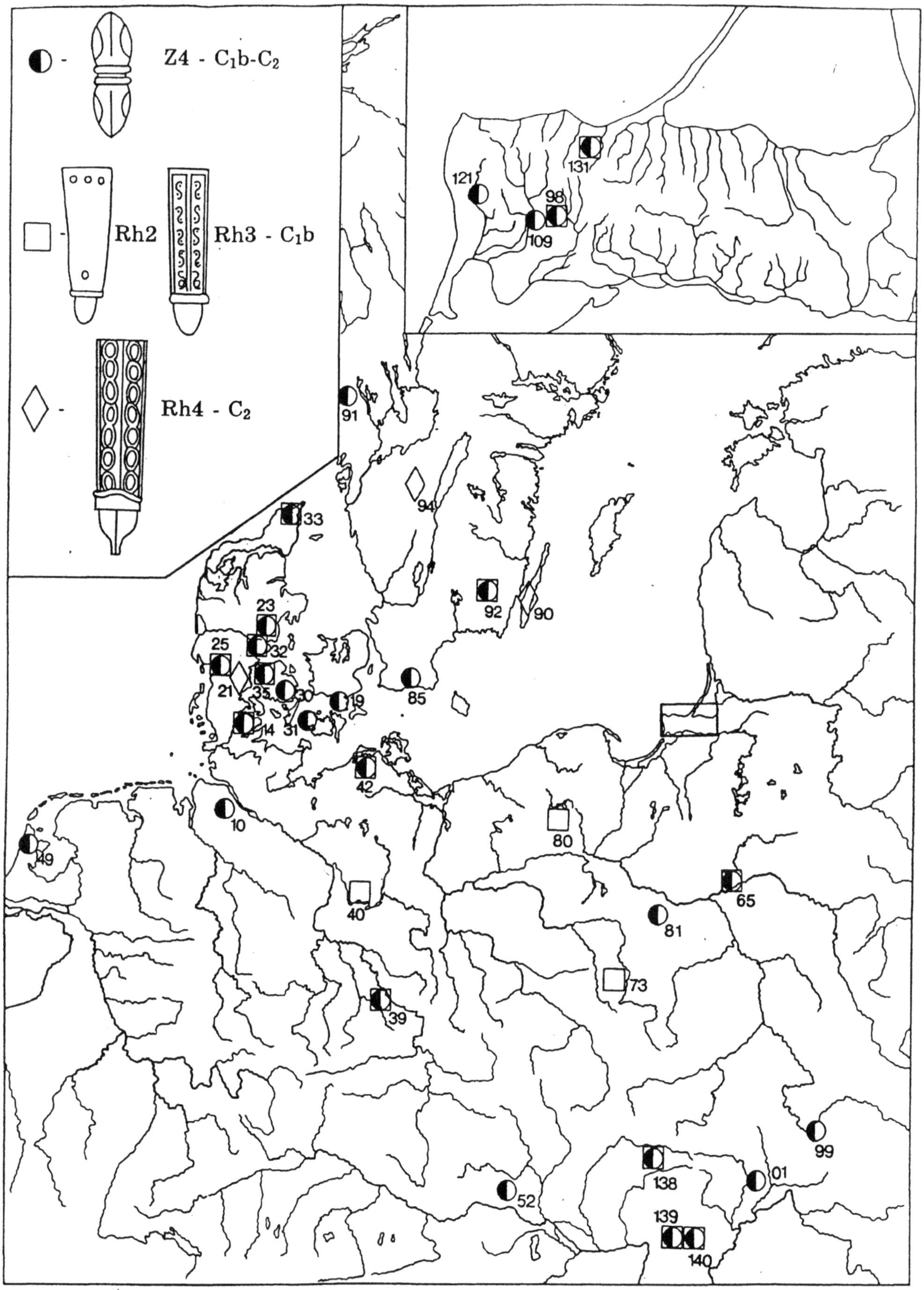

Fig. 69. Distribution of copper-alloy chain links of type Z4 and strap holders of types Rh2-Rh4 in Barbaricum (Wilbers-Rost S., 1994, Karten 3, 3a, 4, 4a, 7, 7a). Inserted map above - finds of these artefacts in Sambia.

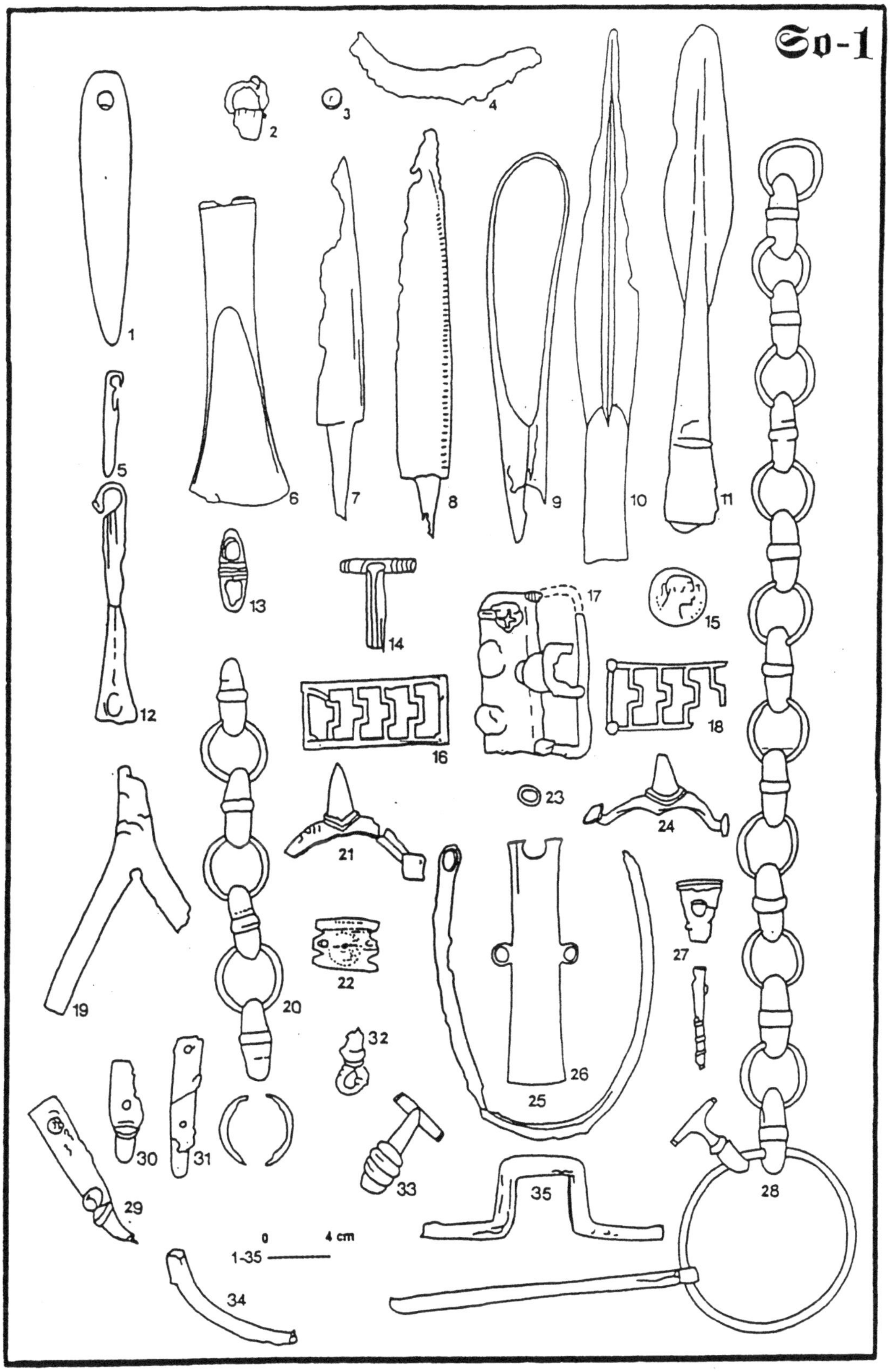

Fig. 70. Kulikovo, Zelenogradsk District (ehem. Sorthenen, Kr. Samland) - links of type Z4 (13, 20), strap holders of type Rh2a (33) and straight snaffle of type Ks1a (28) in composition of headpiece of Vimose type (24, 25, 28, 29-35) from an unnumbered grave, containing grave goods: 1 - whetstone; 2 - fragment of a buckle; 3 - gaming-counter (?); 4 - razor; 5 - fragment of a mount; 6 - socketed axe; 7, 8 - battle knives; 9 - shears; 10, 11 - spearheads; 12 - fire-steel; 14 - brooch of type AV, 137/145, 15 - bronze coin of Lucilla (before 182 AD) (phase B_2/C_1), 16, 18 - copper-alloy belt mounts; 17 - iron buckle of type Madyda-Legutko G36; 19 - mount from a headpiece; 21, 24 - copper-alloy spurs of types Ginalski E3; 22 - copper-alloy spur of type Stuhlsporn; 23-?, - copper-alloy and iron details, 27 - copper-alloy (?) strap-end (Engel C., 1935, Abb. 44).

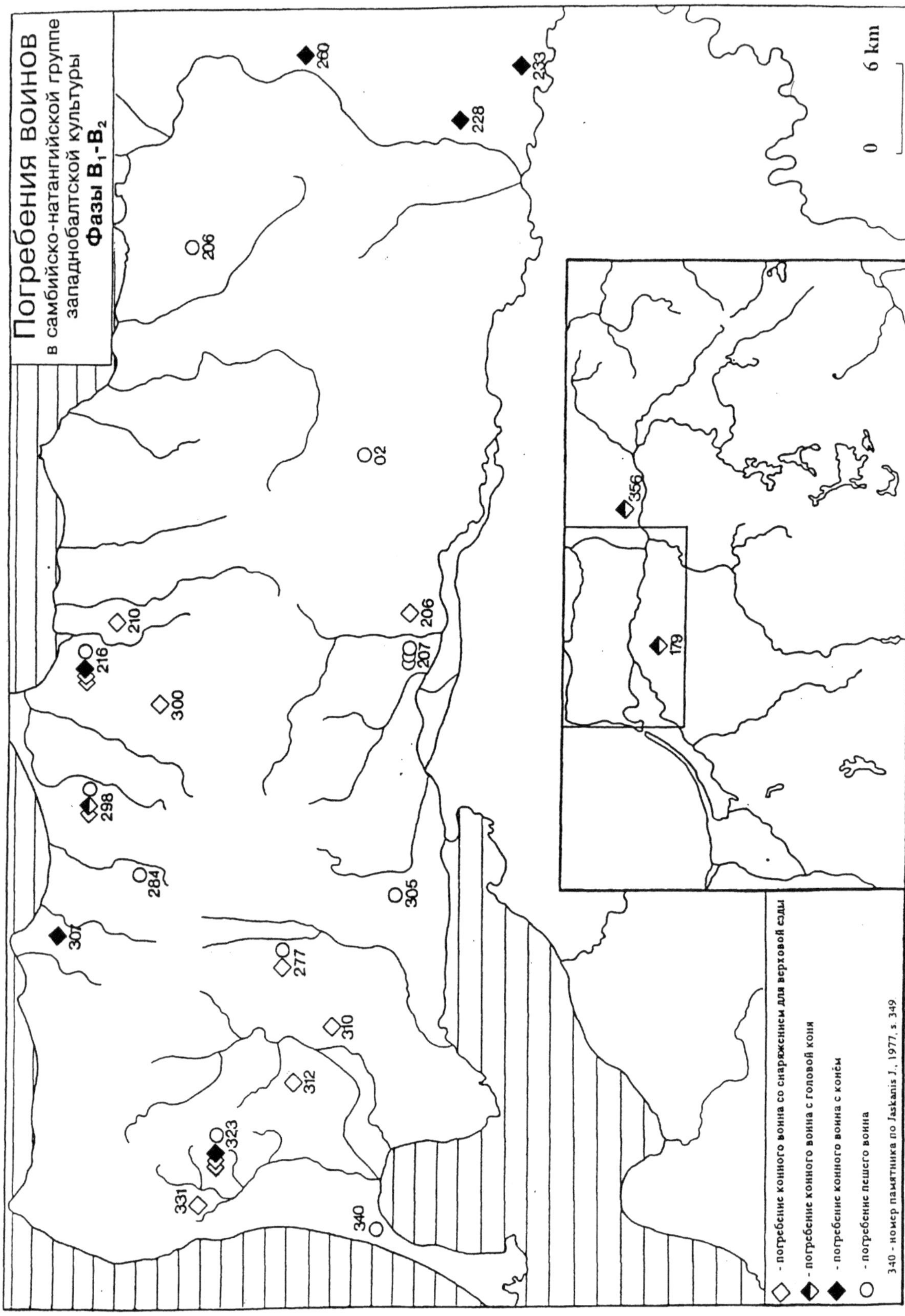

Fig. 71. Burials with horses, horse equipment and weapons in the SNG (phase B_1-B_2): with elements of harness: 179 (hereinafter the numbers of the sites - after Jaskanis J., 1977, s. 349) - Lermontovo (Wogau) - 1 (tooth of a horse); 206 - Zaozerje (Lapsau) - 1; 210 - Lugovskoye (Lobitten) - 1; 216 - Polesye (Gr. Ottenhagen) - 2, 1 (horse); 228 - Soldatovo - 1 (horse); 260 - Izobil'noye (Kl. Fliess) - 1 (horse); 277 - Chrustal'noye (Wiekau) - 1; 298 - Kovrovo - 1; 1 (with a horse skull); 300 - Krasnoflotskoye (Corben) - 1; 307 - Gora Velikanov (Hünenberg) - 1 (horse); 310 - Logvino (Gr. Medenau)-1; 312 - Ramenskoye (Norgau) - 1; 323 - Povarovka - 1 (horse); 1; 1; 1; 331 - Russkoye - find of part of a horse from grave (?); 356 - ehem. Ilischken - 1 (skull of a horse); with equipment of foot-soldier: 207 - Bolshoye Isakovo (Lauth) - 3; 216 - Polesye (Gr. Ottenhagen) - 1; 266 - Sibirskoye (Moritten) - 1; 277 - Chrustal'noye - 1; 298 - Kovrovo - 1; 305 - ehem. Lehndorf - 1; 323 - Povarovka - 1; 335 - Sirenevo (Eisliethen) - 1; 340 - Beregovoye /erroneously - Letnoye/(Tenkietten) - 1; 02 - Prudy (Kadgiehen) - 1.

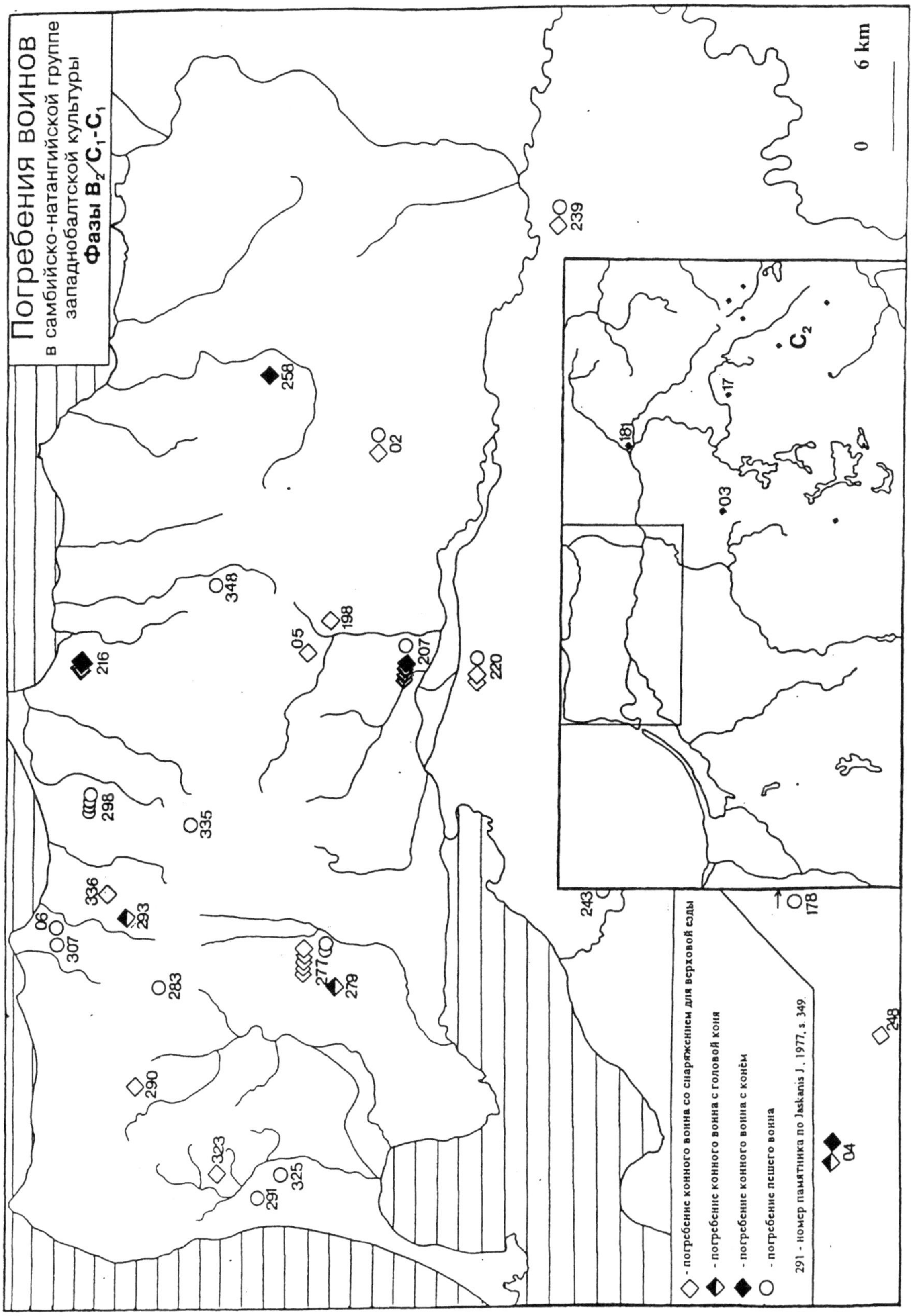

Fig. 72. Burials with horses, horse equipment and weapons in the SNG (phases B_2/C_1 - C_1): with elements of harness: 181 - Chernyakhovsk (Althof) - 1; 198 - Guryevsk - 1; 207 - Bolshoye Isakovo - 1; 216 - Polesye (Gr. Ottenhagen) - 1 (horse); 1 (horse); 220 - Alleya Otvazhnykh, Kaliningrad (Rosenau) - 2; 239 - Suvorovo (Zohpen) - 1; 248 - Uzornoye (Jaknitz) - 1; 258 - Zarechye (Caymen) - 1 (with two horses); 277 - Khrustal'noye - 5; 279 - Pereslavskoye (Cojehnen) - 1 (skull of a horse); 290 - Grachevka (Craam) - 1; 323 - Povarovka - 2; 336 - Kulikovo (Sorthenen)-1; 17 - Grunajki (Gruneiken) - 1; 03 - ehem. Henriettenfeld - 1 (horse); 04 - ehem. Hochschnakeinen - 1; 1 (horse); 1 (skull of a horse); 02 - Prudy (Kadgiehen) - 1; 05 - ehem. Klein Heyde - 1; with equipment of foot-soldier: 178 - Yelanovka (Wackern) - 9; 207 - Bolshoye Isakovo - 7; 220 - Alleya Otvazhnykh, Kaliningrad (Rosenau) - 1 277 - Chrustal'noye - 2; 283 - Dubrovka (Regehnen) - 1; 291 - ehem. Grebieten - 1; 298 - Kovrovo - 4; 307 - Gora Velikanov - 1; 325 - Putilovo (Korjeiten) - 1; 335 - Sirenevo (Eisselbiethen) - 1; 348 - Zaytsevo (Trentitten) - 1; 02 - Prudy (Kadgienen) - 1; 06 - Palwe bei Rantau - 1.

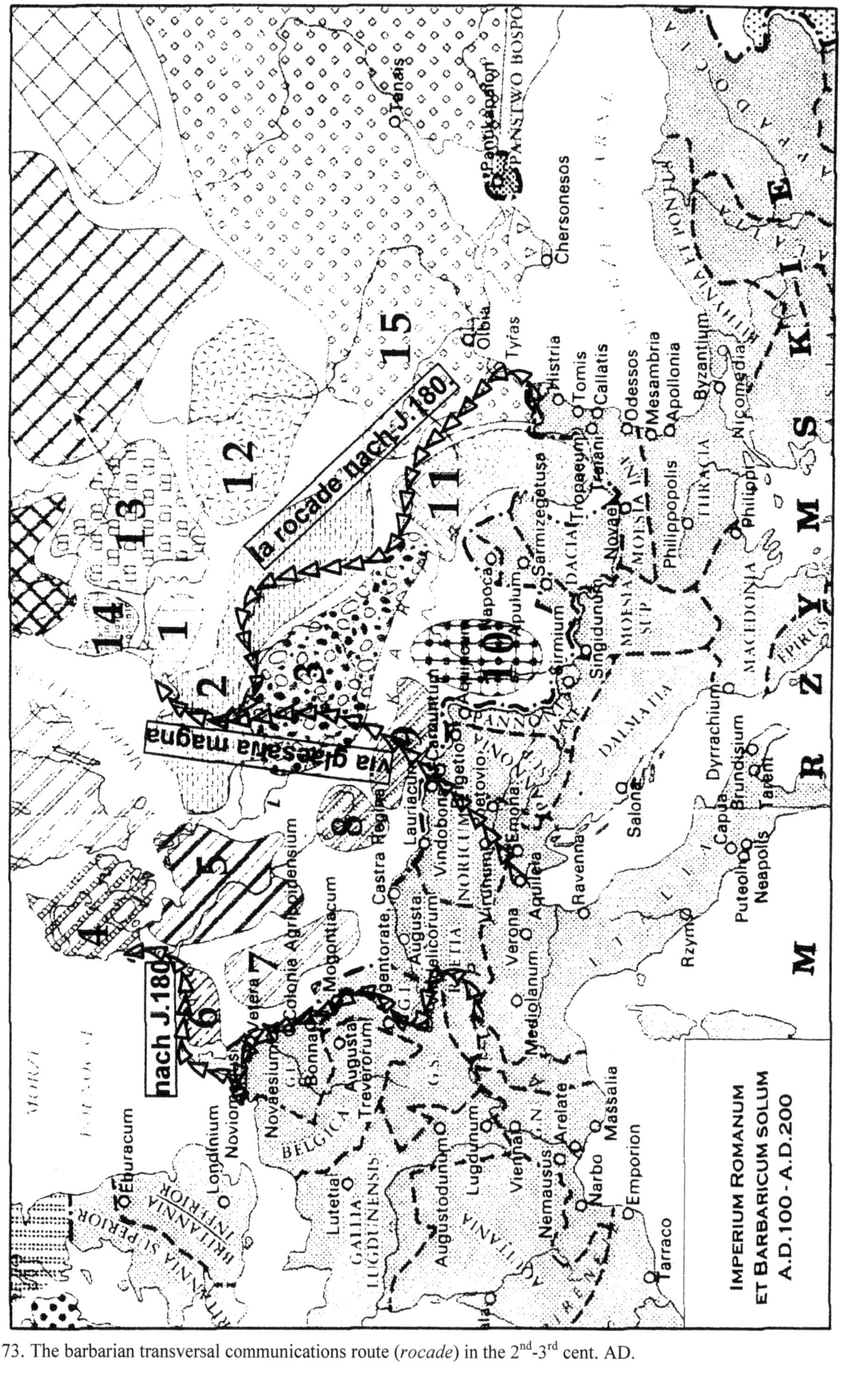

Fig. 73. The barbarian transversal communications route (*rocade*) in the 2nd-3rd cent. AD.

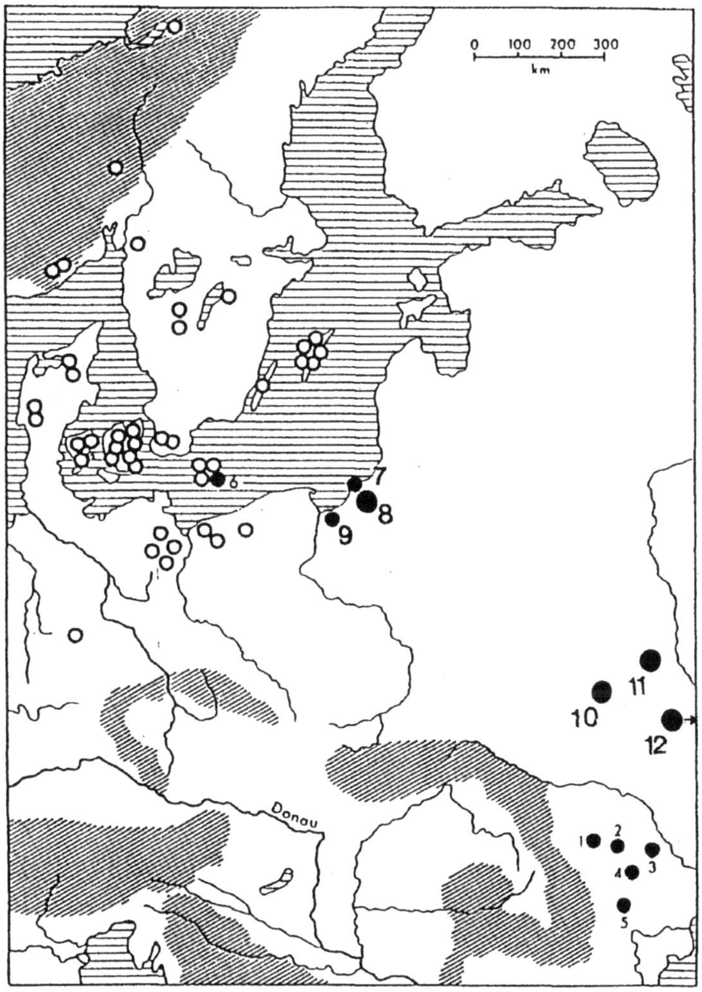

Fig. 74. Distribution of silver brooches of type monstruosa (phases C_1 and C_2): 1 - Todireni; 2 - Vasilica; 3 - Budeşty; 4 - Dančeny; 5 - Chanska; 6 - Slusegaard; 7 - Grebieten; 8 - Lauth, 10 - Petrikivtsy, 11 - Novopolovetskoye, 12 - Medvedovka.

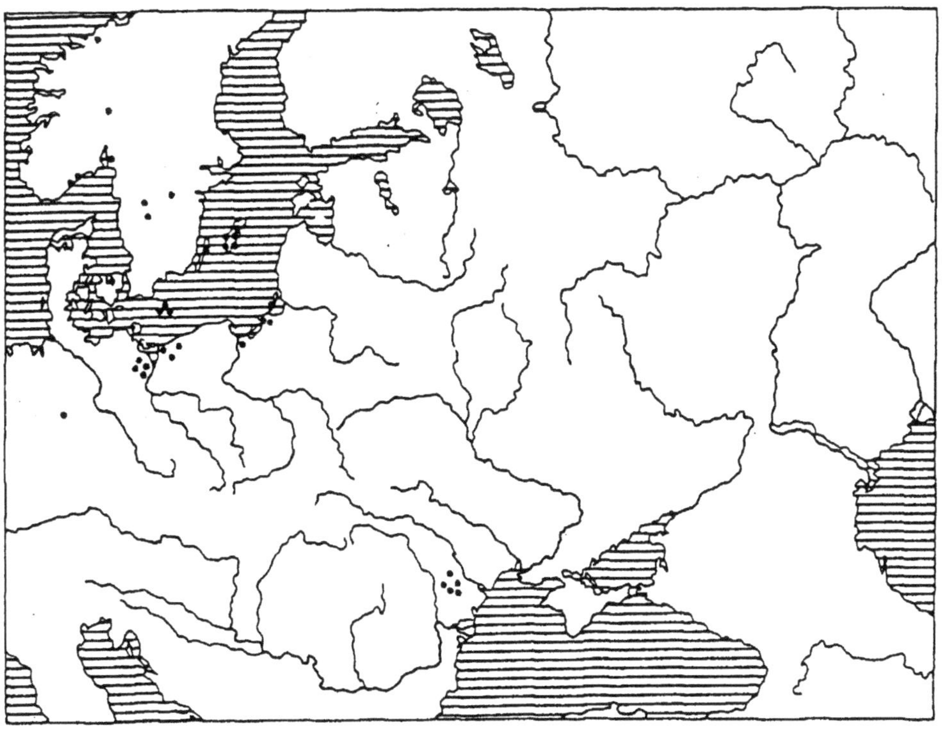

Fig. 75. Elder rune inscriptions of phases C_2-C_3 (?).

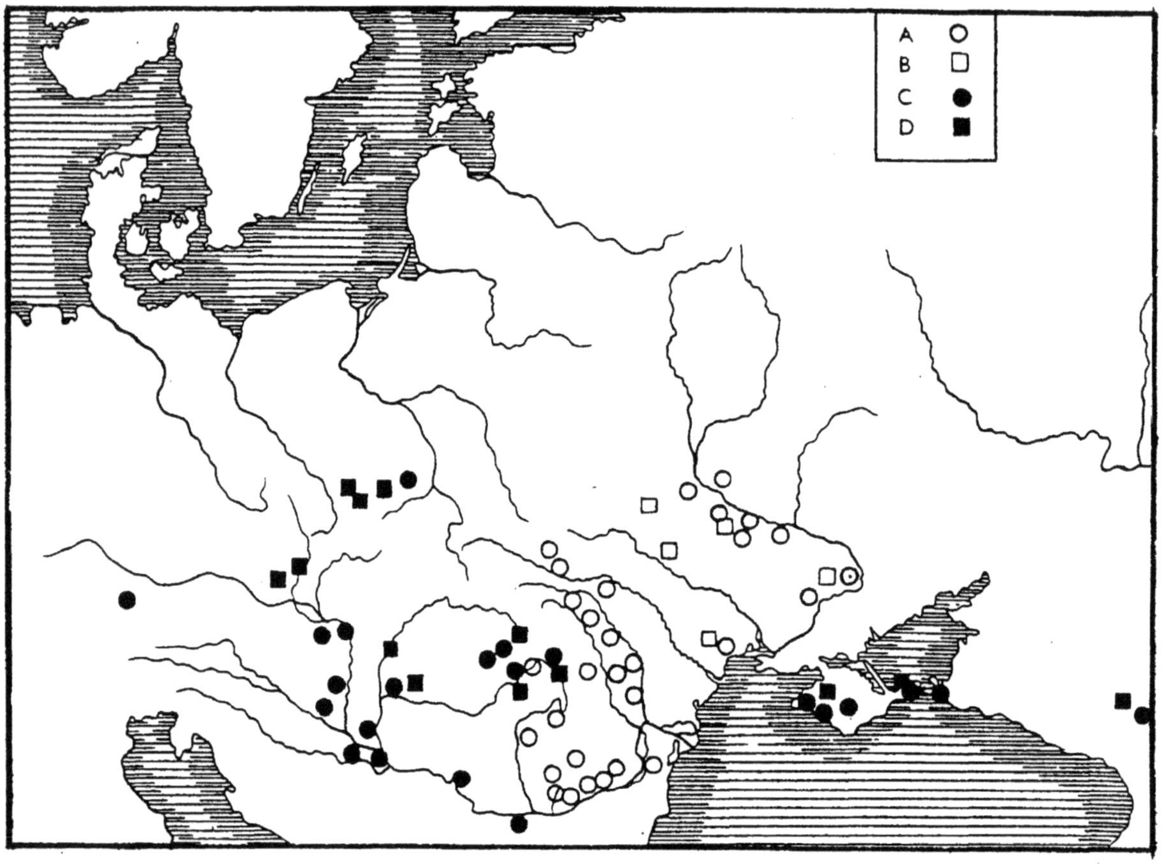

Fig. 76. Distribution of two-plate brooches and buckles in assemblages of phases C_3 (unfilled signs) and D_1 (filled signs): A, C - brooches of subgroup Ambroz I; B, D - large buckles with two or four rivets.

Fig. 77. The appearance of ancient inhabitants of the Amber Coast (Sambia): left - a German of the third quarter of the 2nd cent. AD, right - an inhabitant of Sambia of the first half of the 5th cent. AD.

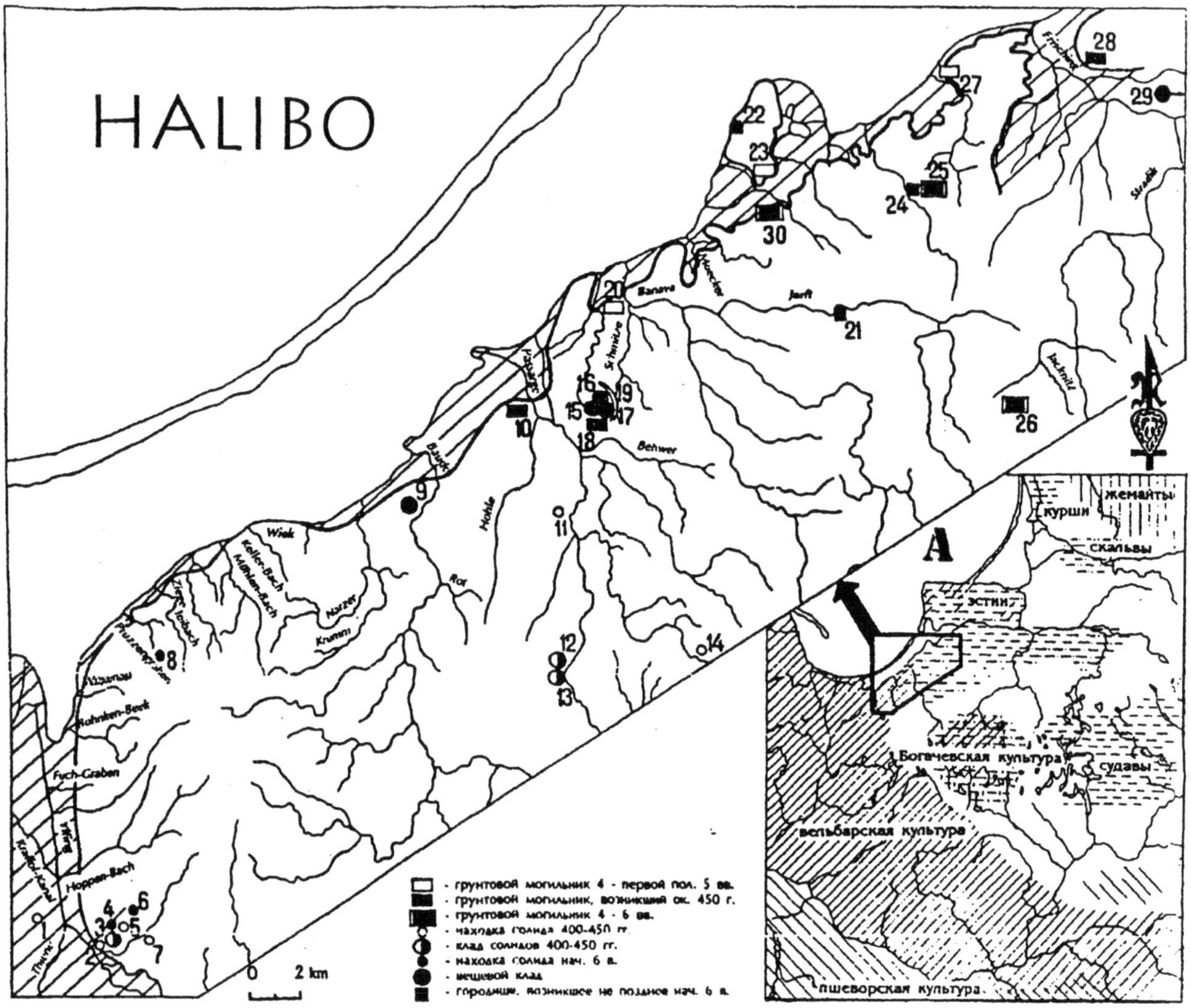

Fig. 78. Halibo. The territory between the rivers Ilfing and Frisching in the 5th cent. AD:
1 - Komorowo, woj. Elbląg - solidus of Theodosius I (361-395); 2 - Dębina, woj. Elbląg - solidus of Honorius (395-423); 3 - Gronowo Górne, woj. Elbląg - 2 (?) hoards of solidii, the latest coins - Livius Severus (461-465); 4 - Elbląg - 11 separate solidii (odd hoard?), the latest ones - Leon I (457-471) and Anastasius (491-518); 5 - Czechowo, woj. Elbląg - 2 solidii, including - Theodosius II (408-450); 6 - Dębice, woj. Elbląg - solidus of Leon I (457-471); 7 - Przezmark, woj. Elbląg - solidus of Valentinian III (424-425); 8 - Lęcze, woj. Elbląg - solidus of Anastasius I (491-518); 9 - Frombork, woj. Elbląg - hoard of goods and solidus of Theodosius I (408-450); 10 - Podgórze, woj. Elbląg (ehem. Huntenberg, Kr. Braunsberg) - cemetery from about 450; 11 - Garbina, woj. Elbląg - solidus of Theodosius II (408-450); 12 - Trąbki, woj. Elbląg (ehem. Kl. Tromp, Kr. Braunsberg) - hoard of solidii, of which 44 were minted under Theodosius II (408-450); 13 - Trąbki, woj. Elbląg - hoard of solidii, of which 11 were minted under Theodosius II (408-450); 14 - Nałaby, woj. Elbląg - solidus of Theodosius II (408-450); 15 - Młotezno, woj. Elbląg (ehem. Hammersdorf, Kr. Heiligenbeil) - hoard with medallion of Constantius II (317-361); 16 - Młotezno, woj. Elbląg - hoard with fragments of dishes; 17 - Młotezno, woj. Elbląg - hoard with two neck-rings; 18 - Młotezno, woj. Elbląg - cemetery from about 450; 19 - Młotezno-Rogity, woj. Elbląg (ehem. Hammersdorf-Regitten, Kr. Heiligenbeil); 20 - Zelenodol'skoye, Bagrationovsk District (ehem. Preußisch Bahnau/Carben, Kr. Heiligenbeil) - cemetery from the 4th - the beginning of the 5th cent.; 21 - Lipovka, Bagrationovsk District (ehem. Grünwalde, Kr. Heiligenbeil) - defended site-sanctuary of Jungfernberg; 22 - Veseloye, Bagrationovsk District (ehem. Balga, Kr. Heiligenbeil) - defended site of Honeda; 23 - Veseloye, Bagrationovsk District (ehem. Scheckenberg, Kr. Heiligenbeil) - cemetery from the 4th/ beginning of the 5th cent.; 24 - Pervomayskoye, Bagrationovsk District (ehem. Warnikam, Kr. Heiligenbeil) - defended site of Pillgarten from about 450; 25 - Pervomayskoye, Bagrationovsk District - cemetery at Kapnieserberg from the 4th-6th cent.; 25 - Uzornoye, Bagrationovsk District (ehem. Jäcknitz/Rosen, Kr. Heiligenbeil) - cemetery from the 4th-5th cent.; 26 - Beregovoe, Bagrationovsk District (ehem. Patersort/Domniksruh, Kr. Heiligenbeil) - cemetery from the 4th 5th cent.; 28 - former Tengen, Kr. Heiligenbeil - cemetery from about 450 AD; 29 - Berezovka, Bagrationovsk District (ehem. Groß Sausgarten, Kr. Heiligenbeil) - hoard of goods; 30 - Krasnodonskoye, Bagrationovsk District (ehem. Keimkallen, Kr. Heiligenbeil) - the 4th-5th cent.
Shaded area shows coastal regions with an altitude of up to 5m, and which might well have flooded in the 5th century as a result of the rising waters in the Baltic (L.A.Zhindarev, V.I. Kulakov 1996. s. 62).

Fig. 79. Młotezno. Hoard with medallion of Constantius II (V.I. Kulakov 1990. fig. 1).

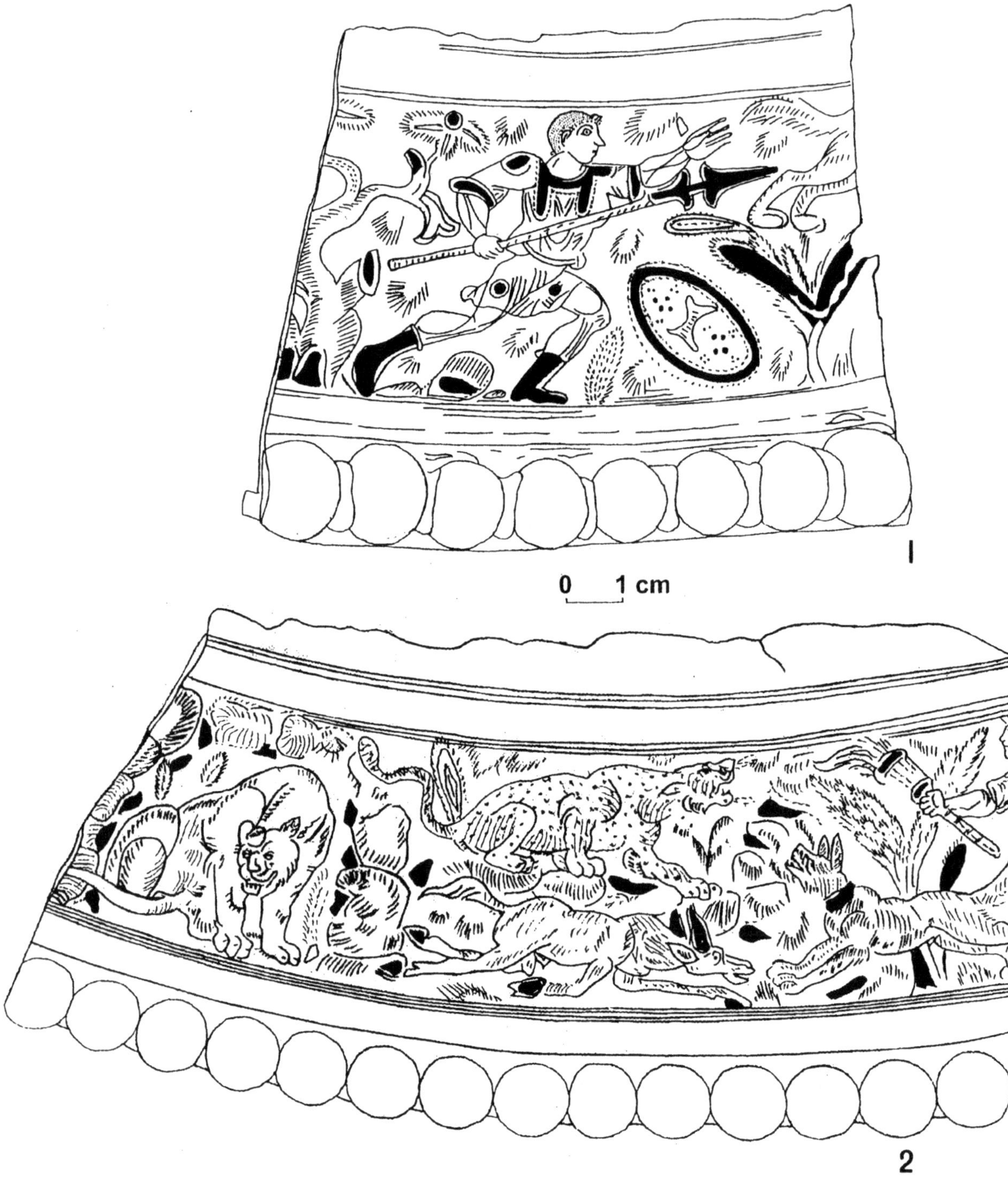

Fig. 80. Młotezno. Hoard with fragments of silver vessels - parts of silver "luterius" with partly gilt and nielloed pictures (1 - W. Gaerte 1929. Abb. 157, 2 - G. Hirschfeld 1886. Taf. VI).

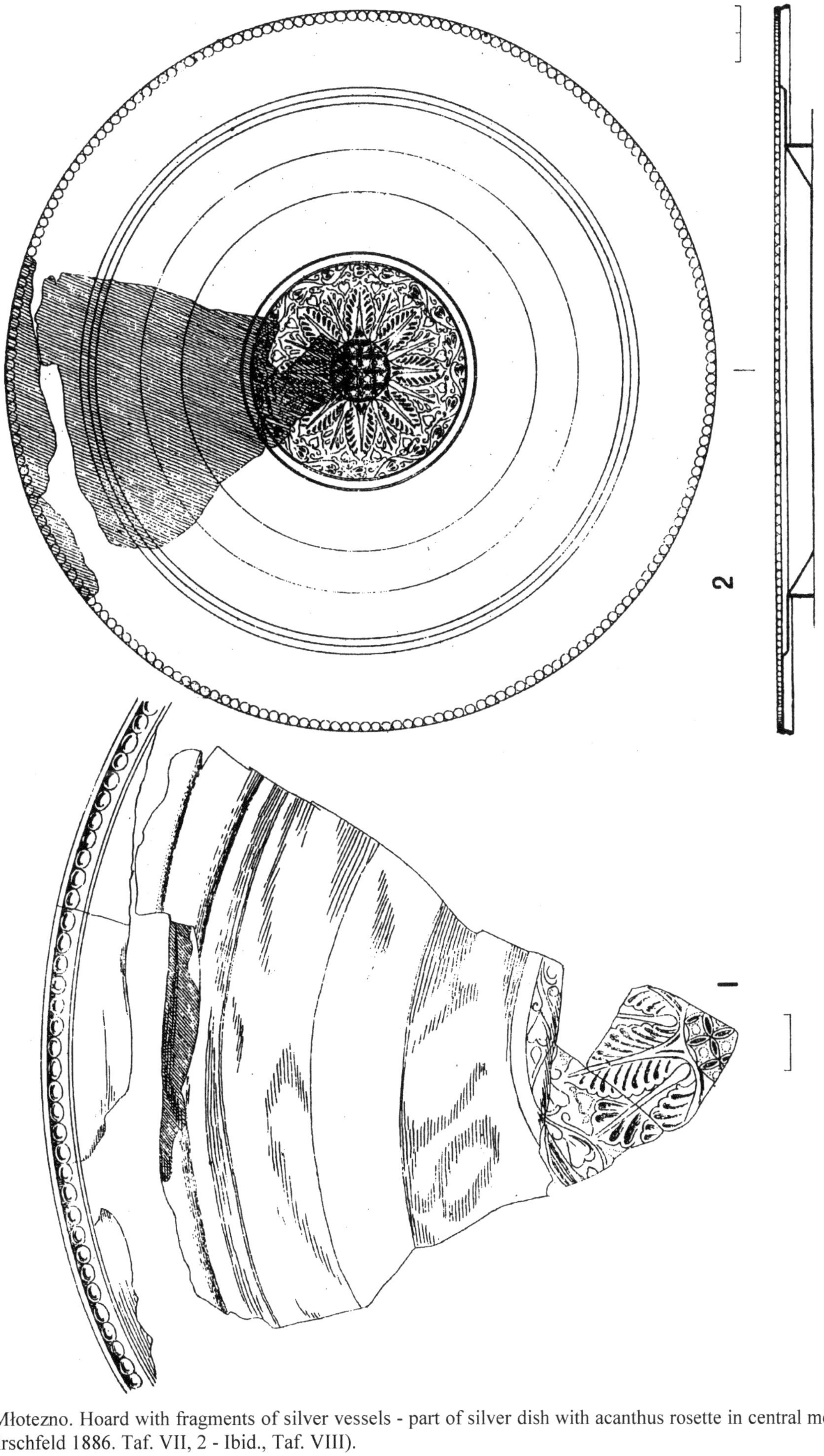

Fig. 81. Młotezno. Hoard with fragments of silver vessels - part of silver dish with acanthus rosette in central medallion (1 - G. Hirschfeld 1886. Taf. VII, 2 - Ibid., Taf. VIII).

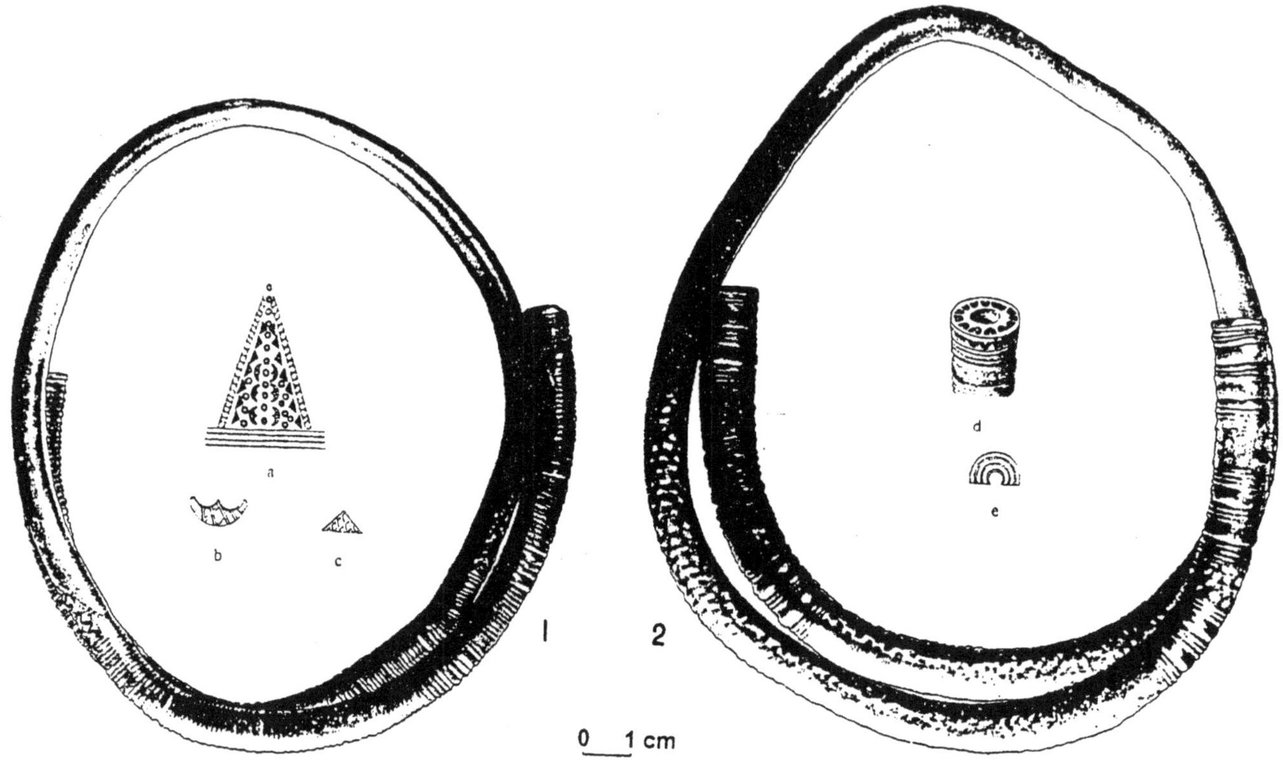

Fig. 82. Młotezno. Hoard with two "neck-rings" (E. Petersen 1939. Abb. 48). Types of prints of stamp for small "neck-ring" - a-c (M. Ebert 1923. Abb. 14, a-c), for large "neck-ring" - d, e (M. Ebert 1923. Abb. 14, d, e).

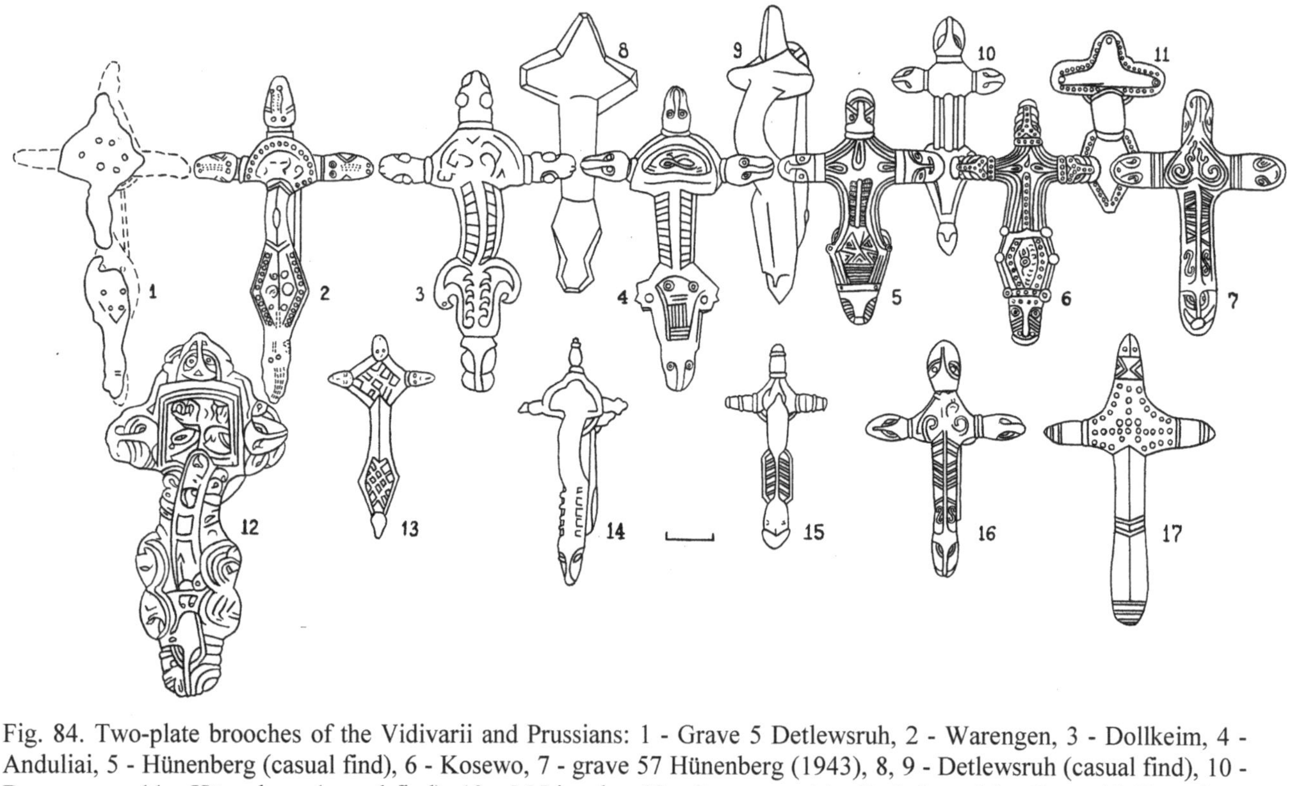

Fig. 84. Two-plate brooches of the Vidivarii and Prussians: 1 - Grave 5 Detlewsruh, 2 - Warengen, 3 - Dollkeim, 4 - Anduliai, 5 - Hünenberg (casual find), 6 - Kosewo, 7 - grave 57 Hünenberg (1943), 8, 9 - Detlewsruh (casual find), 10 - Bogazcewo, 11 - Hünenberg (casual find), 12 - M.Lipovka, 13 - Suvorovo, 14 - Podgórze, 15 - Grave 15 Hünenberg (1987), 16 - Eschenort, 17 - Podgórze.

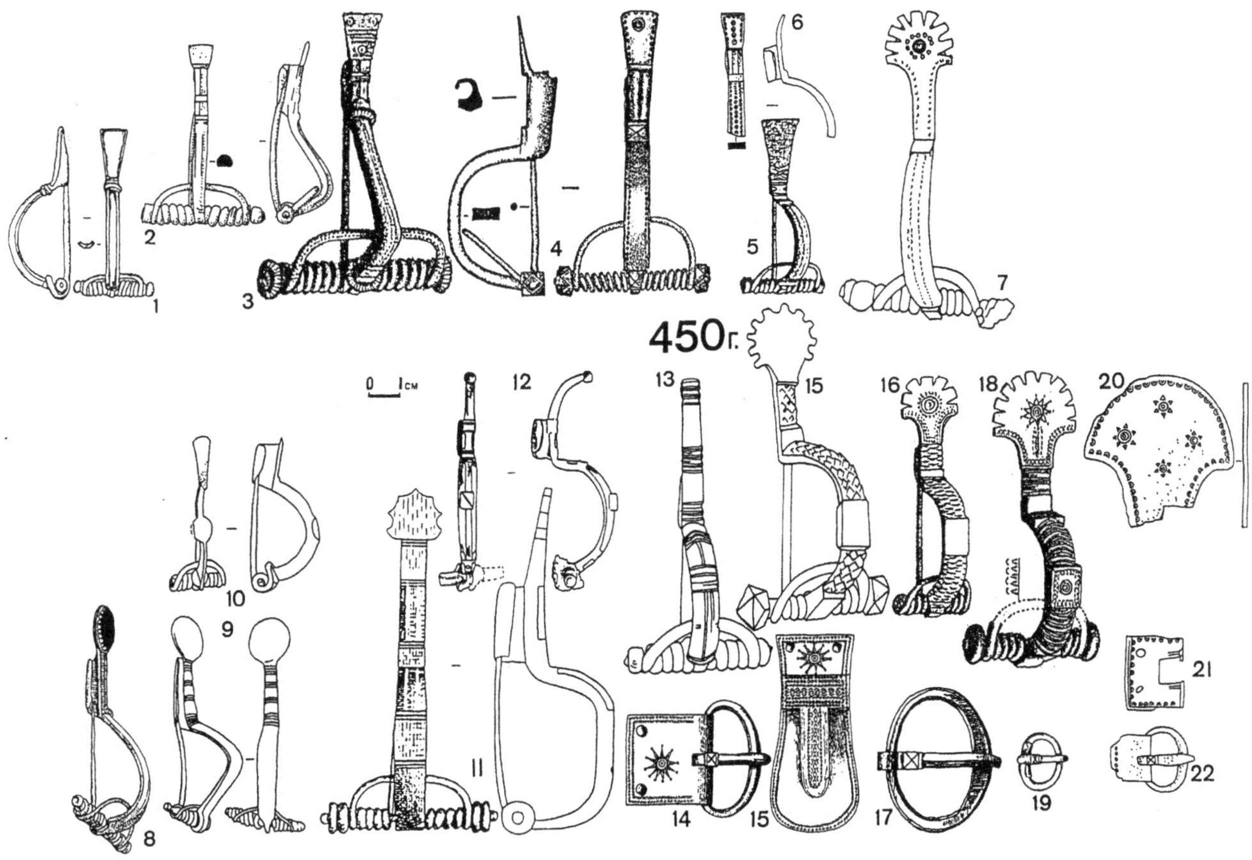

Fig. 83. Brooches of group Almgren VI, 2 (subtypes "with trapezoidal widening of leg" and "star-shaped") from phases C_2 - E from northern zone of Barbaricum (with attraction of accompanying buckles with ornament in Sösdala style): 1 - Grave 8 /AVI, 6 var. 181, stage C_{1a}/ from the cemetery at Gródek nad Bugiem-1c, woj. Zamojskie (T. Borodziej-Mazurek 1988. ryc. 1, b, s. 279); 2 - Grave 223 from the cemetery at Gora Velikanov (ehem. Hünenberg, Kr. Samland) (Archive IA RAS, 1992) /var. Iz Af 2c, phase D_1/D_2/(M. Schulze 1977. Taf. 11); 3 - casual find from the Kovrovo cemetery (ehem. Dollkeim, Kr. Samland) /var. Iz Af 2c with elements of ring set, phase D_1/D_2/ (O. Tischler, H. Kemke 1902. Taf. IV, 18); 4 - Grave 258 at Kovrovo /var. Iz Af 2c, found with pair of brooches with ring set, stage D_1-D_{2a}/(Archive IA RAS, 1996); 5 - casual find from the Gora Velikanov cemetery /derivative var. Iz Af 2c, phase D2/(O. Tischler, H. Kemke 1902. Taf. IV, 23); 6 - Grave 202 at the Gora Velikanov cemetery /derivative var. Iz Af 2c, phase D_2/ (Archive IA RAS, 1992); 7 - Grave 161 from the Kovrovo cemetery /Iz Af 2c with element of star-shaped brooch type II, phase D_2/D_3 (C. Gunter, A. Voss 1880. Taf. 10, Nr 448); 8 - cemetery at Kannikegaard, Bornholm /with colour glass inlay, phase C_2/D_1 (L. Jørgensen 1994. Abb. 121, 7, S. 526); 9 - cemetery at Bjars, Ksp. Larbro, Gotland, a derivative # 8 /phase D_2/(B. Nerman 1935. Taf. 9, Nr 65); 10 - casual find from the Kovrovo cemetery (Archive IA RAS, 1994) /derivative var. Iz Ae 6b, phase D_2/D_3/; 11 - casual find from the Gora Velikanov cemetery (W. Grunert 1944. Abb. 10) /type V according to A. Bitner-Wróblewska, phase D_2/D_3/; 12 - cemetery at Icklingham /West Stow/ type Glaston according to M.Schulze-Dörrlamm, stage E/(M. Schulze-Dörrlamm 1986. S. 633, Abb. 45, 1); 13 - Grave 163 (108) at Kovrovo (C. Günter, A. Voss 1880. Taf. 10, Nr 429) /archetype Duratón, phase D_3/E/(M. Schulze-Dörrlamm 1986. S. 646, Abb. 61, 13); 14 - Grave 163 (108) at Kovrovo (O. Tischler, H. Kemke 1902. Taf. XI, 1) /type 38 of group H, stage D_3/ (R. Madyda-Legutko. 1986, Taf. 20); 15 - Grave 163 (108) at Kovrovo (O. Tischler, H. Kemke 1902. Taf. XI, 2) /type "tongue-shaped" / (A. Bitner-Wróblewska 1989. ryc. 2,5); 15 - cemetery at ehem. Heinriettenfeld, Kr. Gerdauen (Inv. Prussia-Museum Nr 1080, Kopienkatalog, 1880, Nr 13815); 16 - Grave 164 at Kovrovo (O. Tischler, H. Kemke 1902. Taf. IV, 6) /type II according to A.Bitner-Wróblewska, phase D_3/; 17 - Grave 164 at Kovrovo (O. Tischler, H. Kemke 1902. Taf. XI, 18) /type 38 of group H, phase D_3/; 18 - Grave 30 at Warnikam (ehem. Warnikam, Kr. Heiligenbeil (O. Tischler, H. Kemke 1902. Taf. IV, 5, 5a) /type II according to A.Bitner-Wróblewska, phase D3/; 19 - Grave 30 at Warnikam /derivative of type 38 of group H, phase D_3/ (V.I. Kulakov 1997. fig. 8); 20 - Grave 260 at Gora Velikanov (Archive IA RAS, 1993) /derivative of type IIIa according to A. Bitner-Wróblewska, stage E/; 21, 22 - Grave 260 at Gora Velikanov (Archive IA RAS, 1993) /derivatives of type 38 of group H, stage E/.

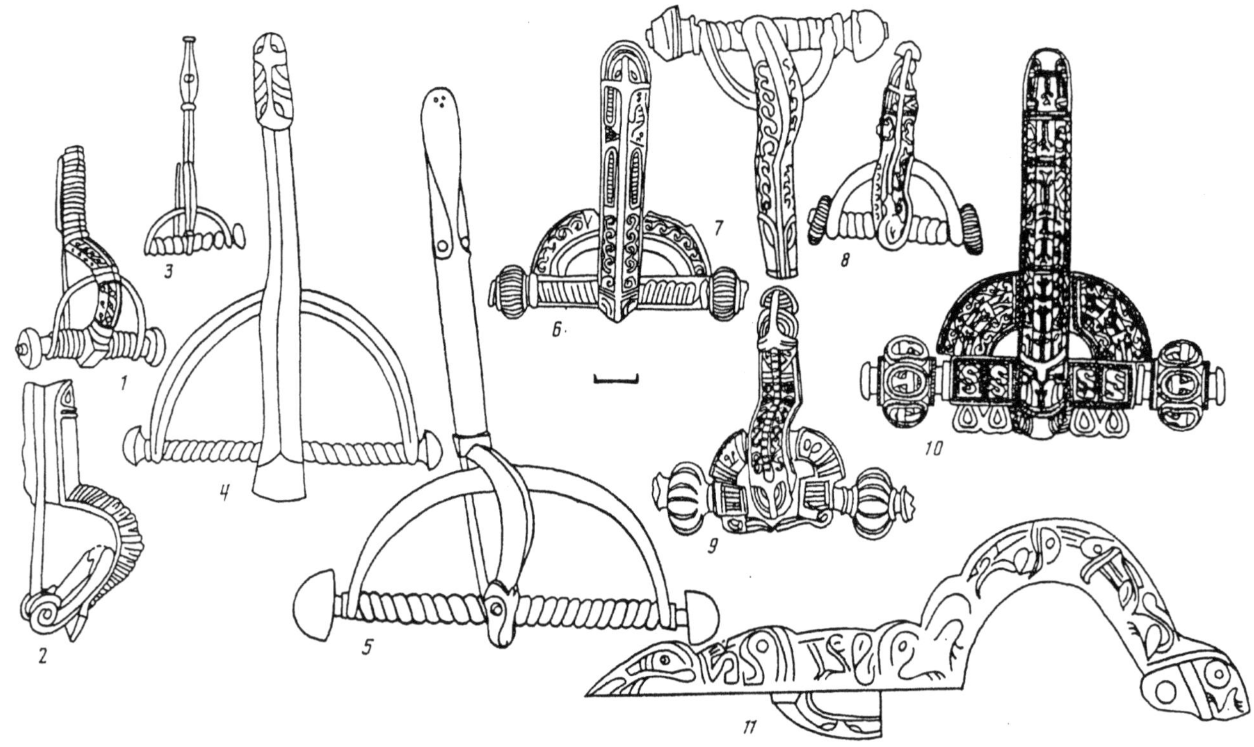

Fig. 85. Crossbow-shaped "animal-headed" brooches of group 2.

Fig. 86. Finds from the Geroyskoye cemetery.

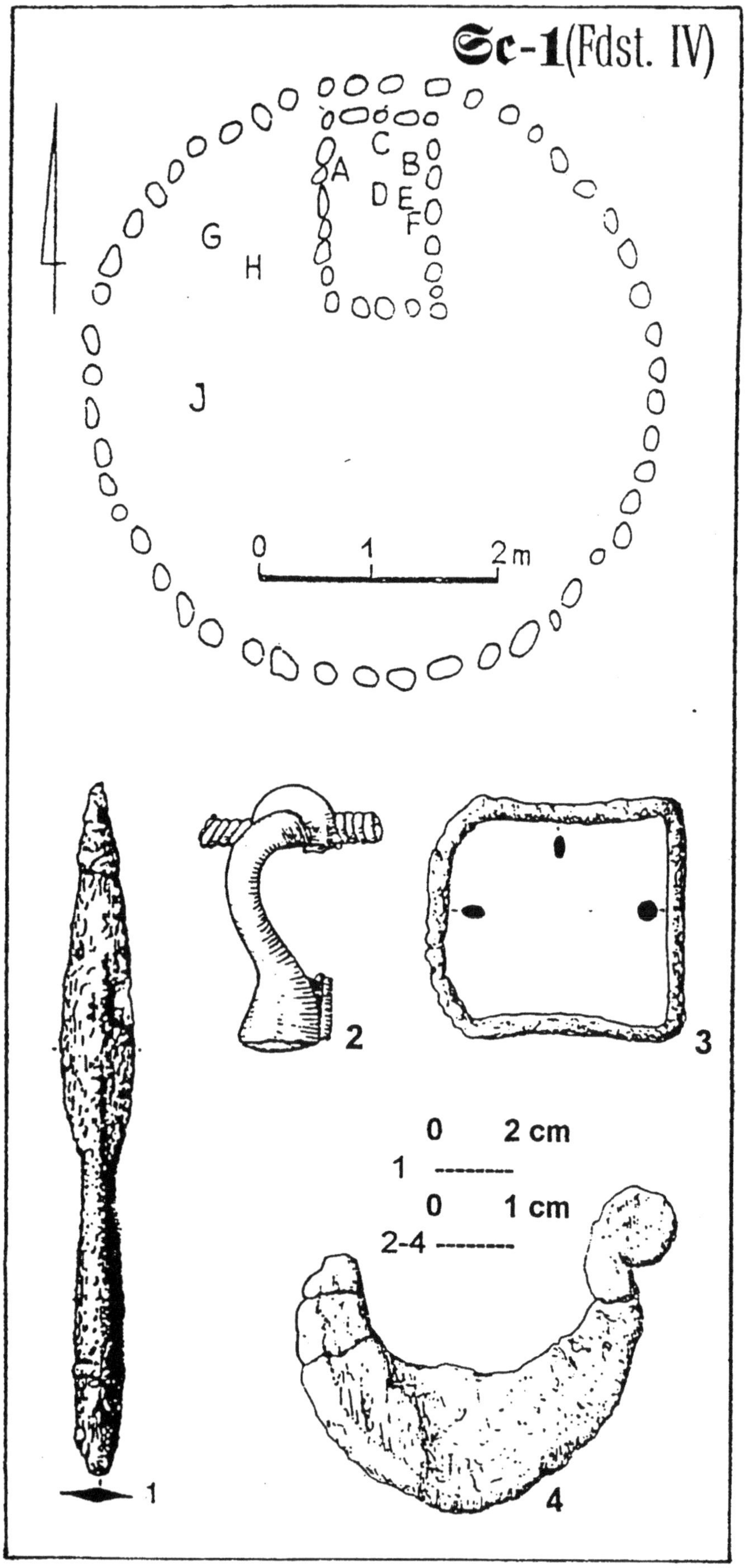

Fig. 87. The plan and assemblage of Grave 1 from the cemetery at Schlakalken-1.

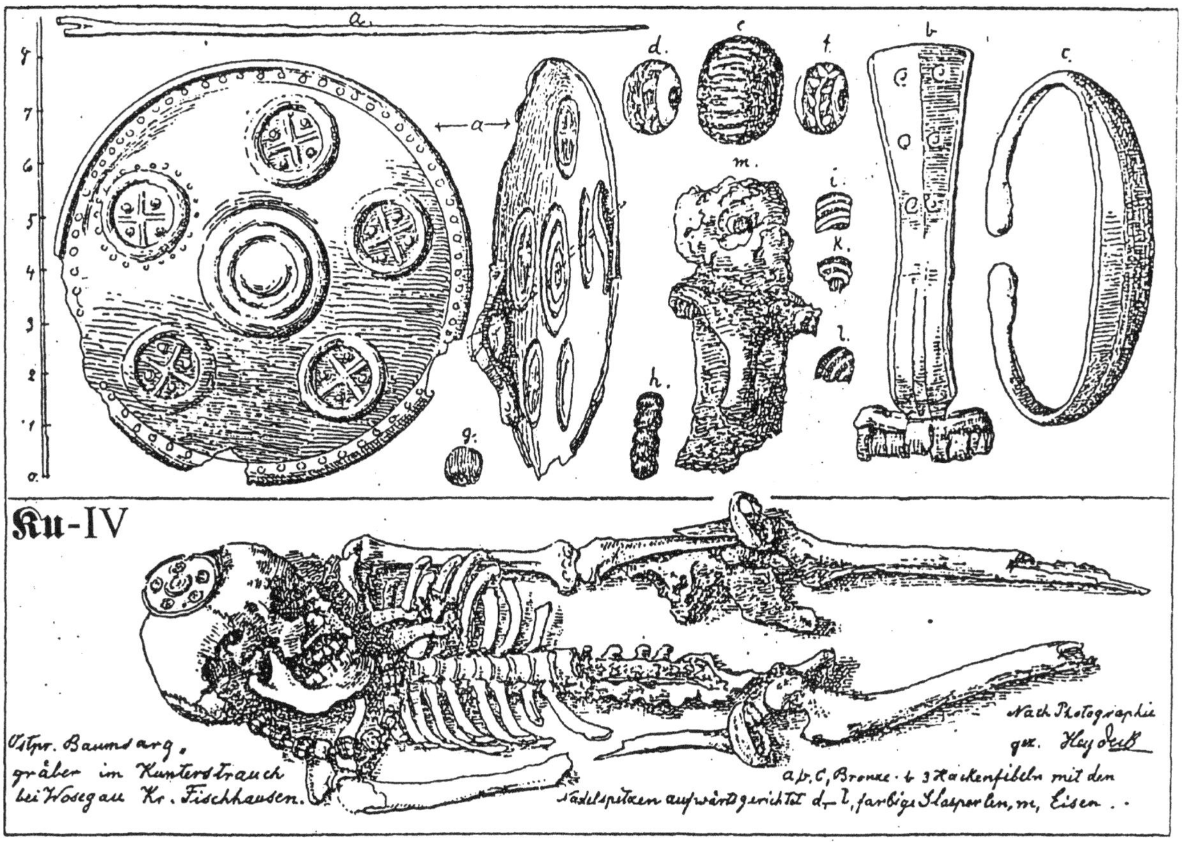

Fig. 88. Sketch and assemblage of Grave IV, Kunterstrauch.

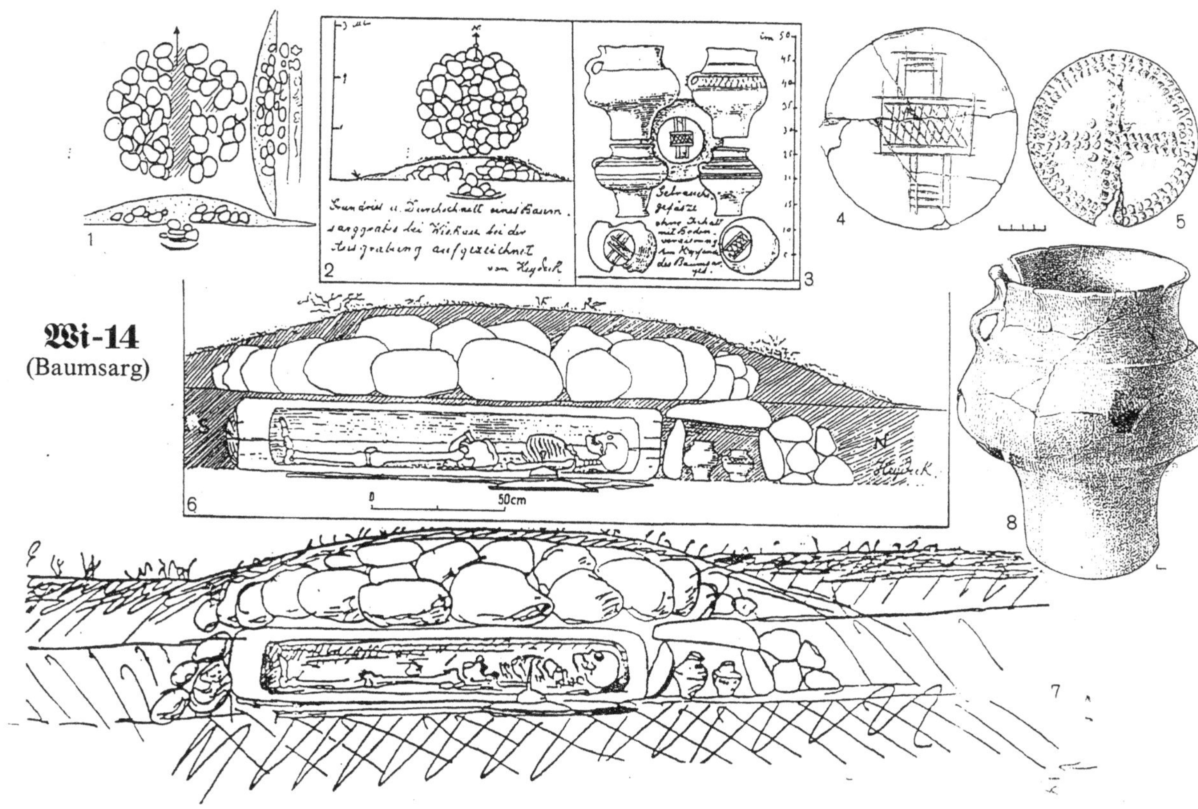

Fig. 89. Grave 14 at the Chrustal'noye cemetery: 1 - field figure from J. Heydeck's archive; 2 - "plan and section of burial in a wooden sarcophagus from Wiekau..."; 3 - "accessory vessel without contents..."; 6 - reconstructed section of Grave 14; 4 - cut-through figure on the base of the accessory vessel; 5 - figure, the contour of which is printed by a punch at the base of urn from the cemetery at Tykrehnen, Kr. Fischhausen; 7 - initial (field) sketch of the reconstructed section of Grave 14; 8 - accessory vessel from Grave 14.

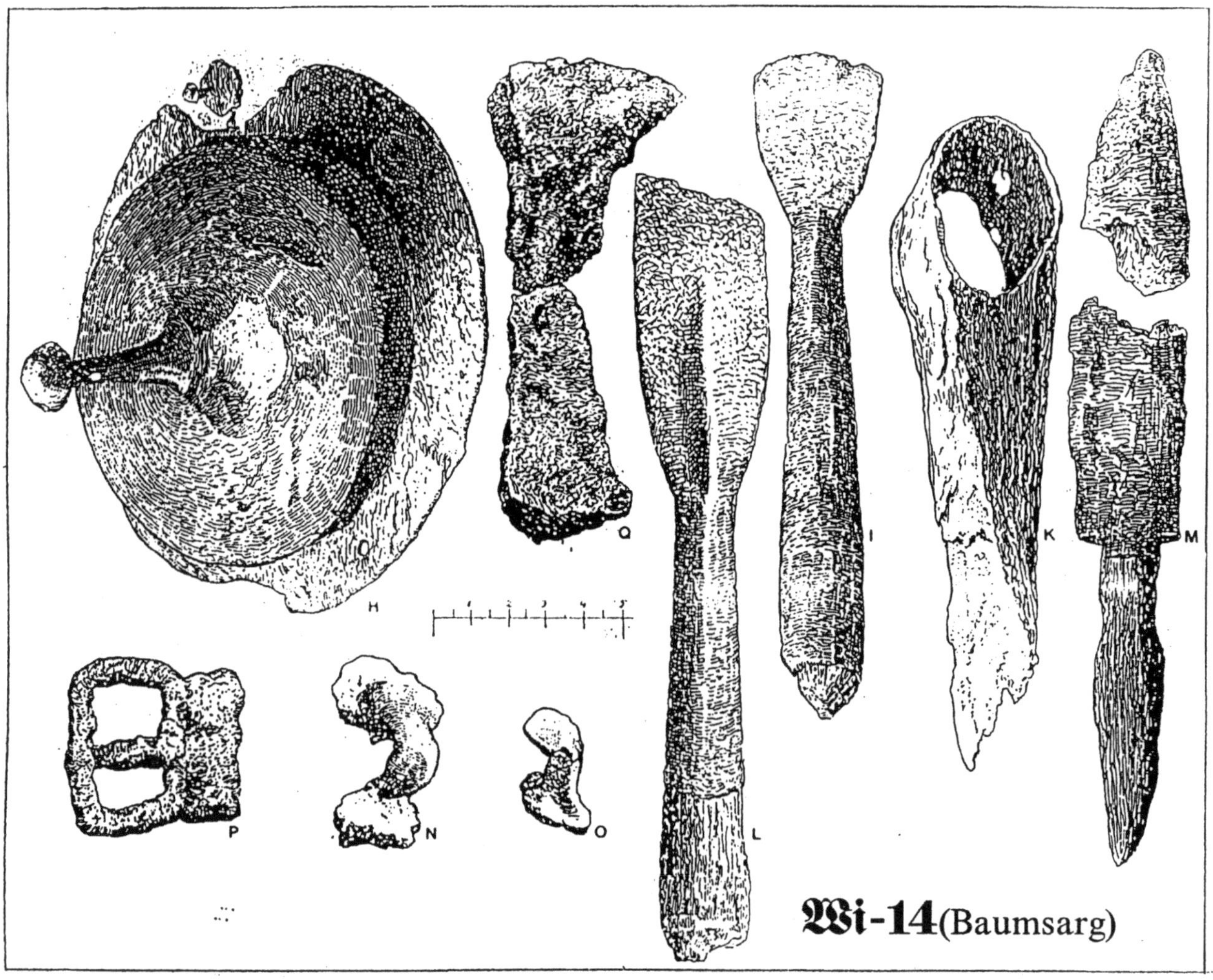

Fig. 90. Metal finds from Grave XIIII at Chrustal'noye with addition of chronological indicators (alphabetic positions - J. Heydeck): h - iron shield-boss with the copper-alloy button on the end of a core; q - iron grip of a shield; l, i - broken spearheads; k – axe; p - iron buckle; n - iron "hat-shaped" (after J. Heydeck) brooch; o - copper-alloy "hat-shaped" brooch.

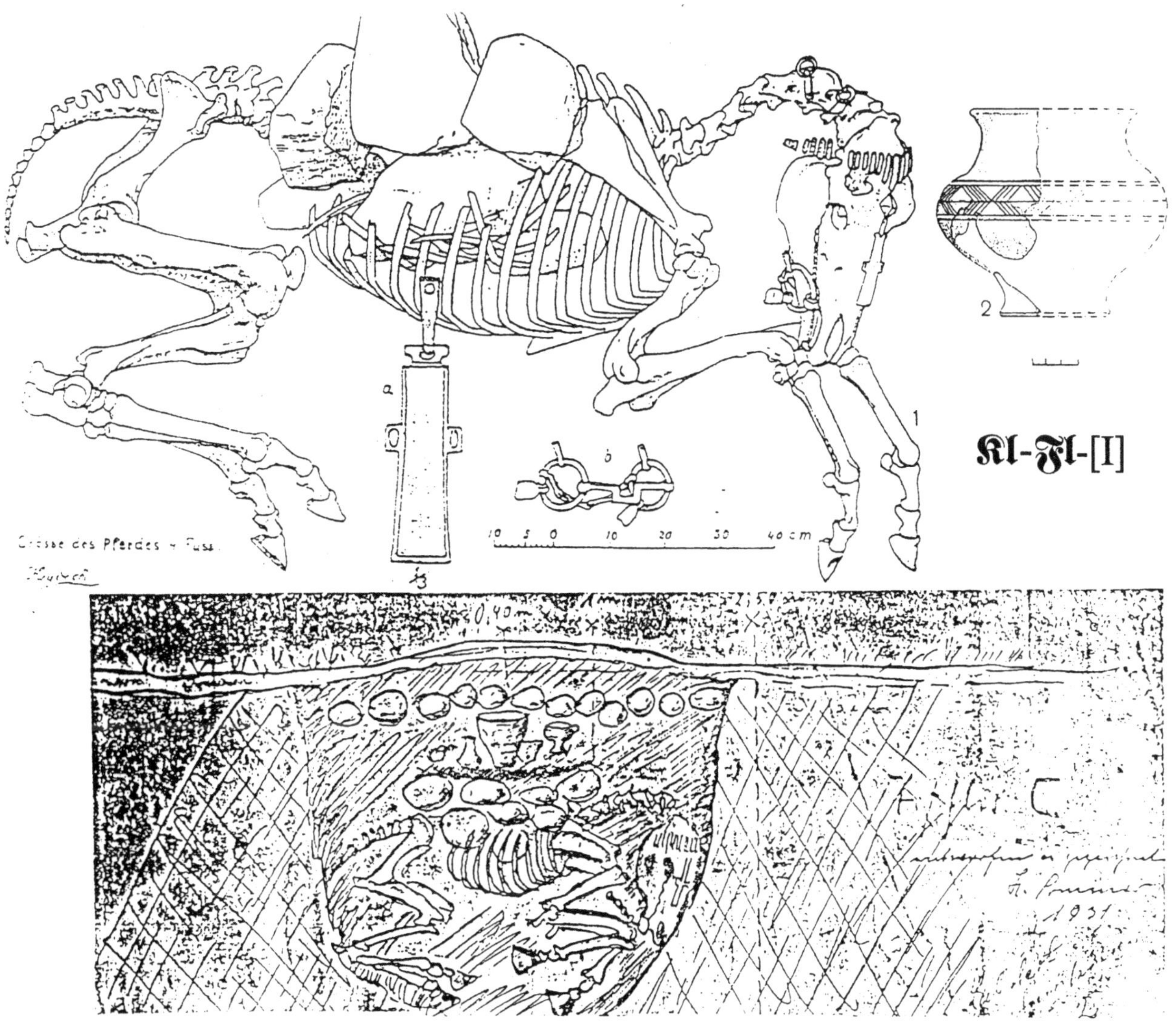

Fig. 91. Burial at the Izobil'noye cemetery: 1 - figure of the horse grave as a separate burial (Heydeck J., 1896-1900, S. 56, Taf. V), 2 - accessory vessel from the layer of funeral pyre, found in the upper level of the grave (Heydeck J., 1896-1900, S. 56, Taf. V, 9), 3 - field figure from the section of split-level grave at Izobil'noye with horse skeleton below.

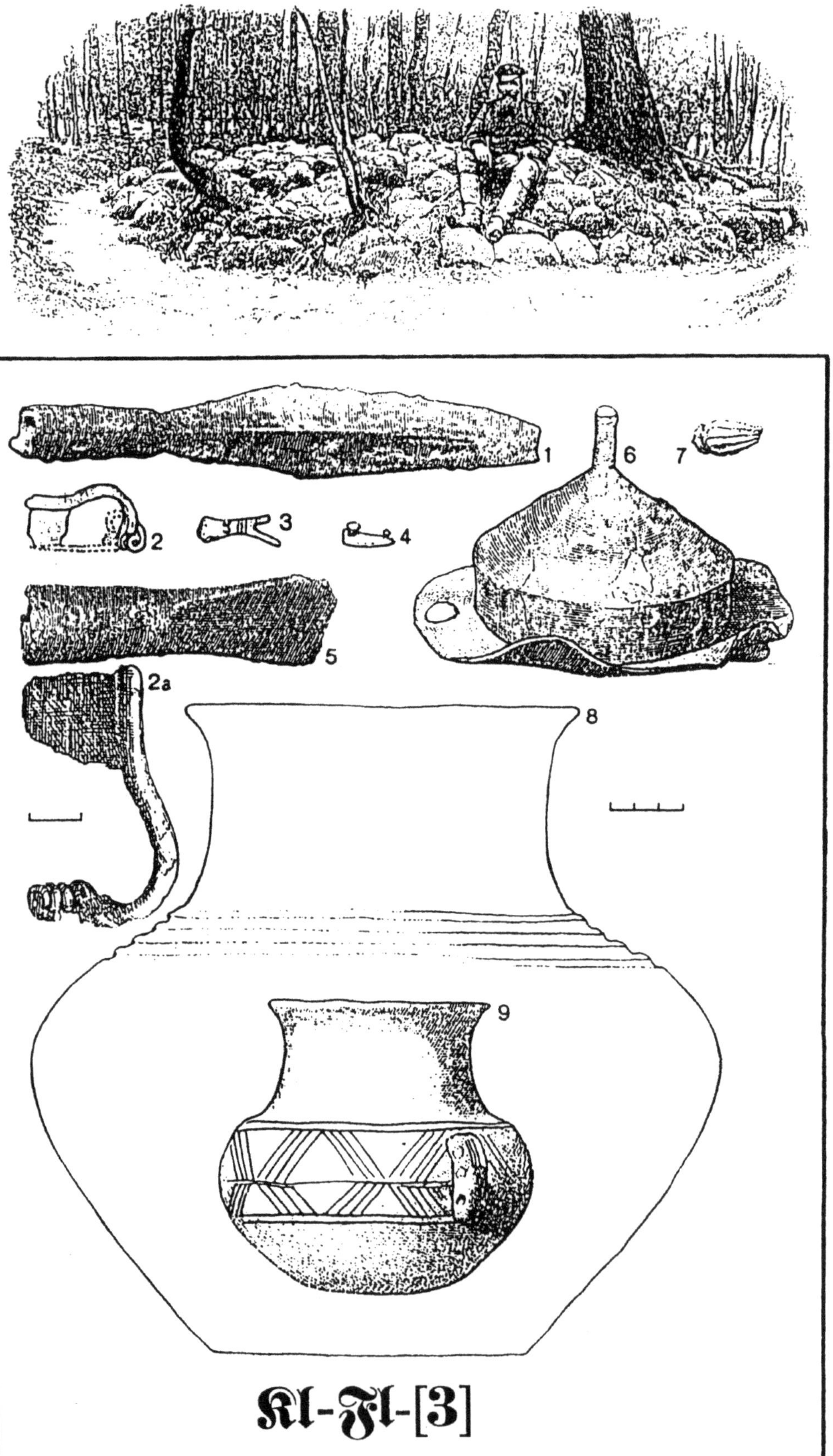

Fig. 92. Appearance of the second burial at Izobil'noye (with forester Bömmel for scale) and finds from this assemblage: 1 - broken spear; 2 - copper-alloy brooch of type AV, 193-203; 3 - fragment of a copper-alloy strap-end; 4 - copper-alloy belt mount; 5 - iron celt; 6 - iron shield-boss; 7 - copper-alloy detail of belt set (?); 8 - urn (?); 9 - accessory vessel; 2a - figure of a brooch Nr 2 (according to Bezzenberger A., 1904, Fig. 107).

www.ingramcontent.com/pod-product-compliance
Lightning Source LLC
LaVergne TN
LVHW071213210926
848482LV00015B/805